PRACTICAL
HOUSEBUILDING

PRACTICAL HOUSEBUILDING

for practically everyone

Frank Jackson

Pencil Drawings by Spike Hendricksen

McGraw-Hill Book Company

New York St. Louis San Francisco Auckland Bogotá Guatemala
Hamburg Lisbon London Madrid Mexico
Montreal New Delhi Panama Paris San Juan São Paulo
Singapore Sydney Tokyo Toronto

PRACTICAL HOUSEBUILDING FOR
PRACTICALLY EVERYONE

McGraw-Hill/VTX Series.

    4 5 6 7 8 9 0 SEM SEM 8

ISBN 0-07-032038-1(PBK)
ISBN 0-07-032035-7(HC)

LIBRARY OF CONGRESS CATALOGING IN PUBLICATION DATA

Jackson, Frank W.
   Practical housebuilding for practically everyone.

   Includes index.
   1. House construction—Amateurs' manuals.
1. Title.
TH4815.J33   1984   690′.837   84-10060
ISBN 0-07-032038-1(PBK)
ISBN 0-07-032035-7(HC)

Editor: Jeff McCartney.
Editing supervisor: Margery Luhrs.

*In memory of Granddad
Frank Goodman
1887–1983*

Contents

Acknowledgments

My grateful thanks to:

Spike Hendricksen, for his optimism and inspired artwork; Don Jackson, for his building know-how, careful review of the manuscript and fatherly advice; Dan Tonge, for typing the original manuscript; Wrench Richard, for countless explanations and inspiration; Steve Starlund and Rex Miller, for their in-depth reviews of the manuscript.

For inspiration, proofreading, and shared building experiences: Carl Blömgren, Mark Giles, Leela Cutter, Georgia Jackson, Don Kneass, Kip Keers, Pat Voght, Nancy Starlund, Chris Miller, Dan Schueler, Jim Garrison, and Steve Abel.

For their interest, as well as their editing and publishing talents: Jeff McCartney, John Aliano, Charlie Buffington, Margery Luhrs, and the other helpful people at McGraw-Hill.

Most of all, the essential ingredient is a jewel named Debby Tonge Jackson, who makes my life such that I can complete an immense task like this.

Introduction

What This Book Will Do for You

The purpose of this book is to provide people who have little or no previous construction experience with:

- A basis on which to decide whether to build your own house

- A primer on how to design and build a house

- Some ideas and inspiration to help carry your project through to a satisfactory conclusion

Recently I heard that less than 10 percent of working people who need a house can qualify for a bank loan to buy one. It's no wonder. The selling price of an average U.S. home is roughly $80,000. At the prevailing 15 percent interest rate, the final price tag with a 30-year mortgage will be over $365,000 by the time a buyer finishes paying for it. And then what do you have? A plain, average house, that's what.

A few years ago Deb and I decided that we should try to get out of the rental life and into our own home. It didn't take us long to discover that we couldn't afford any of the houses on the market, and that most of them didn't appeal to us anyway. But we did meet some people who had built their own houses, spending from $2,000 to $30,000, and they all had very nice homes. These people had designed and built homes that fit their lifestyles. We liked the little surprises—the fold-up stairs, the leaded glass rose in the window, the built-in planter in the corner, and the cozy bunk in the loft. These houses had character. Here was living proof that some imagination and hard work could produce a classier house than money could buy. Eventually we came to a decision; I would build a special home for us.

The only hangup was that I'd never built anything before. Well OK, I built a spoon rack in junior high school and my father and I once built a small sailboat. I *could* pound a nail into a soft piece of wood with a couple of tries. Still, I was determined to build a house. During the next couple of years I bought some books, read and talked about building, sketched some ideas, bought a piece of property, and made the drawings needed to get a building permit. Finally, one day, inspired by Johnny Paycheck singing "Take This Job and Shove It," I up and left my desk job and started building.

Man oh man, what a project! I'd never bitten off anything quite this big before. I learned more about myself and about how things work than I'd learned in the previous 30 years. Unfortunately I learned a lot the hard way. I didn't pour the foundation quite right and spent days chipping away the excess concrete. I put in a lot of unnecessary hours building the floor. Somehow I installed the electrical panel upside down. Things like that. I kept thinking that a few simple illustrations and some words to the wise would surely make things easier.

I needed a book geared to the amateur starting out way behind in terms of construction know-how; a book that outlined all the steps from selecting the property to designing the house to building it in an affordable manner. It wasn't there. Where's the well-illustrated, step-by-step building guide that clearly explained the entire approach in simple, concise terms? I wanted a path through the woods that's easily followed so a person doesn't spend half of his or her time lost in the underbrush.

As I floundered through the brambles, I spent a lot of time paging through books showing an engineer with a $300 transit laying out the house corners, a crew of four framing the roof, and containing dry, expert-sounding text droning on but never getting to the part I needed to know. Nary a mistake ever happened in these books, they were never short of money, and no one was ever in need of a little "divine inspiration."

Fortunately, I had another source for the type of help I needed, not exactly a divine source, but one that most people may not have. Several friends and my father had traversed this path before me and were generous with their advice and encouragement. Between friends, books, and brute persistence, I pieced together the necessary knowledge. Our fantasy of a classic home with a big kitchen and a rooftop sundeck, built with my own hands on a pay-as-you-go basis, slowly came to life. In the end, 95 percent of the actual design and labor emerged from a most unlikely source—me.

The recipe for making any dream come true includes a heavy dose of determination, a dash of ingenuity, a logical approach, and a good source of information. If you can provide the first two ingredients, I think I can provide the last two.

Part One
GETTING STARTED

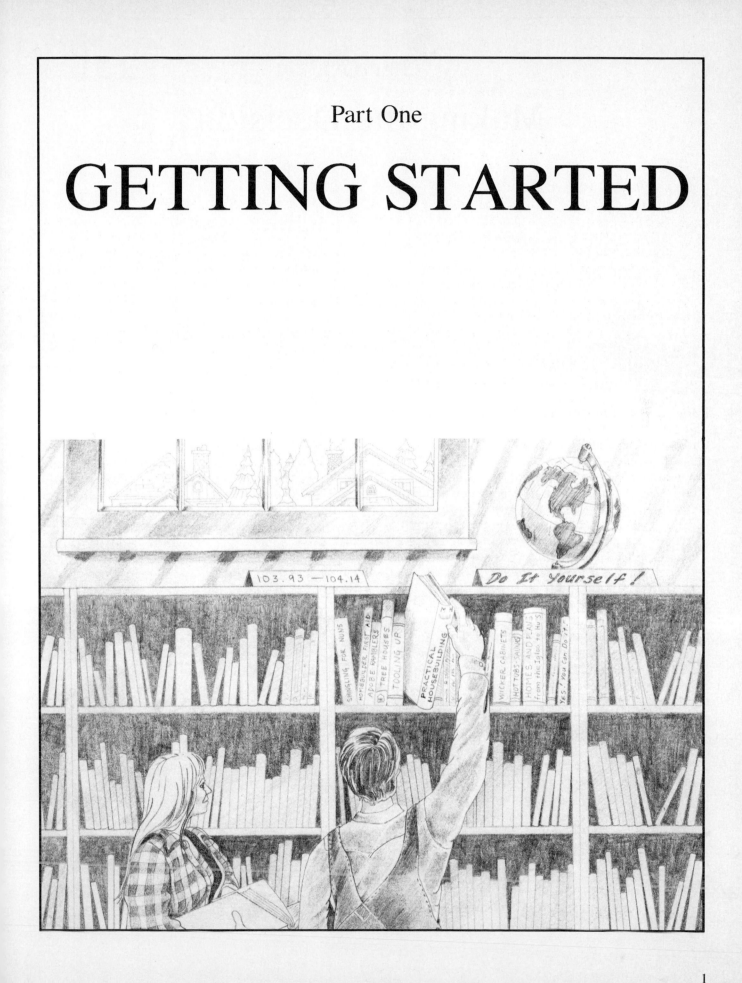

CHAPTER 1

Making the Decision

I don't know how you make decisions that affect your life's direction. But I do it by sort of letting things cook in the back of my mind for awhile. I pick up on facts and opinions, and mix it all into my mental stew and let it simmer. Until there's an obvious need for a decision, I just keep adding to the stew and drifting along until I can work up some enthusiasm about the direction in which I want to go.

Deciding how to get into your own house can be one of those difficult decisions. When Deb and I had a nice place to rent, we sort of drifted along and didn't make any major decisions. Occasionally we visited friends who bought or built a house and that added a few impressions to the mental stew—a carrot here, an onion there. We read about high interest rates and casually looked at home prices. Several sessions with the calculator always gave the same answer—we would probably never be able to buy a house unless I got a drastic pay increase. I wasn't going to hold my breath for that. Mostly we tried not to be depressed and left the stew on the back burner.

When overwhelmed by inflation, high interest rates, and increasingly higher prices, the mainstream of society either goes out and buys a house on a 30-year mortgage or they resign themselves to renting a place to live. There's a smaller side current of society that builds their own houses, scrounging materials and adapting to less than plush living conditions. As our stew cooked, it became apparent that the only way we were going to get where we wanted to go was to build

our own house. So we headed in that direction. We talked to people who had built their own houses. By chance, the people we were renting from had built their own house and were planning to build another. We got a couple of books as gifts that showed some nice photos of handmade houses. We fueled our fantasy by talking and reading and looking. We wanted a place with a tower and somewhere out in the country. During our last rental days, if I saw Wrench (my landlord and mentor) building something, I'd go over and watch and ask lots of questions. Slowly the decision evolved. There was no blinding flash of realization—but the stew was done and our course was set. We were headed for the side current.

It began with a serious savings program, followed by a lot of time spent reading the classified ads and looking at property. We drove around areas we liked looking for "For Sale" signs. After a year and a half of looking and saving, we finally found just the place; 2 acres out in the boondocks. We bought it. We laid plans, schemes, and more plans. We sketched and looked and talked. I built a balsa wood model of our fantasy house, complete with a tower and a sundeck. Weekends were spent clearing blackberries and brush from the lot. A driveway was hacked from the wilderness and a water faucet hooked up. An outhouse was hammered together to take care of nature's calls. None of this happened quickly, you understand. We traveled along our side current from fantasy to outhouse for about 2 years.

Suddenly, there was a stretch of white water and things accelerated. The place we rented was put up for sale at the same time my employer had a business slump. I applied for a building permit and left my job. We bought a trailer and moved out to the boondocks. There was a long stretch of rough water ahead and things looked very scary.

Now, we're 3 years further down the river and things are smoothing out again. We've moved into the house and we love it—rough edges, bare plywood floors, and all. The place is paid for, and that gives us a lot of freedom to pursue some other interests. Things look pretty good for us, compared to our parents who spent 20 and 30 years paying for their homes and compared to people slightly younger than us who are faced with record interest rates and house prices. The radio says that single-family housing sales are lower than any year since World War II. There must be a lot of people out there who want to be in their own house but can't afford it. That is, they can't afford it if they follow the mainstream. Somebody ought to tell them about the side current. It's rougher, but more exciting. It requires guts, determination, ingenuity, and plain old work. It's not for everybody, but it's certainly a way down the river.

To make a decent decision on any subject, you need to be informed. The best way to learn about the trials and triumphs of amateur housebuilding is to talk to someone who has tried it. Now don't pick the most experienced builder you can find; locate a person similar to yourself who has actually done it, or tried and failed. As you're doing now, read a book or two so you can understand the main steps of the process and visualize what it takes.

Following is a short description of the benefits and drawbacks of building your own house, *in my opinion*. We'll talk more about the wild variety and dubious value of opinions in Chapter 4.

An Alternative to a Thirty-Year Mortgage

There's really no way I would consent to spending 30 years of my life working a job for the primary purpose of making house payments, especially when I know what I know today. First of all, there's the amount of money just thrown out the window on interest. Let's look at the actual cost of buying a $60,000 house on a 30-year mortgage. Table 1-1 shows the gory details.

Look at this table carefully. Notice how you've paid the bank $45,520 after 5 years, and how you've bought only $767 worth of house (about one month's payment). Notice how the bank wants over $273,000 to buy a $60,000 house. If you're 32 now, that means you'd pay over $750 every month until you're 62 years old. Thanks, but no thanks!

To add insult to injury, you probably can't sign up for this slave ship unless you have a $20,000 down payment and a hefty income. With the same $20,000 and a year or two of effort, most industrious people with no previous building experience could build a very nice home. Why bother with the extra 28 years of work to pay interest on a mortgage?

I'm afraid I may begin to sound like the weight-loss ads in the Sunday paper, or the "make a million dollars in real estate" people. Don't forget that you have to put in a lot of labor to make this house happen. Therein lies the rub.

TABLE 1-1 The Real Cost of a $60,000 House at 15 Percent Interest*

Time elapsed	You've paid	You own (equity)	You owe the bank (principal)	Fee for using the bank's money (interest)
Day of purchase	$0	$0	$60,000	$0
After 1 year	$9,104	$111	$59,889	$8,993
After 3 years	$27,312	$391	$59,609	$26,921
After 5 years	$45,520	$767	$59,233	$44,753
After 10 years	$91,039	$2,383	$57,617	$88,656
After 20 years	$182,078	$12,966	$47,034	$169,112
After 30 years	$273,118	$60,000	$0	$213,118

* Figured over a 30-year period with a monthly payment of $758.66.

An Invigorating, Productive Experience

Now, my personal slant again. I like doing things that have a tangible result. I spent 5½ years in my previous job and the chief result was a pile of paper that has long ago been incinerated. There is probably nothing so tangible as the house you live in, particularly when you design and create it where there was nothing before but a dream. In addition, your house will reflect your lifestyle and personality and there will be no other place like it. (See Fig. 1-1.)

There's also the self-confidence and feeling of self-reliance gained. I had some serious doubts about my ability to drive a nail, let alone build a house. This project gave me a chance to prove to myself that with enough effort and determination, following a slow methodical approach, I could accomplish a major construction job single handedly. Call it therapy if you will. Therapy for the side effects of a society and economy that cause people to become specialists at some minute mechanism in a vast machine. To prove that I could also be a generalist—carpenter, plumber, electrician, ditchdigger, or most anything—was a personal breakthrough.

Finally, there's nothing like a good building project to get other folks interested in what you're doing. Fellow do-it-yourselfers come out of the woodwork, flooding you with advice, theories, and occasionally free labor. By completion, a lot of people participated in building our house. Establishing new friendships was one of the most enjoyable aspects of building.

Minor Digression and Introduction of Characters

As a way of illustrating some of the crazy things that happen to amateur builders, I've included a generous assortment of real-life stories. (I admit it, I've embellished some of the incidents.) Since some of these folks will be resurfacing throughout the book, I'll introduce them here. Deb is my wife and mainstay; son Garth is a small fry with a penchant for carrying off hammers and tools. The incomparable Wrench will introduce himself. Fellow amateur builders include Carl and Marcia, Ann and Don, Jim and Marnie, Donna and Martin, Marie and Steve, and our old friends and former roommates Steve and Nancy. My number one fallback for tough, "fourth-and-30" situations is good old Dad, retired journeyman carpenter, and ultimate source of resourcefulness. And then there's Uncle Barney, my fictional figment and all-star pessimist (everyone knows an Uncle Barney).

Fig. 1-1 *Comparison of house prices:* (Top) *Price: $60,000. This house has 1000 sq ft and is on a 50 ft × 50 ft suburban lot. With a 15 percent thirty-year mortgage, the total cost will be over $273,000.* (Bottom) *This 1,750-sq ft house cost $27,000 plus about 4,000 hours of work and numerous Band-Aids. It's located on two beautiful acres that cost an additional $12,000. No mortgage was necessary.*

By the way, if you know someone who is also building their own house, keep in contact with them. Steve was working on a log house at the same time I was building and we frequently traded labor when the task demanded two people. We also traded the misery and mistake stories that you don't tell anyone else. And we traded materials, leads on good prices, "trap-line" information, photos, jokes, and cheap wine. It worked out very well. Plus we built that bond of survivorship that made it a foregone conclusion that we both would succeed, hell or high water.

End of Sales Pitch

That's my sales pitch for building your own house. You satisfy a basic living need in less than 5 years, instead of 25 or 30. It's a big step forward if you're aimed toward self-sufficiency. You get a house custom tailored to your needs and desires, and the feeling of tangible accomplishment. The self-confidence gained, skills developed, and friendships made are all rewards for the long, tough journey. You may also open the door to the world of building and a lifetime of enjoyable projects.

And now, lest I be accused of painting an entirely rosy picture, we better look at some of the drawbacks so you can see if the benefits outweigh the negative aspects.

The Dangers and Drawbacks

The granddaddy of all dangers is that you will not succeed. That is, you'll get partly done with the house, run into problems, and in the end, flat-out fail! Now is the time to think about that possibility. I'll let you listen to Uncle Barney for a minute to hear why you can't possibly build your own house, expressed as only Uncle Barney can say it. ("Thirty years in the trades sonny, and I know what I'm talking about!")

"Look at what you do for a living. Do you think the average doofus off the street can come in and do your job satisfactorily? Building a house requires between 10 and 25 professionals who each spent years learning their trade. Willya look at the skills involved: architecture, structural engineering, excavation, foundation (just

wait 'til the forms break and 15 tons of concrete oozes onto your site and then hardens), framing, glazing, roofing, masonry, plumbing, septic system design and installation, wiring (you can burn the place down if you wire it wrong), eh? give up? give up?, furnace and ductwork design and installation, wallboarding (the professionals can do in 3 days what will take you 2 months to do), painting, cabinetry, floor and carpet installation, appliance installation, landscaping, and more! Do you really think you're qualified to do any one of these jobs? The tools alone will cost you a fortune. Without proper tools you'd certainly screw things up beyond all repair. I know you're a little crazy but this is absurd!"

Hmmmm, what else now? Oh yes, "A contractor can build a house in 3 months; it'll take you 2 years. That's 2 years you could be in your house enjoying it, benefiting from the effects of inflation, and doing more pleasant things than slogging away in the mud on a project you know nothing about. You'll probably break your arm the first day on the job. The stress of the situation could cause so much emotional trauma that you'll end up divorced (or *married,* if you're single). What if you run out of money when it's only half completed? You're not really going to go through with this are you? Listen, I've warned you now, so don't come running to me when you're up to your [suspenders] in alligators!"

The Decision Is Yours

Spare me Uncle Barney, spare me! I've got my answers to Uncle Barney's arguments but it makes more sense for you to develop your own. Reading and talking to people who've built their own house are the best ways to determine which particular tasks may be beyond your means. If the house you have in mind is cantilevered out over a clay cliff, has a sweeping gull wing roof, large plate glass windows, and is the ultimate in interior design and cabinetry, listen to Uncle Barney. If your dream house is a modest, comfortable house designed around your lifestyle, you may be the most qualified designer/builder available. Don't forget, you can always subcontract the tough parts to a professional.

The decision is a weighty, time-consuming choice and the implications are sizeable. The good news is that making the decision is inexpensive. Reading, talking, and looking at construction sites comes free. Shopping for land and designing the house require only some gas, pencils, and a few books. Practice projects like sawhorses, a workbench, and a shed are cheap ways to acquire basic building skills. And, if you put yourself on a budget and start saving for a house, even if you abandon the idea you'll be in the best financial shape ever.

Yes, I'm happy I took the plunge. But I can see how other people, a person more committed to a specific profession or a person who has no interest in manual outdoor labor for instance, might feel quite differently. After you've read this book you'll know what's involved and be in a good position to decide. Then, if you get it in your mind that you *can* do it, and you *want* to do it, it'll take more than a team of wild horses and Uncle Barney to stop you.

CHAPTER 2
Financing: An Unconventional Strategy

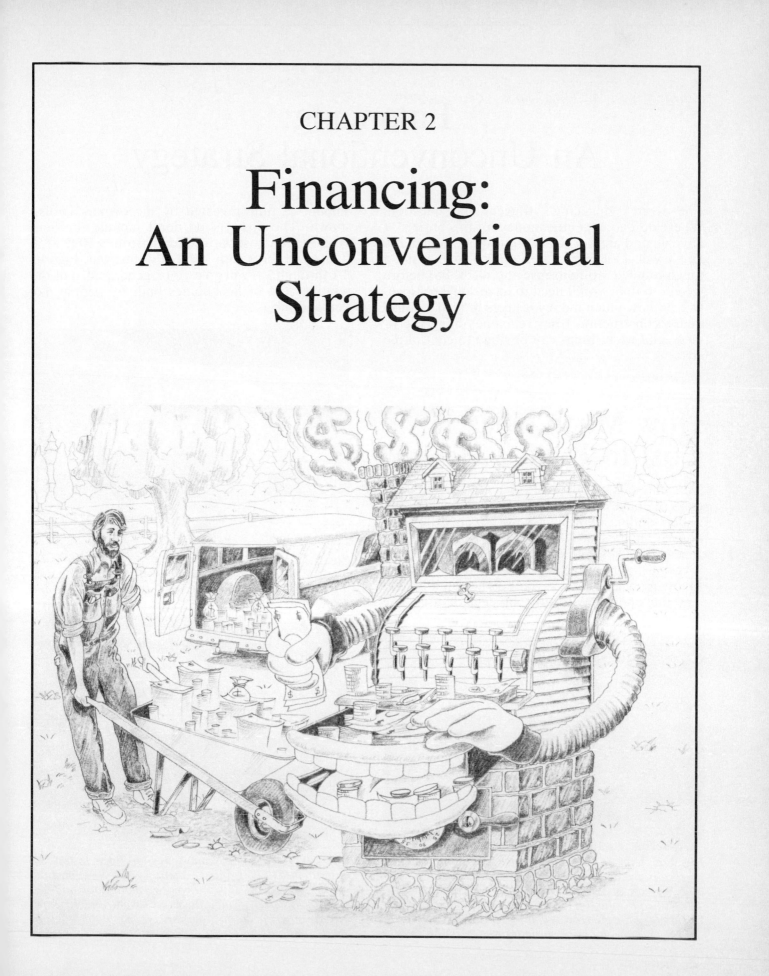

Financing: An Unconventional Strategy

The word "financing" triggers the question: Where do I go to get the money for this project? I have a good answer for that question, though it's an answer a lot of you aren't going to like. It's a tough-minded, no-nonsense approach. But before we get to that, you'll need to have some perspective on how much money it takes to feed a house under construction (they're money-hungry critters), and what things can be done to control the cost.

How Much Does It Cost to Build a House?

You can talk about costs per square foot and class of house all day long and never reach a good understanding of what it will cost to build your house. The best way is to look at some different types of houses with different amounts of owner-invested time (sweat equity). The photos on the following pages show several owner-built houses and the approximate cost of each house. The amount of time invested by the owners is also shown. The costs listed don't include the land because that cost varies greatly from a 50 ft × 50 ft lot to 40 acres with 300 ft of waterfront. Figures 2-1 through 2-6 will give you a general idea of the relative cost of first homes built by people like you and me.

Fig. 2-2 *Jim's house was built on an existing foundation and cost only $2,000. Two people worked full time for 2 months. There are 320 sq ft on the main floor and 200 sq ft of basement storage. Expansion plans are in the works.*

Fig. 2-1 *Carl and Marcia's house was handcrafted of materials from the land. The cost was about $2,000 and 5 years of careful, thoughtful work (part-time). The 380 sq ft, plus 140 sq ft of loft, are nicely finished, complete with cherrywood floors made from an old cherry tree. An addition is under construction.*

Fig. 2-3 *Steve and Nancy's twelve-sided log house cost them $12,000 and 12 months of hard, full-time work. There are 1,200 sq ft on the main floor and 600 sq ft of loft area. Since the house went up, the farm has sprouted a sauna, carport, forge building, root cellar, animal shelters, and now a barn.*

Fig. 2-4 *Deb and I spent $27,000 and 3 years (18 months full-time, 18 months part-time) building our home. Like all the houses shown here, it was our first building project. The 1,750 sq ft of living space is about two-thirds completed on the inside.*

Fig. 2-5 *Ann and Don's house is nicely finished with maple floors, ornate doors, laminated spruce counters, brass faucets, and more. They spent about $30,000 and 1½ years to build it. There are 1,200 sq ft of living space.*

Fig. 2-6 *The final results aren't in yet on Steve and Marie's place. It'll probably cost over $60,000, which includes several parts that were contracted out. There are 3,000 sq ft inside, 500 sq ft of decks, and so far it's taken about 2½ years. You can see that the sky's the limit for amateur builders.*

Where Does the Money Go?

After it's all plumbed and painted, you add up the receipts and wonder how it could really have cost that much. In Fig. 2-7, the actual expenses for our 1,750 sq ft house (the one shown in Fig. 2-4) are summarized.

I had originally estimated the cost at $16,000. While I was fairly close on the foundation, lumber, nails, and structural part, I was way off on the roofing, siding, electrical, windows, plumbing, and furnace. This was partly due to the design improvements made as time progressed. The roofing we selected was twice as expensive as normal asphalt roofing and the roof design required custom-made gutters. The electrical be-

came more elaborate than originally envisioned and a second bathroom and a fireplace were added. Some of the overrun was due to things I forgot to include in the estimate: railings, paint, tar paper, power bills, and permits. Other things just cost a lot more than I had planned. Building materials were experiencing runaway inflation at the time, so inflation alone contributed $6,000 to the cost. If I added hardwood floors, carpets, banisters, millwork, light fixtures, modern appliances, and a solar heating system to the house, I could easily spend another $10,000.

After you have a specific house design, Chapter 8 shows how to go about estimating the materials and labor required. For now, it's sufficient to know that the house structure with a roof, windows, and doors is only half the cost. The house looks complete from the outside but is only half finished financially. The heating, plumbing, and electrical systems are about one-fourth the cost.

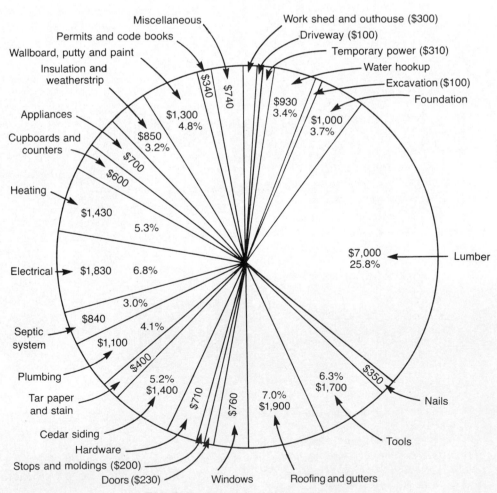

Total Cost: $27,100

Fig. 2-7 *Housebuilding expenses.*

The rest is insulation, wallboard, counters, appliances, tools, and a bunch of smaller costs.

You may also be interested in how fast the money goes. On the average we spent $500 a month when I worked part time on the house and $1,000 a month when I gave the house my full-time effort. This was one advantage to taking nearly 3 years to build; you don't need a whole truckload of money all at once. Five hundred dollars a month keeps you going, gives you time to do the crafty detail work, and is less than most mortgage payments.

Controlling the Cost

How Do You Build a $2,000 House?

I'm not at all sure *I* could build a house for $2,000, but let me make some suggestions. First, I'd design the house small and put it on concrete piers instead of a continuous foundation. I'd have to look very hard for good deals on salvaged lumber, windows, and doors and use logs off the land wherever possible. The roofing cost could be avoided if I could find a cedar log or two and hand split my shakes. These methods could bring the $11,000 I spent on foundation, lumber, nails, windows, and roofing down to $1,000. There could be a lot of trouble getting permits because of code requirements involving foundations, room sizes, kitchen and bathroom facilities, heating systems, electrical and plumbing installations, septic systems, and fire resistant walls. I would also have to try to buy nothing and build everything—counters, cupboards, doors, latches, railings, etc. It would take a heck of a lot of determination, imagination, and ability to build a house for $2,000. But I have living proof that it can be done.

Design

The basic cost of your house will be determined by the design. Naturally, a large house costs more than a small house, equally finished. Large windows cost more than smaller windows. Tile roofing is more expensive than asphalt roofing. A heat pump system is $5,000 compared to $300 for a metal fireplace. Oak floors are about $2 to $3 per square foot and leaded glass windows are at least 10 times as expensive as plain pane glass. A house designed to be a great "party house" will be more expensive to build and heat than a cozy little house designed just for living. To success-

fully control the cost of the house, the most important thing is to *design a house that you can afford*. (More on this in Chapters 4–8.)

Labor

Let's talk about a specific house design. Pick a house, any house. How much is it worth in today's marketplace? $50,000? You can probably build it for half that price if you provide all the labor. The cost of typical houses being built today breaks down to about 50 percent material costs and 50 percent labor and builder's profit. By providing all of the labor, including foundation work, electrical, and plumbing, the price is cut in half. But don't forget that you will somehow have to spend about 2,000 or 3,000 hours of your time to avoid that $25,000 of labor cost. Still, that works out to about $8 to $12 per hour for your effort and a chance to avoid $200,000 of interest payments.

Materials

Generally, the potential for saving money on labor is better than for saving on materials. After all, when was the last time you saw a sale on a truckload of concrete? Or a deal on used roofing? (Don't buy any used roofing.) Although certain materials don't offer much savings potential, a resourceful person can do remarkably well at keeping the cost to a minimum.

Selecting the basic material for the house can be a key decision. Log houses, for example, can be built with the trees from your property. That's how Carl and Marcia did it. Or logs can sometimes be bought relatively cheap. Many of the logs in Steve and Nancy's place came from the shorts of a telephone pole supplier. Excuse me, what I mean is that the logs were too short to be telephone poles, but just right for a twelve-sided log cabin. The nice thing about a log cabin is that the logs are the siding, framing, insulation, and wallboard all in one, which saves a lot of money.

Salvage Materials. Salvage materials can provide major savings. The key factor is to have a discriminating eye to separate the junk from the items of value. For example, salvaged lumber must be carefully examined. Frequently it is split from the salvage operation, or is so hard and dried out that it is nearly impossible to drive a nail into it. Furthermore, unless you get a sufficient quantity to do a whole section of the house, you can have a real problem trying to match old lumber with new because of the difference in sizes. In the early 1900s, 2 × 4's were really 2-in thick and 4-in wide. Then they began surfacing the lumber and 2 × 4's shrank to 1⅝ in × 3⅝ in. Today a standard 2 × 4 measures 1½ in × 3½ in. Trying to use lumber of differing sizes in a floor or wall is not worth the effort.

On the other hand, buying used beams, decking, and hardwood floors often provides valuable savings. The beams in our house came from an old saw mill that was torn down and are far better than any I could have bought new. They were old-growth fir, with over 30 grains per inch. (The slower a tree grows, the stronger the wood.) T & G decking (2 × 6) can be a good salvage deal if the *Tongues* and *Grooves* are intact. The wood has already shrunk as far as it is going to shrink, so you won't end up with gaps between the boards after the lumber dries out. Hardwood floors can be difficult to remove, but are sometimes available at little or no cost, which makes them worth the effort.

Under no circumstance should you use any wood with evidence of termites, carpenter ants, rot, or any type of bugs, even if it's free. You'd be surprised what a small portion of the total cost is due to the lumber that forms the structure of the house; about 20 percent. And, except for the foundation, the structure is the worst place to scrimp.

Appliances, Hardware, Plumbing Fixtures, Windows, Doors, and Tools. These essentials provide the best opportunities for saving money. They are usually available at surplus and second-hand stores, or garage sales. Again, you must develop a discriminating eye or you can end up with a rusted water heater, an old stove missing certain vital irreplaceable parts, or a door that won't quite fit anywhere in your house. Expect to make a few mistakes and learn as you go; I made my share. However, my successes were a $5 toilet, a $10 cast iron bathtub, a $50 refrigerator, some beautiful wood-mullioned windows at $5 each, leaded glass french doors (a "gift" from Deb's parents), a free 12 ft × 18 ft braided rug, and several doors with hinges and knobs for $10 each. I know you'll find this amazing, but you can get just as clean in a $10 tub as in a $600 tub. The extra $590 is for prettiness. For an additional $600 you can get your tushy swizzled as you bathe.

After awhile, you'll develop a "trap-line" of surplus stores, liquidators, Salvation Army outlets, and second-hand stores. Just swing through them every once in awhile to see if there's any good stuff. These places are more than a place to save money, they are a test of your ingenuity, creativity, and the limits of your imagination. Very seldom will you leave the place with the item that you came for, but you almost always leave with something.

Buy in Large Quantities Whenever Possible. A large order reduces a store's overhead. All they have to do is pick up a truckload (of plywood, for instance) and deliver it to you. That is much cheaper than selling the plywood piece by piece out of the store, talking to each person and writing up an invoice for each sale. Don't be afraid to ask for a price quote on large orders.

Support the Small, Local Stores Whenever Possible. Often, the service and advice small, local stores provide is worth the slight extra price. And you help to keep the world from becoming one large corporate conglomerate that you may have to work for some day.

Small stores will often negotiate price if they know you are building a house and will be a regular customer. My local lumber store met the lowest quote of five much larger outfits on my first $4,000 order. That was quite generous I thought, and I made it a point to do most of my follow up business there—another $6,000 to $8,000 worth of smaller orders. I think it worked out well for both of us. In addition, some stores have a homebuilders discount policy, whereby you can set up an account with the store and get a 10 to 15 percent discount.

Take Advantage of Sales. Most of the basic tools, hardware, plumbing, and electrical supplies will be on sale at 20 to 50 percent below regular cost at some time while you are building. This comes from the ability of larger stores to buy in mass quantities. When the price differences between the local store and a discount house reach 25 percent, I compromise my "buy local" ideals. I had to buy very few tools at full price because I planned ahead and bought them before I needed them. I could count on a 30 to 40 percent off sale on every tool I needed at some time during the year. One store had such great prices on wire that I bought it 2 years before I installed it (at one-third the wholesale price 2 years later). In total, I figure I saved about $3,000 by watching the sales.

A Note on Wholesale Prices. Some people go to a lot of trouble to get a contractor's license so they can get wholesale prices. I found that: 1) most discount houses had sale prices less than wholesale prices for the very same quality materials, and 2) some wholesalers were willing to sell to individuals at wholesale prices as long as you had a substantial order for them.

And About My Mistakes . . . I bought an electrical service panel on sale. Two years later, I discovered that I needed a different type of service panel, so I got stuck with something I didn't need. On wallboard, I bought some a few months early and found that it was such a pain to store and keep dry that it's much better to just buy it when you need it. Besides that, the prices started going down on wallboard right after I bought it—yes, *down!*

That came as a real surprise to me. I was so conditioned to prices going up, up, up, that I never thought of them going down. Some of this is seasonal. Most houses get built during the good weather seasons, which makes for less demand and lower prices in the winter. But the main factor is the condition of the housing industry. Right now, the high interest rates and slow market are keeping the professional builders from building. So 2 × 4 studs are cheaper today (99¢ each) than they were four years ago ($1.65 each). I paid $4.50 a sheet for wallboard a year ago and now it is $2.98 a sheet. Building during an economic slump can save you a lot on materials.

Finishing

When you're working with lumber, the costs aren't too bad. You can keep very busy and productive for a month on about $1,000 worth of material. But when you get around to finishing the house, you can spend money like crazy. For example, it only takes about two days to put in 10 light fixtures, an oven and range, a living room rug, and some louvred closet doors. The cost will be in the $3,000 to $5,000 range. You run out of money fast at that rate. You can also settle for 10 porcelain light fixtures (75¢ each), an electric frying pan ($20), a throw rug from the second-hand store ($10), and curtains on the closets ($5). You've still made it into your house, but you'll have to turn to your imagination rather than the bank to make it look homey.

An element of the character-building part of housebuilding is that after you've done your darndest to build a nice house, your house may not be as smooth and elegant as a lot of professionally built homes. This is particularly true if you operate on a pay-as-you-go basis. Face it; it's hard to make a 75¢ porcelain light fixture look as good as a $100 brass fixture. But at least you're not paying interest on the $100, and in time such items are easily upgraded.

How Much Money Do You Need?

You probably have a general idea of the house you want. For example, perhaps it's an 800 sq ft house, functional but unfinished, and built with about one-third salvaged materials. Let's say one full year is the limit of time you can spend building it, which roughly translates to about a $15,000 house. If you know you just couldn't live in a house with plywood floors and bare light bulbs for a year, you're talking about $5,000 more in finishing touches. In addition, there's the property cost. For the sake of example, let's say the land will cost $15,000. Now the primary obstacle is pretty well defined: In this example, you would need to come up with $30,000.

An Unconventional Approach

Face to Face with Reality

The key to financing was explained to me by a hitchhiker. I was down in the mouth, driving home from a beautiful old barn on three acres overlooking a bend in the Green River. We *really* wanted that barn! It would have been a dry place to live, and Deb and I figured we could turn it into something striking and unique. And what a peaceful, picturesque setting. But I had just been outbid by a company that wanted to turn it into a recreational trailer camp. Egad! a trailer camp full of people and litter and noise! What a lousy fate for such a beautiful place. I needed someone to talk with and this hitchhiker looked like he might have a sympathetic ear.

I don't remember most of the conversation. I think I did most of the talking and he was polite and understanding. I remember him lighting a hand-rolled cigarette as I talked, and I was surprised when it smelled like tobacco. Calm and cheerful, he made the point that you don't need to own a place to enjoy it. But if you want to put a lot of blood, sweat, and tears into a place, it's nice to be the legal owner so you can benefit from your efforts and decide when you want to leave, if ever. That means working within the confines of the "supply and demand" economic system. If someone else wants the same place you do and is willing to pay more for it, you lose. As we pulled up to his trail through the woods, his parting words were, "Yeah man, it always seems to get back to the same old thing . . . money talks, hot air walks." I laughed and agreed. He thanked me for the ride and disappeared up a trail through the woods.

It was funny how that catchall expression rang true. Not so much at first, but later. Money talks, hot air walks. You certainly can't talk someone into selling their barn and three acres for $5,000 less than someone else will pay. If you really want something so bad that you spend months or years dreaming about some little place in the woods or by a river, you pretty much have to buckle down and come up with some money to buy such a place. The first step for us was the grim realization that we were never going to have a spot to call home if we didn't figure out a way to get some money. The question that everyone faces is how to come up with the money.

Where Do You Get the Money?

This is where we took an unconventional approach. No need to sell 30 years of your life to the bank. No need to run dope from Colombia. Possibly the need to get a job that is hard work, but no need to work at something you don't believe in (whether it be in a whale rendering plant or in some other environmentally destructive industry). Our secret approach, that so few people utilize, is called "Save your money and pay as you go."

During the 4 years prior to building our house, our combined incomes averaged $17,000 annually. We were able to save an average of $7,000 per year for our "own home" fund, which was enough to buy 2 acres and get the house to the point of being complete on the outside. The inside of the house was paid for from hand to mouth as the money was earned. Instead of a $500 monthly house payment, the money went for plumbing, electrical, insulation, and wallboard. The initial savings plan took a good deal of careful management and some revision of our spending habits. We lived in a modest apartment, drove modest vehicles, and enjoyed inexpensive recreational pursuits. Also, it was financially fortunate that one of our inexpensive recreational pursuits didn't result in any children until recently.

You might think that living on $8,000 (after taxes) a year would make for a terrible penny-scrounging time. Once we adjusted, it wasn't all that bad. In fact, we had a great time cross-country skiing during those winters and spent the summers hiking and enjoying the outdoors. We count those years among the best we have known.

A Financial Philosophy

Now I warned you that you might not like my financing strategy. After all, most people borrow money when they need it. They'll borrow as much as the bank or the store will lend them. From my perspective, the only time I should need to borrow money is when I'm living beyond my means. That is, I'm so eager for immediate materialistic gratification that I have to go out and beg for money from someone else. Yeah, I do it too—credit cards and loans and revolving spending plans. Through interest, the lenders charge me an arm and a leg to rent their money until I can earn some of my own. I don't blame them. I'd expect the same if I made a loan to them.

In the case of a house, most people consume far beyond their current resources. And of course they want a nice flashy house finished with fake plastic beams, texturized plaster, and soft, cushy carpets. It gives other folks the impression that they are doing very well financially, or at least as well as their friends. They want it so badly that they're willing to pay $273,000 over a 30-year period for a $60,000 house; a house they could build themselves for $30,000 if they wanted to work that hard. Salespeople give them a line about increased future earning power and the effects of inflation and before you know it, they've signed on the dotted line. Nice, sensible people like you and me do this every day.

If you think I sound opinionated on this subject, you're right! Lest I get carried away and write this entire book on the virtue of frugality and the sacrifices caused by spending beyond your means, I'll let up right here if you'll do one small favor for me. The favor is this: Take out your income tax forms or W-2 forms for the past 5 years and add up the total amount of money you've earned. Do it now, it'll just take a minute. Then look around and see what you have to show for 5 years of work. Perhaps lots of fine wine and food delighted your tongue, perhaps some sharp looking clothes graced your body, and maybe a lot of time was spent in a shiny car that now has a hole in the upholstery and is burning a little oil. There were many dashing evenings of entertainment, some great memories of that trip to Timbuktu, and of course several wheelbarrows of greenbacks went to various landlords. Any money spent on interest is long gone. Incredibly, there's hardly a trace left of all that dough except possibly a few durable goods—tools, furniture, musical instruments, sporting equipment, maybe a typewriter or camera. If you had put even half of that money into a piece of land or a house on a pay-as-you-go basis, what would it be worth today? I know, I know, I won't rub it in. But you get my drift, don't you? Do you plan to continue along the same trajectory for another 5 years or can you see a way to benefit from your earnings for a longer period of time?

Start Saving Now

Decide how much you are willing to put into a house. Based on your income and a no-frills budget, set up a savings plan and stick to it.

As you save, put your money into high-yielding money market certificates or certificates of deposit whenever possible. These certificates are available from your bank and typically earn 8 to 17 percent interest, depending on prevailing conditions. At 15 percent interest, a sum of money doubles every 5 years. A regular savings account may earn only 5½ percent interest, and it takes your money 13 years to double at that rate. (By the way, don't confuse money market certificates with stocks, bonds or mutual funds—the latter are much riskier investments. Usually the risk isn't worth taking no matter how good the sales pitch is.)

After a couple years of serious saving, you'll have enough money to buy some land and get started on the house. Those 2 years can be spent locating the right piece of property, designing the house, and preparing the building site. If you want to save for an additional 2 or 3 years, you'd be in a much better position to finance yourself to build a fairly exotic home. Granted, 5 years is not a short time if you want a house right now. But I remind you that 30 years is much longer than 5 years.

If you get impatient, you can usually get a bank or a credit union to match your funds with a small mortgage. About half the homes in this book were financed that way, and the owners feel that it was the only way they could have gotten started. The drawback to this approach is that you may have to undergo the scrutiny of the bank officers, meet certain progress inspections, build a fairly conventional house, and work for several years to pay off the mortgage. That is sometimes an acceptable price to pay in order to get started sooner.

No matter what strategy you adopt, the bottom line in the subject of financing is to start saving early. Like right now!

Financing Summary

- Know how much it costs to build a house and where the money goes.

- Set your sights on a house that you can realistically afford.

- Know where you can save money when building.

- Come to grips with the fact that "money talks, hot air walks."

- Adjust your living habits and start following a savings plan so you have some money to talk with.

CHAPTER 3

Selecting Property

CHAPTER 3

Selecting Property

Land. The most basic ingredient. A miniscule corner of Mother Earth to call home. The one place that can be made the way you want your world to be. The very cornerstone of America's social order is that everyone has the opportunity to own a piece of old *terra firma*.

Lest I slip and wax poetical, let me get back to my now hard-nosed, bruised, and practical self. Most folks approach the land question too emotionally, too superficially, and move much too quickly.

"Oh Harold! This is the place. It's beauuutiful! We can build our own cozy house here, raise our kids, and live happily ever after. Oh Harold! Have you ever laid eyes on such a gorgeous spot?"

As only writers can do, I've clamped a 4-lb machinist's vise on Harold's tongue and I will respond for him.

Horsefeathers, Maude! First of all, if the place looks that good already, all cleared and landscaped, you can't afford it. Secondly, you can't know if the property is a good place to live, a reasonable buy, or even a legal house site without a lot of digging for information. The list of pitfalls is endless, and the line of realtors after their 10 percent of selling price is twice as long. Add the smallest dash of enthusiasm from you and the situation is dangerous. I can already hear the agent saying "Trust me. It's a good deal and it won't last long." Finally, the kids could care less about the scenery. They want to know how far it is to the nearest video game parlor.

I'm sorry Maude if I sound so cynical. There are lots of occasions to be warm, trusting, enthusiastic, happy-go-lucky, and uninformed. But buying land is not one of those occasions. Warm and trusting, fine. But also cautious, informed, and aware that the realtor is primarily an agent for the seller.

Now that I've done my duty to prevent the slaughter of innocents, I'll try to stick to the facts with my very best positive attitude.

Finding the Right Place for You

There's plenty of time to select a lot. It's a key decision. In fact, it's the most important decision you'll make on this home-building project. By starting to look at property *before* you finally buy, you can begin to learn the market, the language, and your personal requirements for land. A good place to start looking is in the classified ads of the Sunday newspaper.

Look Before You Buy

Look for a good location and an extra nice piece of property where you will enjoy living for years to come. If it has a rickety little house that you can live in while building or remodeling, great! But the quality of the land and the location are the important things. As the years flow by and you gradually use your time and energy to build or improve the house and grounds, you'll be continually more pleased with the results.

Deb and I spent a lot more on land than we originally intended, which turned out to be one of our best decisions. Several friends have done the same. After the enormous amount of work that goes into a house is done, you'll be happy to have selected a place where you'll be satisfied for a long time.

Shopping for Land Is a Personalized Learning Experience. What aspects of land appeal to you? Notice how a lot is about ten times more appealing on a sunny, summer day than on a rainy, leafless winter day. Notice how a cleared lot with a few fruit trees commands a significantly higher price than a similar, brambly lot that has a better southern exposure. Notice how a 20-mile drive to a place seems trivial on a Sunday afternoon. But imagine what it will feel like at 6:30 AM every work day for the next 10 years.

Learn to Ask the Right Questions. Where are the property boundaries? Is there an available water supply? Is the water company allowing any more connections? If not, are there year-round, good tasting wells in the area? (This helps to predict your chances of drilling a successful well.) Is the sewer company allowing any more connections? If there are no sewers, the percolation of the soil (the rate at which it absorbs water) is crucial to getting a building permit. Have percolation tests ever been done? Would the county building department issue a building permit for this lot? Be sure to include this very important clause in the Earnest Money Agreement: . . "subject to the buyer being able to obtain preliminary approval for a building permit." (I'll explain Earnest Money Agreements in a moment.) Be aware that slide hazards, a high water table, soil that won't perc, building setback requirements, zoning, and other conditions can make an otherwise perfect lot unbuildable.

More questions: How much is the owner asking? Is the price negotiable? Is the owner interested in a cash sale? What is the zoning in the area? Are there any encumbrances to the property? (A glossary of terms like *encumbrances* is coming up.) Does that stream overflow during the spring run-off? Are there any development plans in the neighborhood? Like a new high school next door, tract houses, logging, a superhighway, and so on.

If you don't jump into the first or second deal you see, you'll get pretty good at asking questions.

Learn to Be Observant. What does the soil look like? Is it sandy or gravelly enough to percolate, or does it look like clay? Check the soil along the cut next to the road. How about the vegetation? Any swamp grass? (Swamp grass indicates a high groundwater level, hence no septic system will be allowed.) Is there a view? Do the neighbors have a dirt track for motorcycles? Will the neighboring homes make the neighborhood a good investment? What is the exposure to the sun and to the wind? Pick up a newspaper and see what's going on in the community.

After you spend a month of Sundays looking at lots, you'll start homing in on your likes and dislikes about potential home sites.

Write Down What You Want in a Home Site. Examples might be a view; seclusion; less than 20 miles to work; certain neighborhood characteristics; close to schools, relatives, steelhead fishing, parks; appropriateness for your house design (solar, log, and so on); and space for a horse. If you like to do things scientifically, devise a rating system by giving a point weight to each criterion. Then each piece of property can be rated on a point basis, giving you a quantitative method to help you make the decision.

List the Properties You've Examined. Categorize the properties you've examined by view lots, land with streams, suburban lots, or any way that reflects their comparative value. Include the list price, size, location, and any comments to help you remember the place. With a written list, you'll have a quick reference to determine if an asking price reasonable.

Check It Out

When you think you've found the right place, check it out thoroughly using the authorities given in Table 3-1 before you sign any offers. Also, make sure you spend time at the property and try to talk with the neighbors.

I know. We're talking a good deal of research, time spent on the phone, visiting county offices, and playing detective. A gambler won't bother with it, but any point on the checklist could turn up a very good reason for selling a piece of land at a low price. For example, you might not want property that's about to be flooded by a dam, that's downwind of a smelter or pulp mill, or that borders a loud Nuclear waste generator.

The last item on the checklist is the fun one. Pack a picnic lunch and visit the property some Sunday afternoon. See if you can strike up a friendly conversation with a neighbor about the community and the history of the property. While learning about which wells go dry in the late summer and unusual local conditions, you'll also discover what the neighbor is like. Deb and I met Paul and Sue this way. It turns out that having Paul and Sue as neighbors adds immeasurably to the value of the property. Sue makes chocolate chip cookies that are good enough to kill for, and Paul provides banjo music for all the neighbors within earshot.

TABLE 3-1 Factors to Check Before You Buy Land

Checkpoint	Information source
Water supply	
Availability and cost	Water district/company and well driller
Quality	Health department
Soil percolation	Health department or test it yourself
Property boundaries	County assessor, measure it yourself, or survey records
Assessed value, taxes, zoning, and sales history	County assessor
Building permit possible?	Building department
Impending changes in the area	County/state offices of land development, roads, parks, flood control
Logging operations	Timber companies
Power or gas lines	Power and gas companies
Is property title clear?	Title insurance company
Aerial photos	State department of natural resources
General	Spend time at the property
	Talk with the neighbors

The Real Estate Language

The following is a short treatise on the foreign language called "Real Estate." It's enough to get you started and you'll learn more as you deal with the natives.

Land is located by section, township, and range. A **section** is an area containing 1 sq mi (640 acres). For example, the abbreviation S 16, T 24 N, R 3 E designates the 640 acres in Section 16 (S 16), located on the map by Township 24 North (T 24 N) indicated at each side of the map and by Range 3 East (R 3 E) indicated at the top and bottom of the map. Each set of Township-Range coordinates contains 36 sq mi, thus 36 sections.

Further legal talk describes the particular portion of the section where the property lies. (Fig. 3-1 shows an example.) To locate a property on the map, first find the section it is in, then *read the legal description backwards* locating each successively smaller quarter.

To throw the nonnatives off the track we'll abbreviate the example as S 330 ft of SW ¼ of NE ¼ of NW ¼ of S 16, T 24 N, R 3 E. Irregularly shaped lots often are described by a "metes-and-bounds" description, which you can pa-

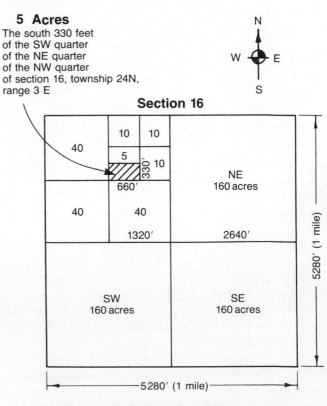

5 Acres
The south 330 feet
of the SW quarter
of the NE quarter
of the NW quarter
of section 16, township 24N,
range 3 E

Section 16

1 Acre = 43,560 sq ft, or an area about 208' × 208'

Fig. 3-1 *Legal description of 5 acres.*

tiently decipher if you remember your high school geometry. Since this system gets cumbersome in cities where the lots are small and irregular, most cities have a platting system. A typical city lot might be described as Lot 7, Farnesworth Plat, S 16, T 24 N, R 3 E. Maps are available to the public at the county assessor's office.

The **title** or **deed** is a piece of paper stating the name of the owner(s) of a piece of property. Some deeds have so many exceptions or conditions to them that they are virtually worthless. A **statutory warranty deed** is the best kind of deed. It says the persons named are the undisputed owners of the land.

Title insurance is an insurance policy written by a company (preferably a well-established company) which guarantees your title to the land or reimbursement for your losses if any dispute over the title arises. The reasons for title insurance are too numerous to cover here but let me be blunt: You're a fool if you don't get title insurance when you buy land. The seller normally provides this and pays the fee. If the title insurance company won't insure the title, don't buy the land.

A **title commitment** or **title report** is a report from a title insurance company in which the company commits itself to provide a title insurance policy, subject to specified terms and exceptions, at the time of closing.

Closing is the actual payment of the purchase price, conveyance of the title by notarizing it in your name with all proper signatures, and recording of the property in your name at the county courthouse.

The **Earnest Money Agreement** is an agreement, usually with very fine print, that you will sign when you make an offer on a piece of property. Get a copy from your realtor and study it. There is a variety of agreement forms which have different terms but they all work basically this way:

> You offer some earnest money to show that you mean business. One hundred dollars is plenty unless you expect competition from another buyer. The earnest money goes toward the purchase price when the deal is complete. If the seller agrees to sell, but you change your mind, you forfeit the money to the seller and the realtor.
>
> You offer to buy the property for a certain price, paid for with a certain amount of cash at closing and with certain financing conditions for any remainder owed.
>
> The seller agrees (upon signing the agreement) to sell the property to you at the price and terms specified. The seller cannot sell to anyone else during the term of the agreement.

Some Precautions

- Add and delete clauses in the contract. *You* are the one taking the big risk here, not the seller or the agent. Don't hesitate to cross out half of the terms on their "standard form" and add your own terms so it reads the way you want it. Better yet, try to locate a form that has the purchaser's interests at heart. You can reduce some of the risk by adding clauses like "agreement subject to an available water supply acceptable to the purchaser." Similar clauses for an "acceptable title report," "acceptable soil percolation," and "contingent upon the purchaser being able to get preliminary approval for a building permit" are common among knowledgable buyers. Beware of sellers who balk at such clauses . . . they may know something that you don't. Be willing to put a time limit on yourself for resolving the water question and finding out if the building department will let you build on the lot.

- Specify a time limit for the sellers to respond to your offer. You'll be on pins and needles waiting for the seller to respond. Specify 10 or 15 days for them to decide. They could hold out for months waiting for a better offer. Meanwhile you are still on the hook until the agreement expires.

- Be sure the seller will provide title insurance.

- Specify a Statutory Warranty Deed to the property. Other deeds have strings attached that may not be acceptable.

- Specify that no liens or encumbrances exist other than those listed and agreed to by the purchaser.

- Cross out any prepayment penalty clauses. Such clauses require you to pay interest over the full term of the contract, even if you pay the contract off years earlier than required.

- Don't leave any blanks on the form. For example, unless you pay cash, the sale will involve a mortgage or real estate contract. Study the contract form to be used and specify the terms and interest rate. A 1 percent hike in the interest rate can cost you thousands of dollars. Verbal agreements mean nothing and there's usually a statement to that affect in the agreement.

- Specify who pays which closing costs.

- If you discover facts that make you decide against the property, go ahead and forfeit the earnest money.

An **encumbrance** is anything that affects or limits the title to the property, such as an easement, mortgage, or restriction of any kind. Examples include a previous owner who has sold a 40-ft easement to a neighbor to drive his logging trucks through the property. Many times the power company has an easement to put power lines across the property.

A **lien** is a type of encumbrance which makes the property security for payment of a debt, mortgage, or taxes. For example, until you pay off the contract or mortgage on the land, there is a lien against it. If you miss a payment, the lender can repossess the place. If you sell the place before it is paid off, you've never really gotten clear title. Sometimes this can lead to a whole chain of owners, each making payments to the previous owner. Get good legal advice on how to buy a property with a chain of owners. In such cases, the title insurance policy is extremely important to you.

There are a lot of fingers in the pot at closing, to the tune of a few thousand dollars. These are the **closing costs.** Typically the seller pays for the title insurance, excise tax, and the realtor's fee. Typically the buyer and seller split the escrow fee and apportion the property taxes according to the calendar date. The buyer pays the fee to have the transaction legally recorded with the county.

Escrow is the deposit of legal papers and funds with a neutral third party who sees that the provisions of a contract or agreement are carried out. For instance, you buy a lot on a 10-year contract that requires payments of $200 a month. The escrow company holds the title to the property, the contract, and takes your $200 payments and forwards them to the seller. When you've paid off the contract, you get the title. By the way, you want **true escrow,** which costs a little more but gives you more protection than a collection account (which might be called escrow).

Make It an Enjoyable Experience

My approach to buying land isn't for everyone. Many folks find it easier just to find a reputable realtor and rely on their judgment. I'm extra cautious because I'm the type of guy the New York cab driver takes for a 3-mile trip to go 6 blocks. Last week the tow truck driver towed Deb's dead VW for a tour of the town before bringing it home and presenting the bill. (Until then, he seemed like a guy you could leave your wallet with.) Those kind of unpleasant surprises I can usually absorb, but in a high-stakes situation like buying land I leave as little as possible to chance.

Some friends and acquaintances haven't been so cautious. One couple was shown a nice piece of property but actually bought (by legal description) the swamp next to it. Countless folks have bought land from someone who didn't really own the land. If they had insisted on title insurance they would have discovered the problem. Many people buy land and then discover that they can't get a building permit for the lot. These mistakes are easily avoided but you have to be concerned enough to learn the basics of real estate and check the facts.

On the flip side, you don't want to be so hard nosed that you alienate the realtor and the seller. After some practice it becomes easy to be friendly and firm at the same time. Then it gets to be really an enjoyable way to spend Sunday afternoons—looking for land and dreaming of locating your ideal homesite. Hang loose and just remember three easy rules:

1. Check out the property carefully

2. Only sign agreements that you fully understand

3. Take your time

Much of the knowledge I have about selecting and buying property I learned from Wrench and Joan, who learned it in the school of hard knocks. Their generous attitude toward sharing information reflects the type of people who build their own homes. A driving force behind this book is to help perpetuate that attitude. Some references that are helpful are listed below.

For More Information

Get the Facts—Before Buying Land, U.S. Department of Housing and Urban Development Pamphlet HUD-183-1, 1978. Washington, D.C. 20402. Check with the real estate division of your state government for a manual published to guide real estate brokers and sales people. Many states provide these at a nominal cost.

Know the Soil You Build On, Agriculture Information Bulletin No. 320, 1967. U.S. Government Printing Office, Washington, D.C. 20402.

Lawrence, James, ed. *The Harrowsmith Reader,* Camden East, Ontario: Camden House Publisher Ltd., 1978.
Includes the articles "The Good, the Bad, and the Swampy," "The Fine Print," "Decisions, Decisions," and "Happiness Is a Rundown Farm."

Part Two
DESIGNING

CHAPTER 4

An Approach to Home Design

Wrench was telling me that we'd probably be happier with our own house design than with any of the standard plans that the magazines offer.

"But how am I gonna design a house?" I complained. "I may have some areas of expertise, but architecture isn't one of them."

"Well there's no time like the present to start learning," he retorted.

"Listen, Wrench, you're an engineer. All this stuff about stresses and strains and heating systems is kid's play to you. We normal folks have to rely on the experts. And to make matters worse, every expert has a different opinion on everything from roof styles to the weight of the hammer you should use. It's really confusing, and there are a lot of things I'm not qualified to decide."

I followed him into the garden where he was setting up a new-fangled irrigation system for his raspberries.

"Baloney Frank. When it comes to your own home, there is only one qualified person to make the decisions, and that's you. You get all the knowledge you can and then you decide what you want based on your own opinion. I've always figured that opinions are like belly buttons; you know, everyone has one, they're all different, and mine is probably just as good as anyone's."

It always took me just a minute to digest Wrench's bizarre brand of logic. While I was contemplating this latest bit of wisdom, this mind-stretching analogy, he disappeared in a row of raspberries. I still hadn't come up with a good answer when he reappeared. In fact, 5 years and a lot of building experience later, I'm pretty sure Wrench was right. It turns out that there are literally millions of ways to design a house and millions of ways to build it. I always started out hoping to find some world renowned expert to tell me exactly how I should do each step. Sometimes I could find an expert to give me an opinion, but as I grew more knowledgeable and more confident it seemed that I was always happier with the results when I'd followed my own feelings and thoughts.

That leads me to this approach to house design. Indeed, the whole approach of this book. Learn all you can from books, friends, and experts each step of the way. Then pluck up your courage and make your own decisions. Be it the shape of the house or the type of paint to use, the successful owner-builder is the one who has spent some time educating himself on the subject and has followed his own tastes and inclinations.

In this book we'll design and build a typical unordinary house like you or I might build. I call it our sample house. It's not a simple sample house, but it's not complicated either. There are a few kinks to it, like all owner-built houses have. After you've seen one house designed and built, you'll be in a pretty good position to embark on your own home. You'll undoubtedly need to do some more specific research in some of the books referenced on various aspects of your special house.

One last warning. Don't spend so much time researching that you never get started on the project. If you start feeling weak-kneed or inadequate about making a decision, just gather a little more information and pull out Wrench's axiom: Your opinion is probably just as good as anyone's.

Suggested Design Procedure

This procedure worked well for us. Feel free to use the parts you like or change them to better suit your needs.

1. Look around. Look at friends' and relatives' houses, home shows, books, magazines, and houses on the street. Learn to be observant. What makes a house look good? The location of the house on the lot, landscaping, window style, entry, roof lines, and so on. Notice how

houses are basically a box or series of boxes with a roof on top. Cut out pictures from magazines and pin them up where you can see them every day. Figure out what house style fits you.

2. Start a list of what you want in your house.

3. Sketch some of your ideas. You don't need to be an artist, just portray the basic shape. Figure 4-1 shows some of our sketches.

Fig. 4-1 Sketches.

4. Select the best home site on your property. Figure 4-2 shows some factors you might think about. Make sure the site is on high ground, is accessible to water and power connections, has driveway access, and has room for a drainfield if there are no sewers to connect with.

5. Draw up some floor plans that make the most of the building site. Make several layouts using all the imagination you can muster. Remember that you'll spend most of your time inside the house, not outside looking at it. Think about view, natural lighting from the windows or skylights, traffic pattern in the house, good use of floor space, and heating and plumbing considerations. For example, high cathedral ceilings make a house difficult to heat. Showers that are a long way from the hot wa-

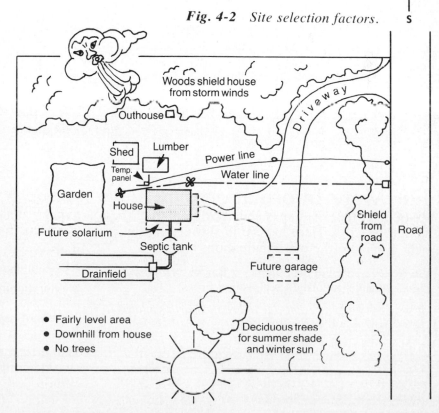

Fig. 4-2 Site selection factors.

ter tank are bad news on cold mornings. All plumbing fixtures have to drain downhill to the septic or sewer connection. Showers, clothes dryers, and stovetops all need exhaust vents and are best put on outside walls. Be sure to plan for future expansion. The easiest way to make floor plans is to use graph paper and let each little square represent 1 sq ft. Figure 4-3 shows a layout for our sample house.

6. Educate yourself on design basics and house construction methods. Then figure out the house structure and build a balsa wood model of it. (See Chapters 5 and 6.)

7. Decide how to heat, plumb, and wire the house. (See Chapter 7.)

8. Draw the final plans. Also make an estimate of the materials, cost, and time required to build the house. (See Chapter 8.)

9. Make continual design modifications as you go along. Of course, sometimes changes will be impossible. But don't overlook any opportunities to make this "the best house ever" as you proceed.

Well, there you have it. A nine-step procedure leading from total ignorance to an excellent house design. It's fun, interesting, and darn near free.

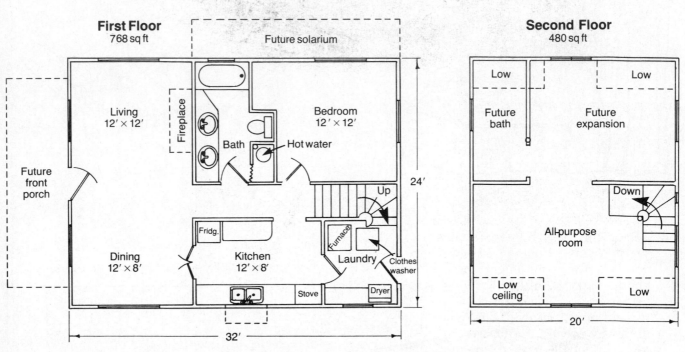

Fig. 4-3 *Floor plan for the sample house.*

For More Information

Remember that reading is an important part of the design process. Here are some sources that will help stimulate the old imagination:

Boericke, Art, and Shapiro, Barry. *Handmade Houses*. San Francisco: Scrimshaw Press, 1973.

Brand, Stewart, ed. *The Next Whole Earth Catalog*. New York: Rand McNally, 1980.

Bruyere, Christian, and Inwood, Robert. *In Harmony With Nature*. New York: Drake Publishers, 1975.

Kern, Ken. *The Owner-Built Home*. New York: Charles Scribner's Sons, 1975.

Leitch, William C. *Hand-Hewn*. San Francisco: Chronicle Books, 1976.

Roberts, Rex. *Your Engineered House*. Philadelphia: J.B. Lippincott Co., 1964.

Shelter Publications. *Shelter*. Santa Barbara: Mountain Books, 1973.
To order this book by mail, write to Mountain Books, Box 4811, Santa Barbara, California.

Sunset Editors, ed. *Kitchens; Planning and Remodeling*. Menlo Park, Calif.: Sunset-Lane Publishing Co., 1976.

CHAPTER 5
Designing the Structure

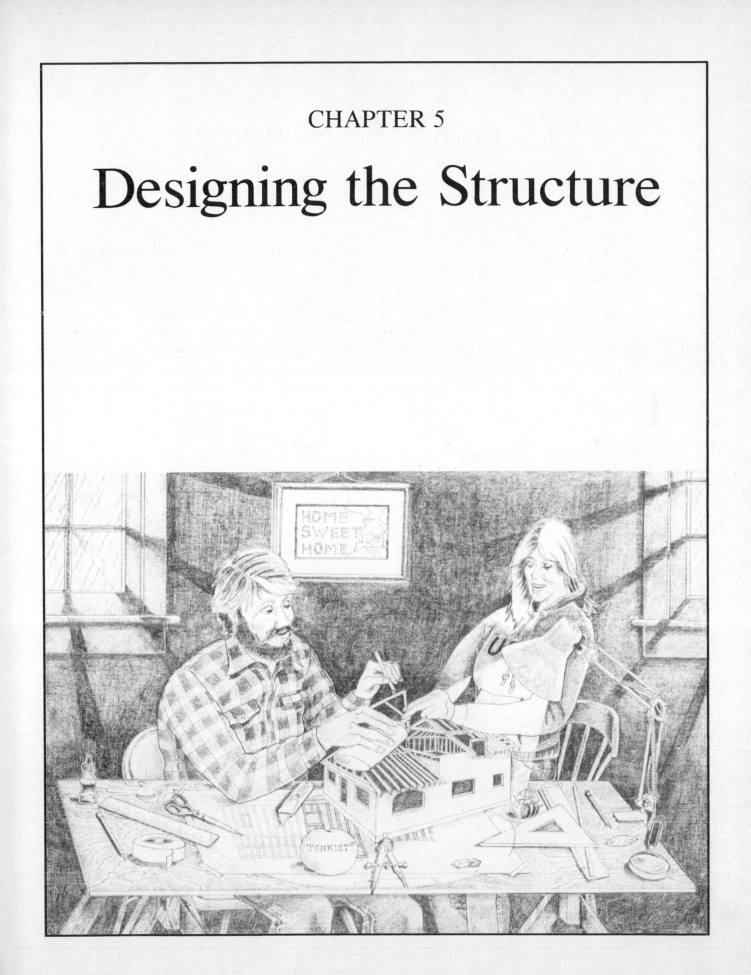

CHAPTER 5

Designing the Structure

With a sketch of our house and a floor plan in our hands, we're ready to start designing the structure. Basically, we'll pour a continuous concrete foundation for the house to sit on so we never have to worry about anything rotting beneath and letting the house sag. Building on stumps or untreated wood posts is guaranteed to cut the life of the house in half. The floor will be made with planks (called joists) set on edge and securely blocked so they can't fall over, and covered with plywood (called the subfloor). Then we'll build exterior walls around the floor and put interior walls wherever necessary to support the second floor and the roof. Once the roof is on, we've got the essential structure of a house.

It may be hard to visualize this if you've never closely observed a house being built. So hop on now for a quick visual tour through the conventional wood frame building process as shown in Fig. 5-1.

Fig. 5-1 The conventional wood-frame building process.

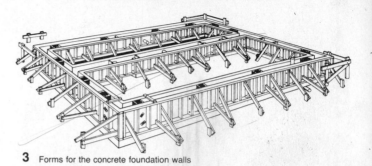

3 Forms for the concrete foundation walls

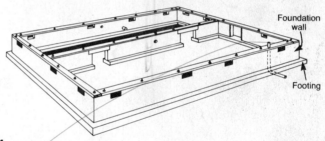

4 Completed foundation

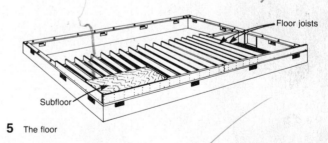

5 The floor

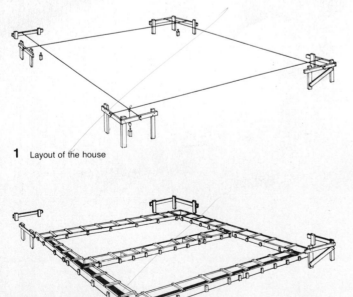

1 Layout of the house

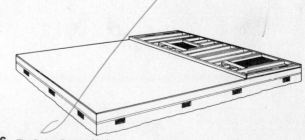

2 Forms for the concrete footing

6 The first wall assembled

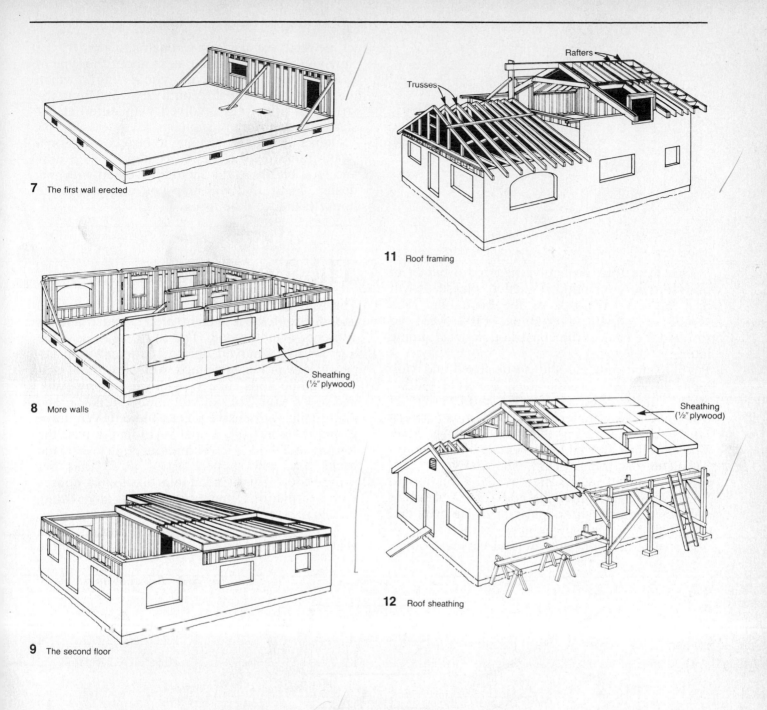

7 The first wall erected

8 More walls

Sheathing
(½″ plywood)

9 The second floor

10 Second-story walls

11 Roof framing

Trusses

Rafters

12 Roof sheathing

Sheathing
(½″ plywood)

13 Roofing

Tar paper

3-Tab asphalt
shingles

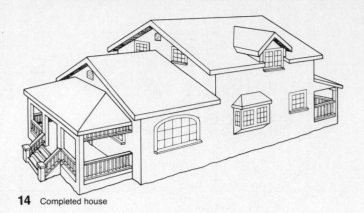

14 Completed house

let me refer you back to Wrench's axiom. By the time you've looked at and read about the type of houses that interest you, you'll know which method fits you best. You'll also find that the structural principles involved with different designs are surprisingly similar.

Our sample house is going to be a conventional wood-frame design. Let's go step by step now and look briefly at the factors involved with this design. Figure 5-2 will help you understand the house plans we'll be using.

This little tour probably triggered a bunch of questions in your mind. Patience, please. It will probably take the rest of the book, and then some, to explain everything. First we'll go through the basics; the construction details come later.

This same house could be designed and built using posts and beams, logs, concrete blocks, adobe, tilt-up concrete panels, and a variety of other ways. For numerous reasons, most modern houses are built using the method shown here. The materials are readily available and economical. The method is fast and relatively easy. There's a lot of information available on how to do it. However, other methods of construction (geodesic domes, yurts, underground houses, and so on) all have their advantages. Rather than give you a sales pitch for one method versus another,

The Foundation

The foundation holds the house off the ground and distributes the weight of the house and its contents over the soil. The topsoil (at least the top 6 in) is removed during the excavation because it won't carry much weight without settling. Some soils, like wet clay, swampy soils, and loose sand are not suitable to build on and may require expensive pilings to be driven down to more stable soil. Usually you'll dig past the topsoil and find a combination of dry dirt and rocks that will support about three tons per square foot. This is adequate to support houses with continuous foundations without any settling or sagging.

A foundation has two parts: a **footing** and a **wall.** The footing gives a wide base to spread the

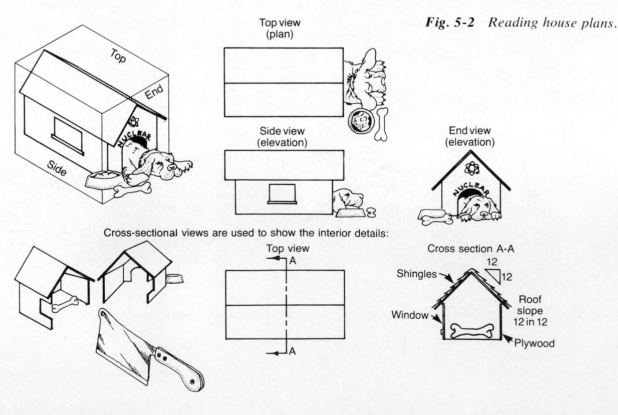

Fig. 5-2 *Reading house plans.*

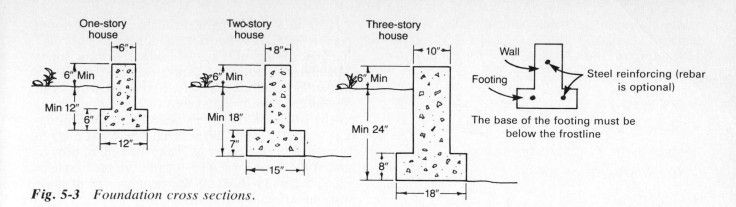

Fig. 5-3 *Foundation cross sections.*

weight of the house over a large surface. The wall raises the house off the ground at least 6 in to keep the wood dry and away from all the bugs, worms, and other wood-eating critters. The base of the footing must be below the frostline to avoid heaving and cracking when the ground freezes up. Figure 5-3 shows the minimum dimensions for foundation footing and walls.

In unusually cold areas, the footing must go much deeper than shown here. Check with your building department. Also, steel reinforcing is required in some areas and is always a good idea. If the concrete cracks during an earthquake for example, the steel "rebar" still holds the concrete in place.

The height of the foundation walls depends on the amount of space desired in the crawl space or basement. If this is going to be your first concrete pour, limit the wall height to three feet or get some experienced help. Most codes (the *Uniform Building Code* is the most universally used) require a minimum of 18 in of crawl space under the floor and 12 in under any girders. Plan on at least

24 in if you are going to be the one who has to slither around down there installing the plumbing, wiring, heat ducts, and insulation. If the lot is sloped, check Fig. 11-2 to see how to make a stepped foundation.

Our sample house will have a foundation per Fig. 5-3 for a two-story house. Long pieces of reinforcing steel will be wired into the forms before we pour the concrete. The site is level and the crawl space is 24 in. The footing will be poured first with short pieces of ½-in steel rebar sticking out of it to connect the footing to the wall. (Ref. Fig. 11-8). After the footing concrete cures, the wall will be poured on top of it. The completed foundations will look like Fig. 5-4.

Notice the vents that allow air to circulate under the floor. There must be at least 1½ sq ft of vent for every 25 ft of perimeter foundation. Hence, our sample house needs about 7 sq ft of vent, or fourteen 6 in × 12 in vents.

Also notice the access hole to the crawl space (min 18 in × 24 in), the J-bolts to anchor the treated planks (called mudsills) to the foundation,

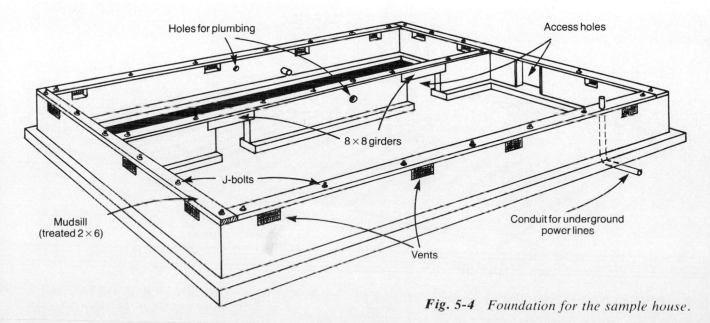

Fig. 5-4 *Foundation for the sample house.*

and the holes for the sewer line and underground electrical service.

In summary, the *key factors for foundation design* are:

- Soil type
- Slope of the site
- One-, two-, or three-story house
- Depth of the frostline
- Crawl space height
- Floor-joist span, which we'll discuss next

The Floor

Floors are designed to carry loads. The load of people, furniture, bathtubs, and the unexpected must all be supported without sagging, creaking, groaning, or complaining.

A floor is designed to withstand a certain load, usually 40 pounds per square foot (abbreviated 40 psf). I use 60 psf as a design load in this book because I like strong floors. You might well wonder if a 60 psf floor would cave in if a 200-lb person stood on 1 sq ft of floor. No—because the 60 psf figure is a rating for the entire floor. For example, a 10 ft × 10 ft room has 100 sq ft of floor, and would withstand a 6,000-lb load without hardly bending. The chances of having more than thirty 200-lb people in such a small room is pretty remote.

How does a nonengineer design a 60 psf floor? Very simple. Just build a conventional floor like the one shown in Fig. 5-5 with the floor joists spaced 16 in center to center (abbreviated 16 in on center or 16 in O.C.) and covered with ¾-in plywood subfloor. The size of the joists depends on the distance they will span between foundation supports, and that size can be looked up in Table D-2 in Appendix D. Take a moment now to look at the floor joist span for the sample house in Fig. 5-5 and to look up the proper size of floor joists in Appendix D.

This is probably a good time to explore the Appendices. Notice that the floor joist Table D-2 qualifies itself by saying that you must use a wood specified as "structural" in Table D-1. That's right, no balsa wood or willow allowed. It also says the wood has to be graded No. 2 or better, which requires a look at Appendix A to discover how lumber is graded. (The grade is stamped on each board.) While we're rummaging around in the Appendices, note in Appendix A how the plywood used for subfloor will be stamped "Underlayment." Appendix C tells us what type of nails to use to nail the floor together and Appendix B describes the different types of nails specified. We'll go through all this in more detail in Chapter 12 when we're building the floor, but it's useful now just to know the Appendices are there and can answer a lot of questions.

The floor for our sample house measures 24 ft × 32 ft. I chose to run the floor joists the shorter direction, with one intermediate support in the middle so the floor span is 12 ft. This allows me to

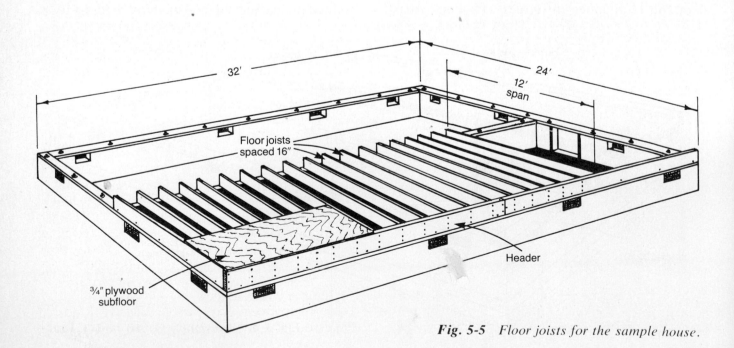

32'
24'
12' span
Floor joists spaced 16"
¾" plywood subfloor
Header

Fig. 5-5 *Floor joists for the sample house.*

use 2 × 10 floor joists. Had I tried to run the floor joists the other way, the only way I could have covered the 32-ft distance would be by designing a foundation with two intermediate supports. Floor joists usually span the shorter distance, in order to keep the size and cost of the joists to a minimum and to avoid unnecessary intermediate supports.

Now take a look at the floor girders in Fig. 5-4. One of these girders supports not only the floor that rests on it but also the wall that rests on the floor, which in turn supports the second floor and ceiling. Table D-3 in Appendix D tells us that if this girder spans 4 ft, it should be a 6 × 8 or larger. The other girder only supports one floor and could be a 4 × 6 per Table D-3, but I specify 8 × 8's in both cases because they rest on an 8-in foundation wall. Incidentally, in carpentry terminology **girders** are usually wood (although sometimes steel) and serve as part of the foundation to support the floor joists. **Beams** are also usually wood and serve to support anything above the foundation. Hence an 8 × 8 installed below the first floor is a girder, but installed elsewhere in the house it will be called a beam.

That pretty well covers floor design. For the sake of making efficient use of materials (particularly plywood), it helps to make the overall floor dimensions a multiple of 4 ft, like 20 ft × 28 ft or 24 ft × 32 ft. If there are any floor openings or porch overhangs, try to run them parallel with the floor joists as shown in Figs. 12-2 and 12-4. (By the way, it's a good idea to read this entire book before you finalize your house design, so there won't be too many surprises when you actually start building.)

In summary, the *key factors for floor design* are:

- The span and size of the floor joists, which are determined by the foundation arrangement

- The spacing of the floor joists, which is usually 16 in O.C.

- The type of subflooring, which is usually ¾-in plywood underlayment

The Walls

Just as the foundation supports the floor and the floor supports the walls, the walls in turn support the second floor, the ceiling, and the roof. The construction of the walls is standard and you'll see how they are built in a minute. Our main problem as designers, however, is to place the walls so they adequately support the roof and second floor, and to make sure there is a foundation under these walls to carry the weight down to the ground.

There are two basic types of walls in a house:

1. A **bearing** wall, which helps to support the structure (by bearing a load).

2. A **nonbearing** wall, which just separates two rooms and has no structural purpose. Nonbearing walls are often called partitions.

The bearing walls are the important ones, so let's look at them first. Look at Fig. 5-6 for a moment and study the way the walls support the second floor, roof, and ceiling. All of the walls shown bear a load. The walls on the first floor carry both the roof load and the second-floor load.

Bearing walls are all built basically the same way. The size of the header varies based on the span and load situation (see Fig. 5-7). The load of the roof or the second floor rests on the double top plate of the wall. The studs spaced every 16 in carry the load through the sole plate to the floor. Where there is a load over a window or door, the header bridges the gap and transmits the load to the trimmer studs that support it. The proper size of the header is determined by looking in Table D-4, Appendix D. For example, a header spanning a 5-ft window and supporting both a floor and a roof would be made of two 2 × 10's or one 4 × 10. For normal construction, plan on all exterior walls being bearing walls as well as those interior walls that carry the load of a roof or upper floor. Exterior bearing walls will be sheathed with ½-in CDX plywood.

Many people are building their walls with 2 × 6 (instead of 2 × 4) studs these days. In fact, your building department may require it. This is an energy conservation measure. The best way to improve house insulation is to make the walls thicker. With the rising cost of energy, the extra lumber cost is paid back in a few years with lower heating bills. In addition, walls of 2 × 6 studs are required on the first floor of three-story homes as a structural requirement. A 2 × 6 stud wall is built just as shown in Fig. 5-7 except with triple planks for headers. At the corners and wall intersections the blocking is doubled.

The nonbearing walls in a house are normally installed last since they have no structural func-

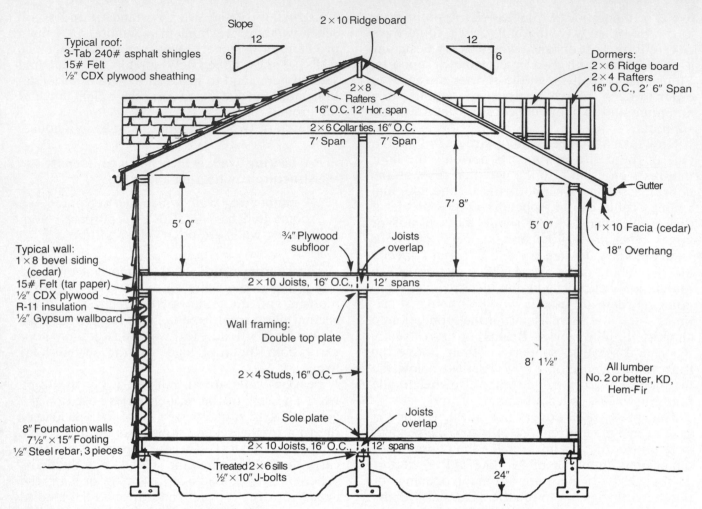

Fig. 5-6 *Cross section of the sample house.*

tion and just get in the way. No headers or trimmer studs are necessary and the stud spacing (16 in O.C.) is determined by where studs are needed to support the wallboard when it is nailed up. A single top plate is adequate, although usually a double top plate is used to aid in connecting the walls at the intersections.

Posts and Beams

A post-and-beam structure functions as a wall, in that it bears a floor or roof load. The beams must be large enough to bridge the spaces between the posts, while the posts carry the entire load. The advantages to posts and beams are that the space under the load is left clear (except for the posts), and often their appearance is striking. Their disadvantages are that they are not as laterally stable as a wall, and they require more muscle to put in place.

Once a designer decides on a post-and-beam structure, the next question is how big do the

beams and the posts need to be? And what is the best spacing of the posts?

First the load carried by the beam must be calculated. For example, we could add a 10 ft × 10 ft sundeck to our sample house accessible from a door in the south dormer. The floor of the sundeck could be supported by a beam and two posts on the outboard side and by the house wall on the other side. The porch beam carries half of the porch load and the house wall carries the other half. The design load on the porch would be 100 sq ft multiplied by 60 psf for a total of 6,000 lb. The beam carries 3,000 lb and spans 10 ft. From Table D-7, Appendix D we see that a 4 × 10 beam would be adequate. In Table D-8 we can determine that 4 × 4 posts, adequately braced and connected to the beam, could support each end. Each post would carry a 1,500-lb load. Furthermore, if the beam were bolted to the posts, (rather than the better method of setting the beam on top of the posts), the number and size of bolts necessary could be determined from Table D-9. Three ⅝-in bolts would carry the 1,500-lb load, although the post would have to be larger than a 4 × 4 in order to space the bolts 3 in apart.

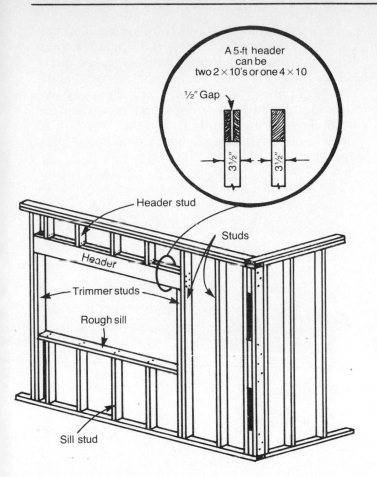

A 5-ft header can be two 2 × 10's or one 4 × 10

½" Gap

3½"   3½"

Header stud

Studs

Header

Trimmer studs

Rough sill

Sill stud

Architecture Isn't So Hard

By now you're probably getting the hang of designing a structure. If you need a floor joist size, you go to the floor joist table, header sizes come from the header table, beam sizes from the beam table, and so on. Of course it's not entirely simple because you have to be sure you don't leave some part of the structure dangling in mid-air without any adequate support. And if you tangle with a normal building code book you'll be faced with an array of lumber species and grades, each with a series of modulus of elasticity values, repetitive- versus engineered-use conditions, maximum extreme fiber bending stresses and horizontal shear stresses, not to mention several hundred pages of fine print that all tends to boggle one's mind.

The tables in this book are for simple and normal housebuilding conditions. Their validity depends on whether you use quality lumber, follow

Fig. 5-7 *Anatomy of a bearing wall.*

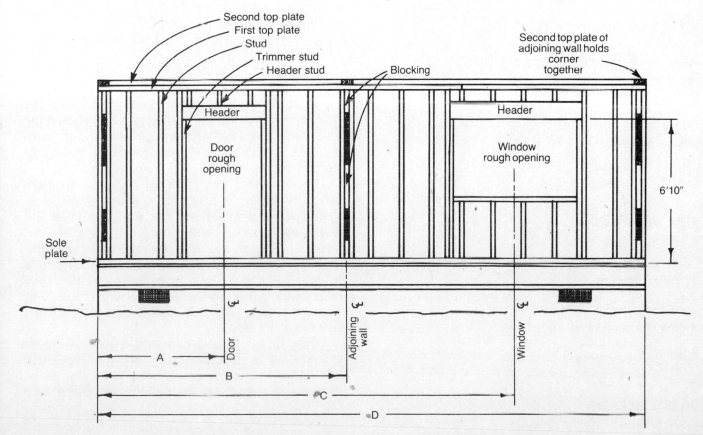

Second top plate
First top plate
Stud
Trimmer stud
Header stud
Blocking

Second top plate of adjoining wall holds corner together

Header

Header

Door rough opening

Window rough opening

6'10"

Sole plate

Door

Adjoining wall

Window

A

B

C

D

the instructions in this book, and above all use common sense. The design loads of 60 psf for floors and 50 psf for roofs are safe, conservative loads under normal conditions. If you live where 3 ft of snow piles up on the roofs, or if you want to drive your car on the floor, you'll have to get the design professionally engineered to withstand those extreme conditions.

Finally, call up your building department and get a summary of the local requirements. When you submit your building plans, there are bound to be some corrections but it helps to know most of the requirements early in the game.

Roof Structures

The number of different types and shapes of roofs is huge. Often the entire character of the house is determined by the roof style. Figure 5-8 shows a few examples.

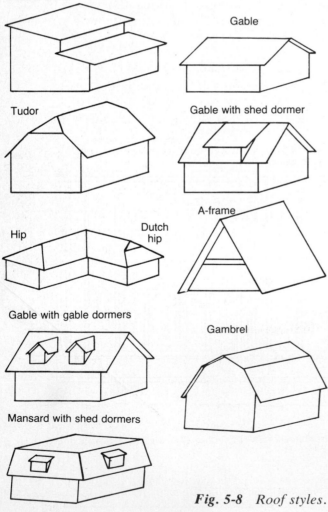

Fig. 5-8   Roof styles.

The easiest, quickest, and often most economical roof structures are made with manufactured trusses. A quick look in the yellow pages under "Roof Structures" or "Building Materials" and a couple of phone calls will net you a catalog or two showing all kinds of trusses and how they are used. Trusses commonly have 4 in 12 or 6 in 12 slopes, although many shapes and slopes are available (see Fig. 5-9).

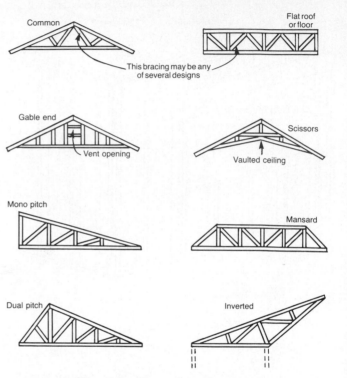

Fig. 5-9   Examples of trusses.

The design of truss roofs is easy. Call up the truss company, tell them the span and roof slope you want, and they'll tell you which truss type to use. As usual, special conditions may require special engineering.

The design of custom-built roofs (roofs built rafter by rafter) is primarily a problem of determining the proper rafter size for a given span and slope. Also, rafter ends must be properly connected to the walls, and ceiling joists (or collar ties) must tie each rafter pair together in a triangle to prevent the rafters from pushing the walls over. Take a look at the drawing with Table D-6, Appendix D to see how rafter and ceiling joist sizes are selected.

While rafter spacing can vary from 12 in to 24 in and roof beams supporting planked roofs can be spaced up to 6 ft apart, we'll limit ourselves to the most common 16 in O.C. situation (see Fig. 5-10).

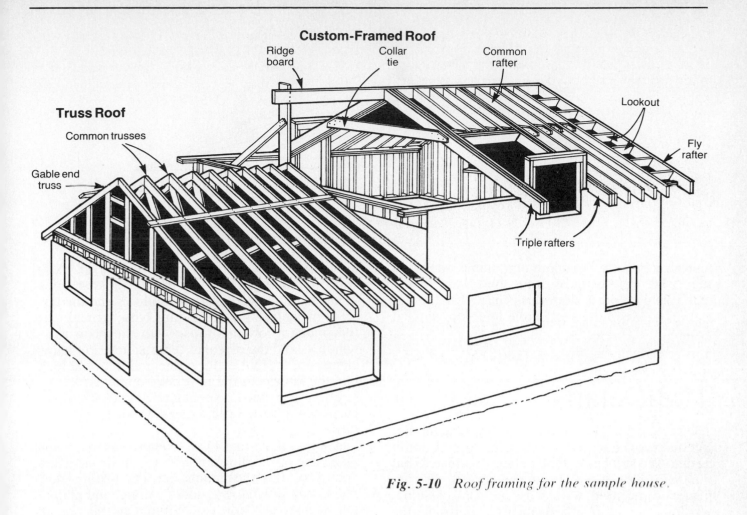

Custom-Framed Roof

Ridge board · Collar tie · Common rafter · Lookout · Fly rafter

Truss Roof

Common trusses · Gable end truss · Triple rafters

Fig. 5-10 *Roof framing for the sample house.*

For our sample house, we'll use trusses over the one-story part. Since the roof over the second floor has dormers, we'll go to a custom-built roof there. I run into some typical design problems here. In Fig. 5-6 you can see that the second-floor walls are only 5 ft tall. That's because an 8-ft wall would make the house taller than suits my taste. Still, I want usable space on the second floor and I'd prefer not to bump my head on the ceiling every time I come upstairs. (Code requires 6½-ft clearance over stairs and landings.) So I go with 5-ft walls and tie the rafters together with collar ties at 7 ft 8 in from the floor. Forget about the standard designs, this house is going to be unique.

The gable dormers are another way I put my personal touch on this house. They provide natural light to the upstairs and they appeal to my anachronistic aesthetic proclivities. (I like them.) Since I'd rather avoid the normal time-consuming roof framing that large dormers entail, I'll limit their width to 5 ft and frame them like a roof opening for a large chimney. The triple rafters on each side function as beams to carry the load of the roof and dormer structure.

The rafters provide another design experience. While Table D-6 will only allow a 2 × 8 to span 11 ft 10 in, by diving into my Uniform Building Code I see that if I use Hem-Fir* rafters I can span 12 ft. Also, I can expect only minor snow loads in my climate and I plan to use asphalt shingles, rather than slate or tile, for roofing. I'm confident that 2 × 8 rafters are plenty strong enough, and I don't want to go to 2 × 10's because they are kind of awkward to handle when working alone on a second floor roof like this.

I select 2 × 6's for collar ties, though Table D-5 shows that 2 × 4's would be adequate to span 7 ft. That's because I need room for 6 in of insulation in the ceiling. The entire roof will be sheathed with ½-in CDX plywood to make it rigid. Though it took a bit of design work, I end up with a roof that is both strong and pleasing to my eye.

Although you'll probably determine the slope of the roof according to the appearance that you want, another constraint is that certain roofing

---

* Hem-Fir is a grading designation for certain species of hemlock and fir, by which lumber strength is indicated. See Appendix D, Table D-1.

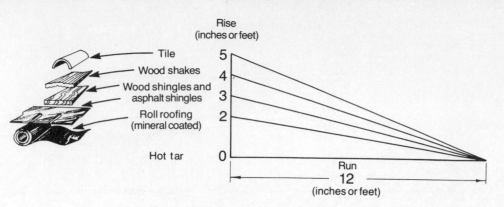

Fig. 5-11 *Minimum slopes for certain roofing materials.*

materials require the minimum slopes shown in Fig. 5-11. For instance, you don't want to use wood shakes on a slope less than 4 in 12. For maximum life of the roof, avoid low slope and flat roofs.

The Details

The sample house is nearly complete structurally. Foundation, floor, walls, and roof. Poof! A house design! We still have some details to attend to but the major part of the design is complete. For appearance and style, we can use any kind of siding, windows, and doors that appeal to us. Inside, the cupboards, counters, wall covering, floor covering, light fixtures, and plumbing fixtures can be worked out at our leisure. Additions like porches, greenhouses, large stone fireplaces, sundecks, and elegant stairways will be planned to fit this basic structure.

I remember finding this part to be somewhat amazing. "You mean to tell me that all these houses are structurally identical?" I asked. Dad and I were looking over a row of houses in the old neighborhood.

"That's right kid. They're not exactly the same of course. You can see how some have the entry located on the side and with others it's on the front. The location of the large living room window varies. The floor plans may be different, but the roof structures, bearing wall positions, and foundations are the same. The white house and the brick one are mirror images of the others."

"How can a brick house be built like a wood house?" I wanted to know, visualizing Practical Pig building his wolf-proof house in my childhood storybook.

"Just stack up bricks instead of nailing on siding. The bricks seldom are structural, they're just a facing on a wood wall supported by an extra wide foundation. See that rustic looking brick house with the leaded glass windows and the turret at the entry? It has the same basic structure as that modern looking house with the plate glass windows and the diagonal cedar siding. Put a low slope roof on that cedar home and it would look like the houses in the new developments."

For all the time I'd spent living in houses, I was surprised at how little I knew about their structure.

Another thing that I had to learn was that when designing or building a house, the basic structure goes fast. But the details like the turrets, dormers, stairs, windows, doors, siding, and plumbing, seem to take forever. Without getting into all the gory details, the stairway is a good example of a time-consuming design detail.

Stairs

First, let's talk about normal stairs. Section 3305 of the *Uniform Building Code* specifies that, among other things, stairways must be at least 30 in wide, have 6 ft 6 in of headroom, maximum step rise is 8 in and minimum step run is 9 in. Winding and spiral stairs have special requirements. As you look at various flights of stairs you'll develop your own guidelines. For example, I prefer stairs to be at least 36 in wide and no steeper than a 8 in rise and a 10 in run (see Fig. 5-12). With my guidelines, no matter where I put stairs in a house I need about 40 sq ft on the first floor and 36 sq ft on the second floor. The floor opening must be a minimum of about 3 ft × 9 ft.

For the sample house I have this picture in my mind of a nice stairway with a bend in it and a pretty stained glass window partway up the stairs. This arrangement raises havoc because it causes more floor joists to be severed than can be accommodated with a standard floor opening run-

Typical Stairs

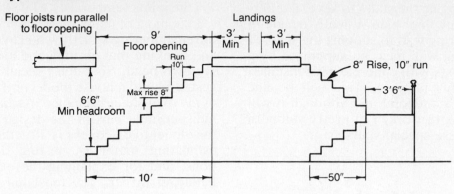

Stairs for Sample House

Stair plan

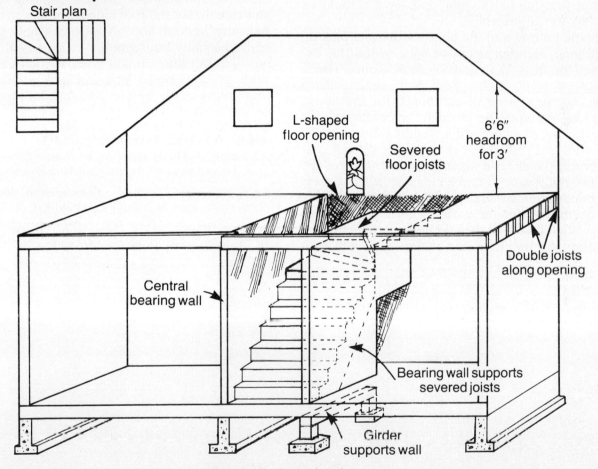

Fig. 5-12 Stair details.

ning parallel to the floor joists. After a great deal of thought and moving the stairs to several different locations on my floor plan, I finally decide to put in a short bearing wall to support the severed joists. The bearing wall must be supported by a girder and two posts with concrete pads installed specifically for this purpose. This will be fine from a structural standpoint and although it will cause me some extra work, I'm pretty stubborn about wanting these special stairs.

Building a Model

Now for the fun part! We'll build a scale model of the house using balsa wood and see if it looks as good as we've imagined.

Hustle on down to the hobby store and pick up some long, slender pieces of balsa wood (for example, $1/16$ in $\times$ $1/8$ in and $1/16$ in $\times$ $1/4$ in). You'll also need some white glue, pins, and a sharp knife. Using a piece of cardboard for the house floor, push the pins up through the cardboard and into the balsa wood walls, holding them erect. The wall and roof surfaces are made to scale using graph paper (one square represents 1 sq ft). It's not necessary to make every stud and rafter, but be sure the general proportions of the house are accurate. Get the major wall and roof lines established first, and then fill in the details. Model building requires care and patience. It will take several evenings because the glue dries slowly. Pin the joints temporarily while the glue is drying.

Like most of mankind, I suffer from the malady of remembering exactly who gave me bad advice and thinking that all the good ideas originated in my own head. So I don't recall if someone suggested I build a house model or if it was a product of my own sheer brilliance. At any rate, it was an indispensable part of the design process. I used the model to discover my structural mistakes, for estimating materials, testing design modifications, and for explaining the excavation to the bulldozer driver. The most important use however, was to get a good idea of what the house was going to look like. Deb and I spent a lot of time out in the yard holding the model up against the developing house frame, squinting our eyes and visualizing the roof slopes and porch designs. Ask any builder how many times he's gotten a house partially built and the owner says "Hmmm boy, I didn't know it was going to look like this!" With a scale model you can look at the house from every angle before you start building it.

For More Information

Anderson, L.O. *Wood Frame House Construction*. U.S. Department of Agriculture Handbook 73, 1970.

Ching, Francis. *Building Construction Illustrated*. New York: Van Nostrand Reinhold Co., 1975.

Wagner, Willis. *Modern Carpentry*. South Holland, Ill.: Goodheart-Willcox Co., 1976.

CHAPTER 6
Structural Analysis: A Primer

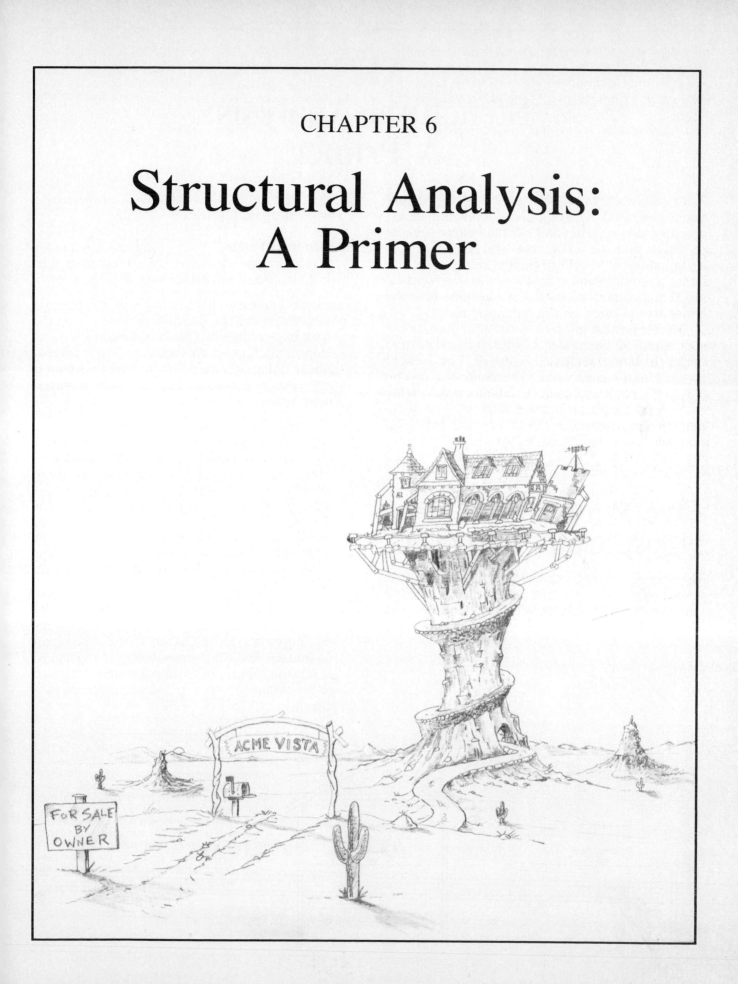

CHAPTER 6

Structural Analysis: A Primer

A lot of blood, sweat and tears will go into any house you build. To see a house with good ideas, careful craftsmanship and overall beauty sagging or tilting because of a structural oversight is a crying shame. If every builder spent just a few hours to understand a few structural principles, the thousands of hours he or she puts into the house would often end in a happier result.

Trouble is, though, most structural literature is such a put-off because it is written in exhaustive (and exhausting) technical language. The average owner-builder lasts about 15 minutes before he closes the book and decides to figure it out on his own. What follows is a translation of "Structure-talk" geared for the nonengineer and heavy on practical examples.

Five Ways Your House Can Fall Down

The way I see it, there are five main ways that your house, or a part of your house, can fail structurally. A structural failure is when any part of the house cannot hold its shape and sags, breaks, squishes, snaps, falls over, or otherwise gives up. Let me illustrate:

1. **Bending Failure.** Any long horizontal or slanted member* that is not strong enough to hold its load† will either sag or break.

2. **Shear Failure** (whacked off). A member or connector may be literally sheared off by the load imposed on it. This is often caused by an impact such as an earthquake. To visualize a shear failure, think of what would happen if the bolts and nails in Fig. 6-2 were replaced with pencils or toothpicks.

3. **Compression Failure** (crumpling or squishing). A member can be cracked, crushed, crumpled, or squished by the weight it carries. To visualize a compression failure, think of what would happen if part of the foundation in Fig. 6-3 was replaced with a pencil or an egg. To witness a compression failure, stand on an upright aluminum beer can with one foot and tap the side of the can with your other foot. (Careful now, don't sprain an ankle!)

4. **Tension Failure** (snapping). A member can be pulled apart or snapped by opposing forces pulling on it. Tension failures are not common in houses, but they are possible. To visualize a tension failure, think of what would happen if the ceiling joist in Fig. 6-4 was replaced by a string.

---

* A **member** is a structural piece of the house. Usually it is a beam, board, or post. It can also be a steel I-beam, a concrete structure, a cable, or a piece of plywood. Even a pencil, an egg shell or a string can be thought of as a structural member, though they are much too weak to be used in a house.
† A **load** is a weight or force put on a structure (see Fig. 6-1). Ten 200-lb people put a 2,000-lb load on a floor. If the floor itself weighs 500 lb, the total bending load on the supporting structure is 2,500 lb.

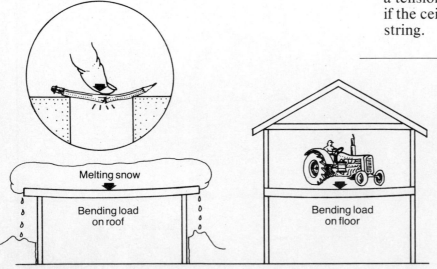

Fig. 6-1 Bending load.

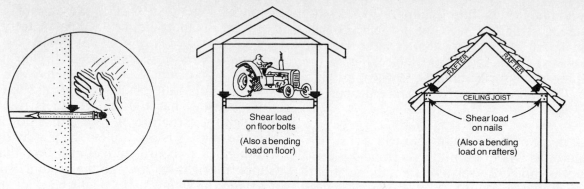

Shear load
on floor bolts

(Also a bending
load on floor)

RAFTER   RAFTER

CEILING JOIST

Shear load
on nails

(Also a bending
load on rafters)

Fig. 6-2   Shear load.

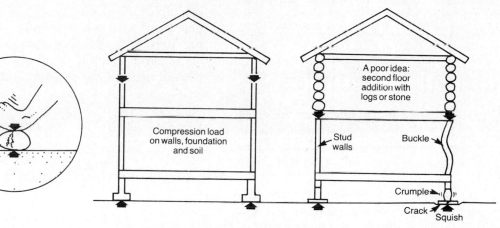

Compression load
on walls, foundation
and soil

A poor idea:
second floor
addition with
logs or stone

Stud
walls

Buckle

Crumple

Crack

Squish

Fig. 6-3   Compression load.

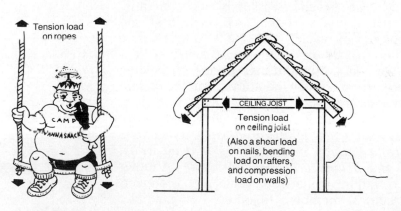

Tension load
on ropes

SNAP!

CAMP

I WANNA SNACK

CEILING JOIST

Tension load
on ceiling joist

(Also a shear load
on nails, bending
load on rafters,
and compression
load on walls)

Fig. 6-4   Tension load.

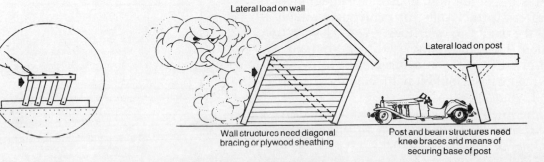

Lateral load on wall

Lateral load on post

Wall structures need diagonal
bracing or plywood sheathing

Post and beam structures need
knee braces and means of
securing base of post

Fig. 6-5   Lateral stability.

5. **Lateral Stability Failure** (falling over). A wall or similar structure may not be rigid enough to maintain its shape when a large lateral (sideways) force is put on it (see Fig. 6-5). Nothing necessarily breaks, it may just fall over. Or it may distort, twist, and pull apart at the joints.

Yes, a nuclear blast would also cause a structural failure, as would termites, a fire, and possibly an unrestrained 2-year old. I'm talking about the more predictable stuff. Now if you've got a feel for the most common types of structural failures, we can discuss ways to avoid such problems in your house design.

Structural Principles

I believe that a small amount of engineering knowledge and a good supply of common sense is all that's required to design a solid house. Here are four common sense principles that will guide you.

Principle 1: Carefully evaluate the possible loads that each part of the house may have to carry and design accordingly. A major cause of structural failure is loads that were not anticipated by the designer. Examples are melting snow loads, heavy roofing material, cast-iron bathtubs full of water, water beds, things stored on a floor (wallboard, lumber, magazines, car parts), remodeling additions, groups of dancing people, earthquakes, windstorms, and the weight of the structure itself.

Principle 2: Mentally go through your house design making sure all the joints and connections are strong enough and checking that all walls, posts, and beams are laterally stable. Another cause of failure is inadequate means of connecting structural members together. For example, omitting the knee braces on post and beam structures, inadequate connection of floor members, not using diagonals or plywood to prevent walls from distorting or leaning, inadequate attachment of the rafters to the walls, or inadequate attachment of the walls to the floor. A good way to think about lateral stability is to ask yourself if this part of the house could survive the force of a car backing into it. Could it withstand the force of a 70 mph wind?

Principle 3: When in doubt, design the structure about twice as strong as you think it needs to be. The cost of materials to make a house strong is only a fractional part of the cost of a house. "Strong" is more important than "pretty."

Principle 4: Learn to "see" how forces are transmitted through a structure. It helps to exaggerate. For example, mentally envision a giant 100-ton grapefruit placed smack dab on top of your house. That 100-ton weight is transmitted through the structure of your house down to the dirt the house sits on. Follow the path of that load through each board and connection that must support it right on down to the ground. Take special note of where there might be a lot of stress and predict which parts of the structure may fail.

If you can understand and abide by these four principles, you can design a strong house. It's best to stick to conventional, tried-and-true building methods. Be thorough and use that good sense your mama was always talking about: "Walk *around* the puddles, not through them!"

Fig. 6-6 *The 100-ton grapefruit test on a weak structure.*

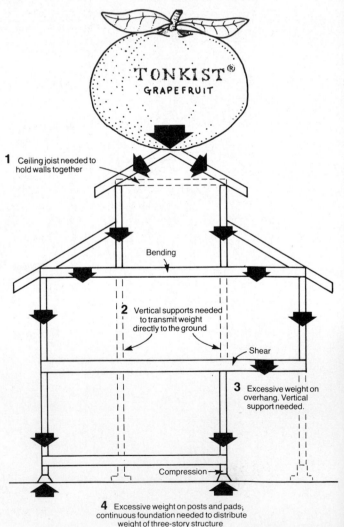

1 Ceiling joist needed to hold walls together

Bending

2 Vertical supports needed to transmit weight directly to the ground

Shear

3 Excessive weight on overhang. Vertical support needed.

Compression

4 Excessive weight on posts and pads; continuous foundation needed to distribute weight of three-story structure

To give you some practice in applying these principles, let's look at some specific weak points in a poorly designed house.

In Fig. 6-6 the rafters would want to separate from the walls first. If the rafters were well connected to the walls, then the top of the walls would begin to be pushed out, pushing the walls over. A ceiling joist or a collar beam would help to hold the walls together. Then the third story walls would groan under the compression load. If the walls didn't crumple, the third floor would bend, sag and break next. If the third floor was stiff enough, the weight would be transmitted down the walls to the second floor. Then the second-floor overhang would sag or collapse. If by some miracle the upper structure survived, the foundation posts would crumple under the weight. Finally the foundation pads would crack and the dirt under them would compress, causing the whole structure to sag or tilt.

Most 100-ton grapefruit find their way to the ground eventually. If you have trouble imagining a 100-ton grapefruit, think of 50 teenagers on your roof watching the hydroplane races. Actually, most types of roofing weigh more than 50 teenagers. It may help to take a look at some old, crumbling houses. When you build something new it seems real strong but it is surprising how time and Mother Nature can point out the faulty places in a house design.

A good way to see how your particular structure will react to a heavy load is to test the balsa wood model you made. Put a load on it and see what happens. The design of the house I built was tested at the breakfast table one Sunday morning. By subjecting the model to the load of, you guessed it, one large grapefruit, I discovered some interesting things. For instance, the rafters tended to push the walls over, which was corrected by lowering the collar ties. While it wasn't exactly a space-age laboratory testing technique, it did tell me what I wanted to know. You can probably devise even better testing techniques. Use a small bag full of sand instead of a grapefruit, for example.

How to Design a Strong House

We've talked about the types of loads and failures your house may have to resist. Now how do you make it strong enough?

For example, suppose you are concerned about how much the floor will bend when the room is full of people. From Chapter 5 you learned how to go to a table to look up the proper size of floor joists. The size of floor joists in Table D-2 are calculated so the floor will bend only $1/360$ of the span when carrying its design load of 60 psf. For example, a 10 ft × 10 ft floor with a 6,000-lb load (for example, thirty 200-lb fullbacks), would bend 10 ft ÷ 360 or less than $3/8$ in. That's so slight you probably wouldn't notice it.

To answer the question, how do we make the various parts of the house strong enough: We make the floor, roof, ceiling, window/door openings, beams, and girders strong enough by using the size of lumber indicated by the tables in Appendix D. All floor joists, rafters, ceiling joists, window/door headers, beams and girders will be subject mainly to bending loads. The walls and foundation will be subject to compression, shear loads, and lateral loads and we'll make them strong enough by using standard stud walls sheathed with plywood and by building a continuous concrete foundation reinforced with steel. We'll resist shear and lateral loads at the joints by using the nailing schedule in Appendix C to nail all the lumber together.

These tables and construction methods are tried and true guidelines that meet or exceed the requirements of the Uniform Building Code. You can rely on them. Wherever you expect your house to be subject to normal living loads, which include groups of people in a room, normal snow loads, and normal use of the house, you can play it safe by using the tables and standard construction methods.

For more unusual methods of construction or unusual load conditions (like underground houses or houseboats), you will have to do some research beyond what is covered here. By the way, you can build a very unconventional, kinky house using standard, conventional building methods.

So why all this structural information, you ask? Because if you understand structural loads and failures, how to identify weak points in a house design, and how to strengthen weak points, you'll be a lot more comfortable with *why* you will be doing certain things a certain way. You'll know why it is important not to cut big notches in the floor joists when plumbing, why the walls should be sheathed with plywood, and why there should be three nails holding each rafter end to the ceiling joist. And if you do venture into some unconventional building methods, you'll be less likely to make any mistakes.

Resisting Bending Loads

The normal way to reduce the amount a structure will bend is to build it with stiffer members.* For example, to make a floor stiffer, you would build it with stiffer floor joists. Figure 6-7 illustrates several ways to span 10 ft, each progressively stiffer. Please study them.

Notice that a 2 × 12 is as stiff as a 6 × 8, even through a 6 × 8 has twice as much wood in it. A 6 × 8 also weighs twice as much and usually costs twice as much as a 2 × 12.

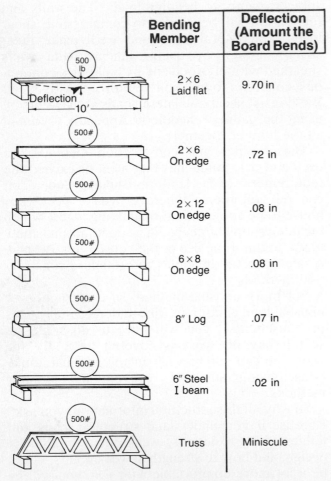

Bending Member	Deflection (Amount the Board Bends)
2 × 6 Laid flat	9.70 in
2 × 6 On edge	.72 in
2 × 12 On edge	.08 in
6 × 8 On edge	.08 in
8″ Log	.07 in
6″ Steel I beam	.02 in
Truss	Miniscule

Fig. 6-7 *Comparison of bending members.*

Bending Exercise

The floor shown in Fig. 6-8 spans 14 ft and is 10 ft wide. Refer to Fig. 6-8 to answer these questions:

---

* Another way to stiffen a structure is to space the members closer together.

1. What size floor joists would you use and how many? (Use Table D-2 in Appendix D.)

2. How large is the load that each crossbeam may have to carry? (Use a design load of 60 psf.) What size beam would you use? (See Table D-7, Appendix D.)

3. What do you think about lacing a 3 ft × 9 ft floor opening for the stairway as indicated by the dashed lines?

4. Would this floor be strong enough to support a 6-ft-diameter hot tub?

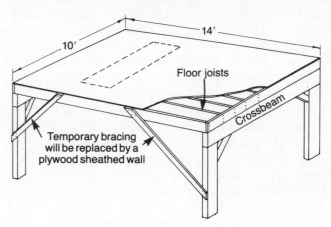

Fig. 6-8 *Bending exercise.*

My answers:

1. I would use nine 2 × 12's for floor joists spaced every 16 in. The last spacing would be less than 16 in. If the floor supported any walls running parallel to the joists, double joists would be needed under the walls.

2. 140 sq ft × 20 psf equals 8,400 lb total load. Each crossbeam carries half of the load, or 4,200 lb, and spans 10 ft. Using Table D-7, Appendix D, I would select 6 × 10 beams.

3. Placing a stairway as shown would sever darn near every floor joist in the floor. The stairway should run parallel to the floor joists and the joists next to the opening must be doubled up as shown in Fig. 12-2, Chapter 12.

4. Well, let's see now. If the water in the tub is 3 ft deep, the 6-ft-diameter (3-ft-radius) tub would hold: Volume of a cylinder = area of the circular base × height equals $\pi r^2 \times 3$ equals $3.14 \times (3)^2 \times 3$ equals 84.8 cu ft of water. A cubic foot of water weighs about 64 lb, so the water alone would weigh $84.8 \times 64 = 5,426$ lb. Since there would also be the possible 8,400-lb load of people, the pump and

equipment, and the weight of the deck itself, this floor wouldn't be strong enough. For one thing, the tub is a concentrated heavy weight that puts all the stress in one spot. (Most floor loads are distributed over the whole floor and are less dangerous.) If you put the tub right in the middle of the floor (where the bending stress is the greatest) you would have a dangerous bending situation. Even if you put the tub right over one of the 6 × 10 crossbeams, the beam is only rated for 4,700 lb *distributed* over the length of the beam. (See Table D-7 in Appendix D.) With the tub and the normal floor load we'd have about 10,000 lb on the beam which is far too much. If I just had to have the tub up on the deck instead of down on the ground, I would support the floor under the tub with a couple of short 4 × 12's running perpendicular to the floor joists and supported by four additional posts to transmit the weight down to the ground. In general though, I'd feel more comfortable with that 3-ton tub sitting on a concrete pad on the ground.

Note: This is what engineers do when they design a house. It involves a lot of figuring and calculating. You can't read this stuff as fast as you can read a good paperback novel, you know, so take your time and think *about things. A chapter every two nights is a good reading speed for engineering material.*

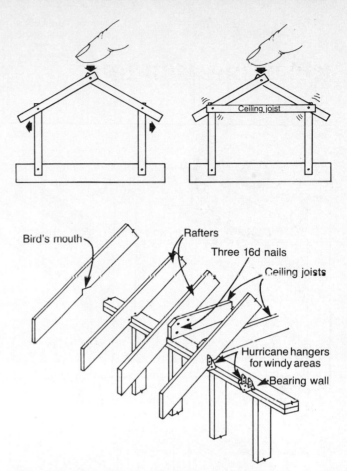

Fig. 6-9 *The rafter/ceiling joist/wall joint.*

Resisting Shear and Tension Loads

One common place that shear loads occur in a conventionally built house is at the junction of the rafter, the wall, and the ceiling joist. The most common place for tension loads is in the ceiling joist.

To understand this rafter/wall/ceiling joist joint, take a pair of scissors, some cardboard (or popsicle sticks), and some pins (or thumb tacks) and make the model shown in Fig. 6-9. Then push the peak of the roof with your finger and watch how the walls are pushed out and fall over. Your finger represents the weight of the roof.

Now add a ceiling joist to the structure and push again. Vive la difference! The structure now holds the weight as long as you push directly downward and put no lateral (sideways) load on the structure. You can see the stress at the joints, trying to shear off the pins and putting tension on the ceiling joist. You can use a rubber band for the ceiling joist if you want to see the tension stress better. Keep this model. We'll use it again in the section on lateral stability.

The tension load on the ceiling joist is not so great that we have to worry about the ceiling joist snapping. But you can see the problem that will arise if you neglect to put in the ceiling joists altogether. Normally you'll want ceiling joists in most rooms anyway so you have something on which to nail the ceiling wallboard (sheetrock). The main stress on the ceiling joists is a bending stress from the weight of the wallboard. The proper size of joist can be determined from Table D-5, Appendix D.

Now, about that shear load: How do you know how many nails are needed to hold this joint together? The easiest way is to follow conventional building methods. The nailing schedule in Appendix C shows that the ceiling joists must be face nailed to the rafters with three 16d nails. This is based on the requirements of the Uniform Building Code, which are based on the shear strength of nails and the normal loads in a house.

Cutting bird's mouths in the rafters as shown in Fig. 6-9 provides a larger surface to support the

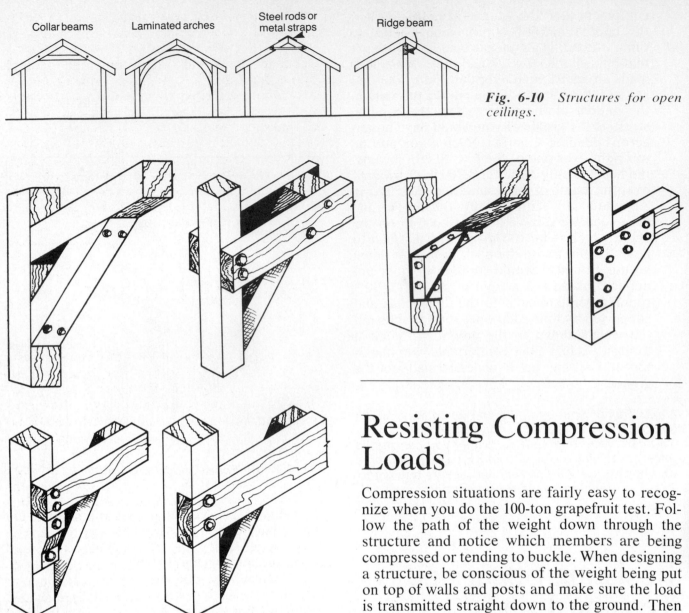

Collar beams Laminated arches Steel rods or metal straps Ridge beam

Fig. 6-10 *Structures for open ceilings.*

Fig. 6-11 *Post-and-beam connections.*

weight of the roof on the supporting (bearing) wall. It also allows enough room to toenail the rafters to the top plate of the wall so the roof won't blow off during a windstorm. Metal connectors called hurricane hangers should be used in areas where high winds are expected.

If your house design includes open or cathedral ceilings, there are no ceiling joists to hold the walls together so the roof load must be resisted in some other way. (Fig. 6-10 shows some examples.)

In post-and-beam construction (Fig. 6-11), the large shear loads are resisted by bolts and steel connectors at the joints. Knee braces are added for lateral stability.

Resisting Compression Loads

Compression situations are fairly easy to recognize when you do the 100-ton grapefruit test. Follow the path of the weight down through the structure and notice which members are being compressed or tending to buckle. When designing a structure, be conscious of the weight being put on top of walls and posts and make sure the load is transmitted straight down to the ground. Then select members that are strong enough to carry the load.

In general, if the **bearing walls** are of the standard construction described in Chapter 5 there won't be any problems. For three-story homes, the walls on the first floor must be built with 2 × 6 studs, rather than 2 × 4 studs. For walls and other structural parts of the house, use only lumber graded "Standard and Better." "Stud," "No. 1," "No. 2," "Construction," and "Select" are also acceptable grades. Do *not* use "Utility," "Economy," "No. 3," or "No. 4" lumber for any part of the house that is structural. Any lumber that helps to hold up the house is considered structural. For example, a bearing wall is structural while a nonbearing wall (or partition) is not.

Post strength is primarily dependent on the size and length of the post and the species and grade of the wood used. The longer a post is, the more likely it is to buckle. This is particularly true when the post stands alone and is not constrained by walls or braces.

Table D-8 in Appendix D shows acceptable loads for different post situations. Note that the table specifies certain species of No. 1 or better, seasoned wood. The same is true of the beam table. Both posts and beams carry large loads and are susceptible to cracking and splitting, which leads ultimately to their failure. I use No. 1 Douglas-Fir for posts and beams whenever possible.

Posts must be set squarely on a very solid surface. The end must be protected from moisture (by an asphalt roofing shingle or a post base) to avoid the moisture that is transmitted through concrete. Also protect posts from drip splashes and other sources of moisture. A rotted post won't carry much weight.

A **bearing surface** is a surface that bears a load. Use your common sense to make bearing surfaces large enough to transmit loads from one member to the next. Figure 6-12 shows two situations where the size of the bearing surface is important.

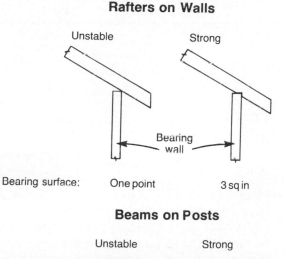

Rafters on Walls

Unstable   Strong

Bearing
wall

Bearing surface: One point   3 sq in

Beams on Posts

Unstable   Strong

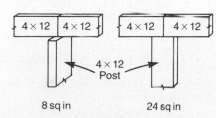

4 × 12   4 × 12   4 × 12   4 × 12

4 × 12
Post

8 sq in   24 sq in

Fig. 6-12 Comparison of bearing surfaces.

A house is only as good as its foundation. The foundation and the soil under it are subjected to the largest compression loads. By using a continuous foundation of the type described in Chapter 5, the weight of the house (approximately 100 tons) is spread along the foundation and over the ground in a safe manner. The stone castles that you see in the travel brochures for England and Scotland are usually built on solid bedrock. Those that are still standing are on bedrock anyway. Imagine that you've set a large boulder on your finger, then multiply that pressure by several hundred boulders and you can get an idea of the compression load under one of those castles, or under the Great Wall of China. That compression load is the reason why some heavily built structures don't last as long as lighter structures. This is particularly true of structures built on steep, slide-prone or eroding slopes.

Resisting Lateral Loads

So far our discussion of how to design a strong house has centered on what has to be done to make the structure strong enough to support the weight of the house and its contents. Most bending, shear, tension, and compression stresses occur from this downward force of weight on the structure.

Now we need to talk about how to resist lateral (sideways) forces on the house. Examples include such loads as windstorms, falling trees, or a car backing up against the house. A lateral force can cause a combination of bending, shear, tension, compression, and twisting stresses on different parts of the house. The many different directions from which the force can come makes it difficult to prepare for lateral loads. Nonetheless, here is an exercise that will teach you a great deal about how to build a house to resist lateral loads.

Lateral Stability Exercise

Get the cardboard (or popsicle stick) and pin model that you used to make the rafter/wall/ceiling joist joint. You'll also need some more sticks and pins (or thumb tacks). Build the three models A, B, and C shown in Fig. 6-13. It helps to use a short board as a base on which to pin the sticks.

Apply a lateral force with your finger to each of the structures. What happens? Don't just nod your head—get some sticks and tacks and *do this!* It only takes a few minutes and it is very enlightening.

Let's see what it takes to stiffen up these structures so they won't fall over. Add to the first three models so they look like D, E, and F, then push them again.

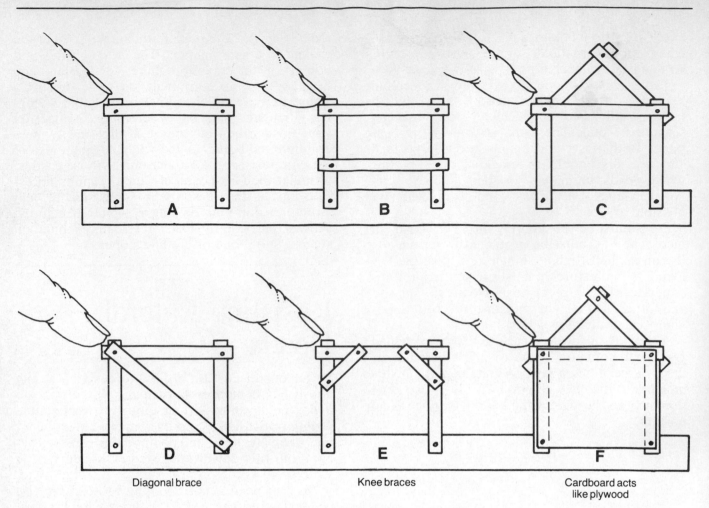

A

B

C

Diagonal brace

D

Knee braces

E

Cardboard acts
like plywood

F

Fig. 6-13 *Lateral stability exercise.*

Structural Knickknacks

Overhangs

Overhangs present a complicated combination of both bending and shear loads. The engineering books have a field day here with free body diagrams, shear and bending equations, and load diagrams that won't quit. Depending on the way an overhang is connected to the main structure. it may be considered a cantilever, which calls for several more chapters of technical analysis. It gives most of us a major headache trying to decipher all that stuff. So let's just keep overhangs safe and simple by observing the following rules:

1. Keep the load on the overhang minimal. The 100-ton grapefruit drawing (Fig. 6-6) shows an unsafe load on an overhang.

2. Limit the amount of overhang according to the chart in Fig. 6-14. If you want a larger overhang, support it with posts resting on pier blocks.

3. Frame overhangs as described in Chapter 12. (See Fig. 12-4.) The overhanging joists/rafters/beams must extend inward (into the floor or roof) twice as far as they overhang the supporting wall. For example, a 3-ft overhang requires the member to extend 6 ft into the floor.

The Effect of Notches

Any piece of lumber can be severely weakened by notching, knots, or other imperfections. This weakening is most serious when the notch occurs on the bottom side of a bending member and in the middle of the span. (See Fig. 6-15.)

For practical purposes, a notch along the edge of a bending member 2 in deep effectively reduces the strength of that member to the strength of the next smaller lumber size. For example, a 2 × 10 with a 2-in notch in it is only as strong

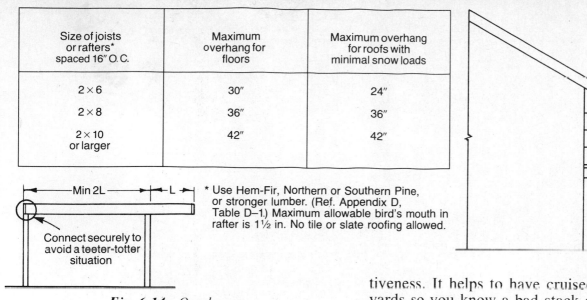

Size of joists or rafters* spaced 16" O.C.	Maximum overhang for floors	Maximum overhang for roofs with minimal snow loads
2 × 6	30"	24"
2 × 8	36"	36"
2 × 10 or larger	42"	42"

Min 2L ——— L

Connect securely to avoid a teeter-totter situation

* Use Hem-Fir, Northern or Southern Pine, or stronger lumber. (Ref. Appendix D, Table D–1.) Maximum allowable bird's mouth in rafter is 1½ in. No tile or slate roofing allowed.

2 × 8 Rafters

3'

2 × 8 Joists

Fig. 6-14 *Overhangs.*

as a 2 × 8. A 2-in hole in the central portion of a 2 × 10 reduces the strength of the board about 10 percent or less.

When drilling holes for the plumbing and the wiring, always drill near the center of the face of a joist or rafter. Avoid notching, particularly avoid midspan notches. Design the plumbing and wiring so major structural members are undisturbed.

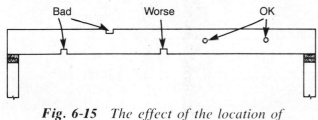

Bad   Worse   OK

Fig. 6-15 *The effect of the location of notches, holes, and knots on structural integrity.*

Lumber Grades and Species

Always specify lumber grades when you order lumber. In most cases, "Standard and Better" or "No. 2 and Better" is the grade you should specify. (See Appendix A, Table A–1.) For beams and posts, specify No. 1 or Select lumber of the strongest species that is locally available. The lumber grading standards are published by the American Lumber Standards Committee and are applied by various lumber producer associations such as the Western Wood Products Association. Some crooked outfits use their own grading systems though, so look at the lumber before they unload it and have them take it back if you don't like it. It's a good chance to practice your asser-

tiveness. It helps to have cruised a few lumber yards so you know a bad stack of lumber when you see one. Particularly avoid lumber with large splits, checks, knots on the edges, soggy wet, warped, or with inconsistent dimensions.

In addition, always specify kiln-dried (KD) lumber. Green lumber usually warps and cracks as it dries out. KD lumber is dried slowly in a big oven under carefully controlled conditions.

Avoid rough cut lumber, unless you specifically want rough cut siding or beams. Rough cut lumber is not surfaced or graded and many inspectors will require that all lumber be one size larger than code requires if it is rough cut.

The strength of a particular piece of lumber is fairly predictable if you know its species and grade. Each species of wood is rated to withstand a certain amount of stress, measured in pounds per square inch (psi), before it's in danger of breaking. The amount of stress the wood can take depends on whether the load is a bending, shear, tension, or compression load. Table D-1 in Appendix D shows the relative bending strengths of some common types of wood and indicates which should be used to structural purposes.

The strength of a species of wood is an important property. Another property that you will see listed in books is the stiffness of a wood, as measured by its modulus of elasticity, E. The higher the E value, the stiffer the wood. A dry, brittle twig is stiff, but not strong. A green, supple twig may be strong, but not stiff. The wood species indicated as "structural" in Table D-1 are both stiff enough and strong enough to be used as structural members for your house.

Steel Connectors

There is a large assortment of manufactured steel connectors that can be very helpful when you are

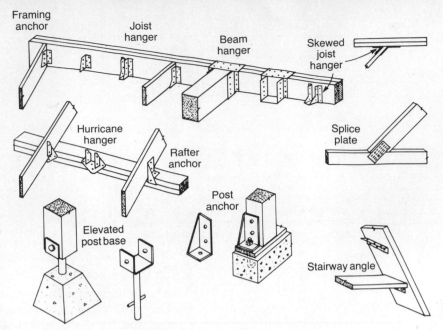

Fig. 6-16 *Examples of steel connectors.*

building. (See Fig. 6-16.) You can get a brochure from your lumber yard. Some common types of connectors are shown below. Most come with special stubby nails that you usually forget to pick up. Then you have to make another trip to the store or cut off some 16d nails with the side cutters on your pliers. The nails holding most of these connectors will be subject to some large stresses, so be sure they are well nailed.

In Conclusion

So there you have it. These last two chapters covered a lot of territory but if you stuck it out, you now know how to:

- Recognize bending, shear, tension, and compression stresses in a house structure.

- Calculate the load on a structure (floors 60 psf; roofs 50 psf).

- Determine the proper size of bending members by using tables (floor joists, rafters, ceiling joists, beams, and so on).

- Decide how to locate stairways and hot tubs, and similar openings and special loads.

- Determine the proper type and size of structures that are being compressed (walls, posts, foundations).

- Connect members properly with nails (Appendix C) and with bolts.

- Make a structure laterally stable.

- Design porch and roof overhangs, use the right grades and species of lumber, select steel connectors, and know the effect of notches and knots.

I hereby confer on thee thy Engineering Merit Badge.

For More Information

Dwelling Construction Under the Uniform Building Code. Whittier, Calif.: International Conference of Building Officials, 1979.
Also see *Uniform Building Code* published by the International Conference of Building Officials. You can write to them at 5360 S. Workman Mill Road, Whittier, Calif., 90601.

McNeese, Donald G. and Hoag, Albert L. *Engineering and Technical Handbook.* Englewood Cliffs, N.J.: Prentice-Hall, 1957.

Nash, William A. *Schaum's Outline of Theory and Problems of Strength of Materials.* New York: McGraw-Hill, 1972.

Salvadori, Mario. *Why Buildings Stand Up.* New York: McGraw-Hill, 1982.

Western Woods Use Book. Portland, Ore.: Western Wood Products Association, 1979.
You can write to the Western Wood Products Association at the Yeon Building, Portland, Oregon, 97204 and, for $2, receive a handy span computer.

CHAPTER 7

Designing the Systems: Plumbing, Electrical, and Heating

Designing the Systems: Plumbing, Electrical, and Heating

"Plumbing is a cinch. All you've got to know is that a toilet won't flush uphill and payday is Friday. Since you won't be getting paid it'll be a double cinch."

Wrench sounded pretty convincing, as usual. But I knew better. It probably *was* a cinch for him . . . the guy who was reputed to be born with a monkey wrench in his fist and to have rearranged his mother's plumbing into a methane digester during the last weeks of the pregnancy. But it wasn't going to be a cinch for me. The only thing I knew about plumbing came from some hazy recollections of Dad disappearing down into the crawl space with cast iron pipes, oakum, and a lead pot. The part I can remember is that the task seemed to require a lot of words I hadn't heard before, all delivered with gusto.

Despite Wrench's assertions that plastic and copper plumbing have brought the profession into the easy do-it-yourself range, I didn't really believe him until I tried it. And now it's my turn to tell people how easy plumbing can be. Wiring a house isn't much harder. Of course it doesn't matter if the task is scrambling eggs or grafting a fruit tree; it always looks simply terrible before you try it and terribly simple afterwards.

Designing the Plumbing

House plumbing is actually two separate systems. The **supply system** brings clean water from the water main in the street (or from the well) and distributes that water to the sinks, shower, clothes washer, dishwasher, and toilet tank. The **drain-waste-vent** (DWV) system takes the used water from the sinks, tubs, toilet, and appliances and carries it off to the sewer or the septic system.

The supply system for our sample house is shown in Fig. 7-1. This system could be installed with any of several types of pipe. My preference is copper pipe because of its easy installation and durability. Plastic pipe (the CPVC variety) is gradually winning acceptance by the plumbing codes and is also easy to install. The galvanized steel pipes used in the houses prior to 1965 are difficult to cut and assemble, require special threading tools, and are prone to clog with rust over the years. If you've ever lived in one of those old houses where the water barely piddles out of the shower you know about the disadvantages of clogged pipes.

Since the water in the supply system is pressurized, all you do is route a pipe to the point where you want water. There are very few code rules applying to the supply system, but there are some common sense things to know:

- Locate the water heater close to the shower and sinks. It takes about 30 long seconds for hot water to travel 20 ft.

- Avoid excessive bends and turns. It reduces the flow rate of the water.

- Provide shut off valves where the water enters the house and at each fixture and appliance. This makes repairs easy.

- Locate the pipes so installation is easy. Hanging the pipes from the underside of the floor joists in the crawl space works well, though you may have to insulate them in cold climates to avoid freezing. Running pipes between and parallel to studs and joists makes the installation easy; running pipe perpendicularly through the studs and joists will give you difficulties.

- By sloping all the pipes so they drain toward an outside faucet (hose bib), you can shut off the water to the house and easily drain the system. This is handy for repairs, extensions, and long winter vacations when the house is left unheated.

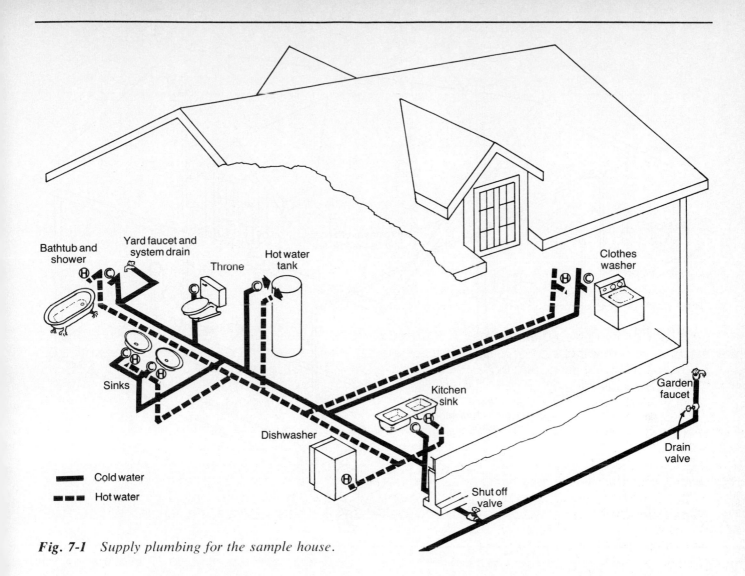

Fig. 7-1 *Supply plumbing for the sample house.*

Sketch the route of the pipes on a copy of the floor plan. Draw a line coming into the house under the foundation and to the water heater. Draw red lines (representing hot water) to the sinks, shower, clothes washer, and dishwasher. Cold water lines branch off the incoming line. The toilet uses only cold water and the dishwasher uses only hot water. My preference is to route the pipes under the floor (where there are few obstacles or drilling to be done) and then up through the floor at the desired location. I use ¾-in pipe nearly everywhere, except for the last few feet to the sinks, toilet, and dishwasher where I use ½-in pipe. While some designers try for a layout that places most of the plumbing in one location, I think that liveability is a more important layout factor and well warrants a few extra feet of pipe. The drain-waste-vent (DWV) system for our sample house is illustrated in Fig. 7-2.

The best way to plumb a DWV system is with plastic pipe. The plastic is called Acrylonitrite-Butadiene-Styrene (ABS). Saw it to length with a handsaw and glue the pieces together. ABS pipe is cheap, durable and easy to install. No molten lead, oakum, and $50 cast iron fittings for me, thank you. If you happen to be in an archaic locality where the code still requires cast iron pipe, use "no-hub" cast iron and remember to vote your city fathers out of office at the next opportunity. They've been there too long.

As Wrench so eloquently pointed out, the drains from all fixtures and appliances must be sloped toward the septic tank or sewer connection. Use a slope of ¼ in per foot. You can plan on needing a septic tank and drainfield system in most rural areas. The septic tank holds the sludge while it degrades and the drainfield lines distribute the effluent to be absorbed by the soil. If the soil doesn't absorb the effluent (percolate) adequately, you'll end up with a smelly yard and the run-off will pollute neighborhood streams and lakes. Hence my emphasis in Chapter 3 on buying a lot that either "percs" well or has a sewer line that can be connected to.

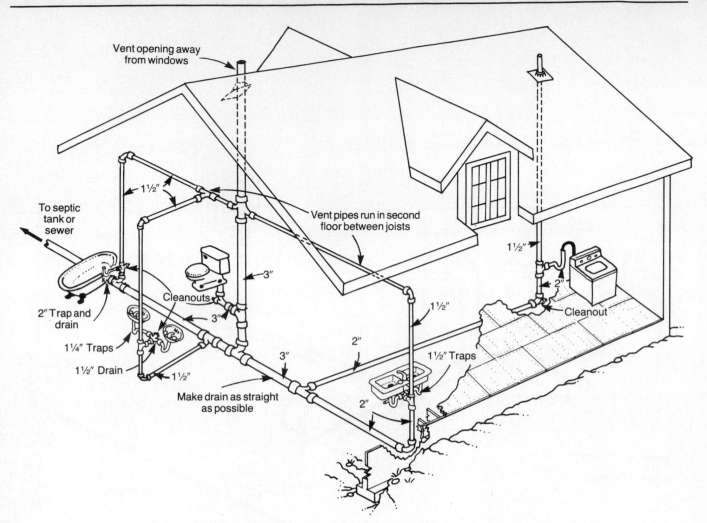

Vent opening away from windows

Vent pipes run in second floor between joists

To septic tank or sewer

1½"

3"

1½"

2"

Cleanouts

2" Trap and drain

1¼" Traps

1½" Drain

1½"

3"

3"

Make drain as straight as possible

2"

2"

1½" Traps

1½"

2"

Cleanout

Fig. 7-2 *Drain/waste/vent plumbing for the sample house.*

TABLE 7-1 Fixture Trap and Drain Size Requirements and Maximum Distances between Traps and Vents

Fixture	Minimum trap size (inches)	Drain pipe size (inches)	Maximum distance, trap to vent
Bathroom sinks	1¼	1½	2′6″
Kitchen sinks	1½	2	3′6″
Dishwashers	1½	2	3′6″
Kitchen sink and dishwasher on the same trap and drain	1½	2	3′6″
Bathtubs	1½	1½	3′6″
Shower	2	2	5′0″
Clothes washers	2	2	5′0″
Floor drains	2	2	5′0″
Toilets	3*	3	6′0″

* The toilet itself is a trap.

The vent part of the DWV system has a double purpose. (See Figs. 7-3 and 7-4.) If you've ever poured liquid from a can (for example, oil, juice, or beer), you know how it glugs if you don't punch a pour hole and a vent hole. That's the first purpose of vents, to allow a good flow in the system. The second purpose is to allow the sewer gases to escape somewhere. For this the rooftop is usually the least offensive place.

Traps are installed at each sink, tub, and appliance to prevent the sewer gases from escaping into the house. The U-shape traps water and holds it there, providing an effective barrier to the gases. Toilets are designed so the trap is part of the fixture. That's why there is always some water in the toilet bowl.

The size of the drains and the location of the vents are the most important parts of the DWV design. The Uniform Plumbing Code gets pretty involved here. My space is limited here so I'll show you a simplified procedure that avoids the most complicated rules, although it may cost you an extra $20 for pipe. Be aware that your local code may differ here.

First, select the proper drain pipe sizes from Table 7-1.

Fig. 7-3 *Septic system.*

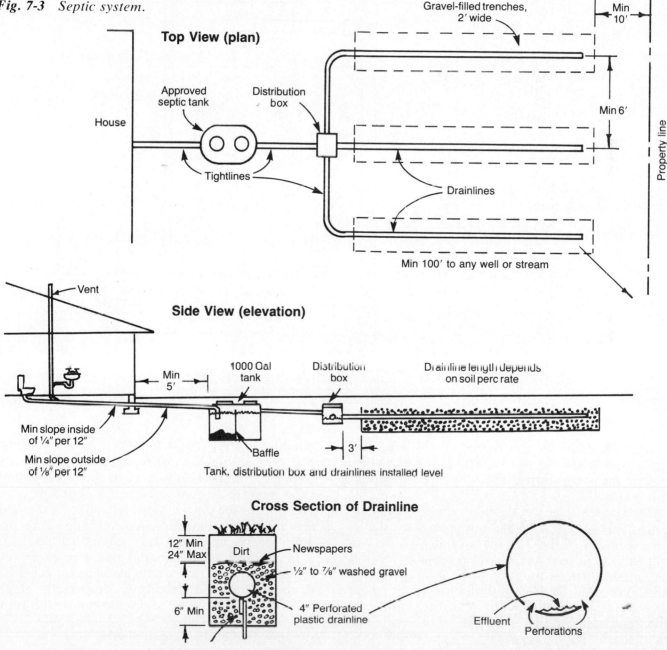

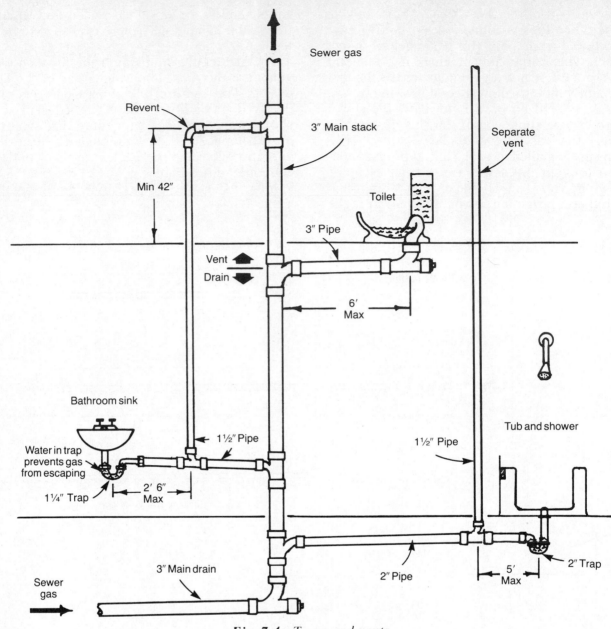

Fig. 7-4 *Traps and vents.*

Second, slope all the drains into a large 3-in drain that leads out to the septic tank or sewer. Avoid any sharp turns. The straighter the drain, the better.

Third, locate the vents and revents. This is the tough part. There'll be the 3-in main stack that is an extension of the 3-in main drain. It's best to put this stack near the toilet, in a place where it can go up through the roof without hindering the structure, and its smelly end shouldn't be near any windows or sundecks. With the drains and the main stack located, the rest of the vents can be placed.

In accordance with my simplified procedure here, only the highest drain on the stack can drain directly into the stack without a revent or separate vent. In Fig. 7-4 I've marked where the stack changes from a drain to a vent. All lower drains must have their own revent or a separate vent. The reason for this is that if someone flushes the upstairs toilet while the downstairs shower is running, for example, the shower won't drain right.

In addition, the distance between any trap and its vent may be no more than indicated in Table 7-1 or there is a good chance the drain will "glug" and the water may be siphoned out of the trap. Take a minute to study Fig. 7-4 and Table 7-1 to see how these rules are applied.

You'll probably end up with one or two extra revents than are absolutely required by the code using this procedure. An alternative is to get a copy of the plumbing code and get into the subject more deeply. When you've completed the DWV design, submit it to the plumbing inspector so he can indicate any corrections that are necessary. In my case, the inspector was friendly and helpful.

For More Plumbing Information

Planning Bathrooms. U.S. Department of Agriculture Home and Garden Bulletin 99, 1967.

Step-by-Step Guide book on Home Plumbing. Salt Lake City: Step-by-Step Book Co., 1975.
For more information you can write to them at Box 2583, Salt Lake City, Utah.

Sunset Editors, ed. *Basic Plumbing Illustrated*. Menlo Park, Calif.: Sunset-Lane Publishing Co., 1975.

Time-Life Books. *Plumbing*. Alexandria, Va.: Time-Life Books Inc., 1979.

Uniform Plumbing Code. International Association of Plumbing and Mechanical Officials, Los Angeles, California.

Electrical Design

Electricity, queen of the mystic sciences. Invisible energy that sparks and flashes. One wonders if even Gandalf the Wizard really understands electricity. Yet 250 years ago old silk-stockinged, bespectacled Ben Franklin reasoned that there was some energy in those lightning bolts and set out to harness it. I get the feeling that Ben was really just a regular sort of guy. After all, anyone who could write "Fish and house guests stink in three days" and go to bat to make the turkey the national emblem couldn't be too highfalutin. It's simply that the logic circuits in his brain were less cluttered than yours and mine. The same goes for Tom Edison, who figured out how to put electricity to work to light his workshop 100 years ago. Our good fortune is being in line behind Ben,

	Outlet (duplex receptacle)
Ⓛ Light	S₃ 3-Way switch
S Switch	▶ 220-Volt connection

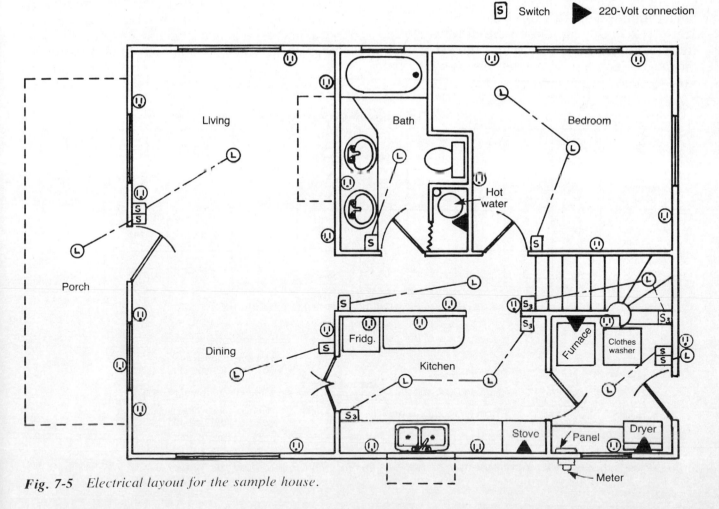

Fig. 7-5 *Electrical layout for the sample house.*

Tom, and a bunch of other people who conquered electricity and made it relatively simple for us to deal with.

Now we can set aside our structural engineer and plumber hats and become an electrical designer. In a way, the electrical system is a lot like the supply plumbing system. Here our goal is to route wires, instead of pipes, to the places in the house that need juice.

The first and most important design job is to locate all the lights and outlets in the house. This takes some thought, since everyone has different needs and desires. Figure 7-5 shows an electrical layout for our sample house. The three-way switches shown allow lights to be turned on or off from two locations, from both the top or the bottom of the stairs for example.

The next design task is to arrange all the lights and outlets on circuits. A normal house **circuit** is nothing more than a plastic-coated cable, containing three wires, that connects a number of lights and/or outlets. The circuit originates from the **panel**, which is a metal cabinet where the large wires come into the house. The panel distributes the electricity to all the house circuits. The size of a circuit depends on the amount of current that will flow through it. For instance, the oven requires a big 50-amp circuit while the lights are on 15- or 20-amp circuits. Before we go further, let me define some terms in the electrician's language. (You'll be a regular linguist by the time you're done designing your house.)

Current is the amount of electricity flowing through the wires. It's measured in **amperes** (or **amps**). If a circuit carries a lot of current to appliances like the oven or electric furnace, large wires are required to prevent the wires from getting hot. (See Fig. 7-6.)

Most circuits in your house will require "12-2 with ground Type NM cable," which is called **housewire** at your local store. The term **cable** refers to two, three, or four wires bundled together with a covering. The covering is plastic (that is, nonmetallic) on **Type NM** cables. The individual wires in the cable include three No. 12 (12-gauge) wires, one covered with black insulation (the **hot wire**), one with white insulation (the **neutral wire**), and one bare **ground wire**. I'll explain how to hook up these wires in Chapter 16. All you need to know to design the general lighting and outlet circuits is that one housewire (a "12-2 with ground" cable) must be routed from the panel to about 10 to 12 lights and/or outlets. (See Fig. 7-7.) (Although 14–2 cable is adequate for lighting circuits, if you ever want to expand such a circuit you'll kick yourself for using 14-2.)

For design purposes, think of electricity flow like water flow. The current flows from the panel, through the circuit breaker, along the housewire to each outlet, switch, and light. A switch acts like a faucet, turning the current on or off. If there are too many **devices** (light bulbs, toasters, refrigerators, and so on) on the circuit, the current drawn will exceed 20 amps and the **circuit breaker**

Ampacities of Copper Wires

American wire gauge number (AWG)	14	12	10	8	6	4	2	1/0	2/0	4/0
Amp rating, types T and W	15	20	30	40	55	70	95	125	145	195
Amp rating, types RH, RHW and THW	15	20	30	40	65	85	115	150	175*	230

*2/0 allowed for 200 amp residential service entrance (ref. NEC 310-15, Note 3 to Table 310-16).

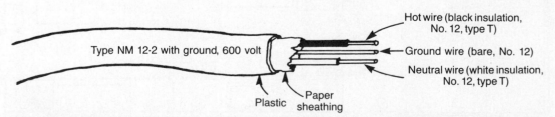

Fig. 7-6 *Electrical wire sizes and amp ratings (ampacities) for copper wire.*

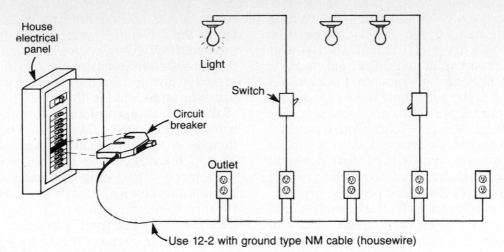

Fig. 7-7 *A typical house circuit.*

will trip, shutting off the current before the wires get hot.

Circuit breakers are available with ratings from 10 to 200 amps, depending upon the size of wire they are protecting from overheating. For example, by looking at Fig. 7-6 you can see that a No. 10 wire is rated to carry a maximum of 30 amps. Hence, a circuit wired with No. 10 wire must be protected by a 30-amp (or smaller) circuit breaker. This way the circuit breaker will trip when the 30-amp limit is reached. In older houses, fuses perform this function.

In order to design a circuit that won't continually be tripping the circuit breaker, we need to spread the **load** (the amount of current being drawn) evenly among the circuits. Table 7-2 shows the load of some common household devices.

TABLE 7-2 Power Consumption of Typical Household
Appliances, Devices, and Tools

Appliance	Watts	Amps	Appliance	Watts	Amps
Kitchen and			*Bedroom*		
Dining Room			Electric blanket	200	1.7
Blender	800	7.0	Clock-radio	10	0.1
Can opener	150	1.3	Baseboard heater	1,500 (230V)	6.5
Clock	3	0.03			
Dishwasher	1,800	15.7	*Living Room*		
Food processor	200	1.7	Lamp	100	0.9
Freezer	1,000	8.7	Large stereo	100	0.9
Microwave oven	1,000	8.7	TV	300	2.6
Mixer	200	1.7	*Shop*		
Oven and range	10,000 (230V)	43.5	Table saw, 1-hp	1,600	13.9
Range hood	200	1.7	Portable saw	1,300	11.3
Refrigerator	1,000	8.7	Portable drill	600	5.2
Toaster	1,000	8.7			
			Miscellaneous		
Laundry			Vacuum cleaner	500	4.3
Clothes washer	1,000	8.7	Sewing machine	100	0.9
Clothes dryer	5,500 (230V)	23.9	Steam iron	1,000	8.7
			Light bulbs	100	0.9
Bathroom			Fluorescent lights	40	0.3
Hot water heater	5,000 (230V)	21.7			
Hair dryer	1,000	8.7	*Furnace*		
Electric shaver	15	0.1	Electric, forced air	18,000 (230V)	78.3
Wall heater	1,000	8.7			

Look at the amps column in Table 7-2. With a 20-amp circuit (regular No. 12 housewire), we could have a hair dryer (8.7 amps), six 100-watt light bulbs (5.4 amps), a clock (0.1 amp), a large stereo (0.9 amp), and a vacuum cleaner (4.3 amps) all going at the same time for a total of 19.4 amps. If we added six lights (5.4 amps) to the circuit, the total possible load would be 24.8 amps. It's unlikely that all these devices would be on at once however, and if they were it would simply trip the breaker and you could turn off the vacuum and a couple lights and reset the breaker. The designer's problem is to intelligently guess how much load will be on each circuit and try to avoid overloads. For example, shop outlets for large power equipment and kitchen circuits with toasters, microwave ovens, and refrigerators should be limited to three or four outlets per circuit. If the lights dim when an appliance is turned on, there's too much load on that circuit.

Now how about the real power eaters like the clothes dryer and the hot water heater? By looking at the UL label, which all electrical devices are required to have, you can determine their maximum load. Some are listed in amps and some in watts just to confuse you a bit. To convert watts to amps, divide the watts by the voltage. (Take the example of our clothes dryer: 4,500 watts ÷ 230 volts = 24 amps). There's no such thing as a 25-amp circuit breaker, so we'll use a 30-amp breaker. Since it's a 220-volt device,* we'll need "10-3 with ground" wire (called "clothes dryer wire" at the store). The larger No. 10 wire is required to handle the 24-amp load. (See Fig. 7-6.) The three wires are required for a 220-volt dryer because the heating elements run on the two hot wires (220V) and the timer (110V) runs on one of the hot wires and one neutral wire. The ground wire (the fourth wire) is for safety.

* The actual voltage averages about 230 volts, but everyone refers to it as 220 volts. Use the 230V figure for all calculations.

TABLE 7-3 Circuits for the Sample House

Circuit	Volts	Breaker (amps)	Copper wire size with ground (AWG per NEC Sec. 310-16)	Code reference (NEC)
1 Oven and/or range	230	50	#6/6/8 w/g	220-19
2 Small appliances, kitchen and dining room	115	20	#12-2 w/g	220-3b
3 Small appliances, kitchen and dining room	115	20	#12-2 w/g	220-3b
4 Clothes washer	115	20	#12-2 w/g	220-3c, 210-25
5 Electric clothes dryer	230	30	#10-3 w/g	220-18
6 Electric water heater	230	20	#10-2 w/g	422
7 Dishwasher	115	20	#12-2 w/g	422
8 Electric furnace (60-kBtu, 18-kW)	230	100	#1-2 w/g*	220-15, 424
9 Bathroom outlets (GFI protected)	115	20	#12-2 w/g	210-8, 210-25b
10 Garage and outdoor outlets (GFI)	115	20	#12-2 w/g	210-8a2, 210-25b
11 12 13 14 15 16 17 } General lighting and outlets	115	20	#12-2 w/g	210-23, 220-11

With a #8 ground wire.

Table 7-3 shows a list of the circuits I would install in the sample house listing the voltages, breaker size, and wire size for each. Most of these circuits are required by the National Electric Code (NEC).

Plan for a panel that will hold 20 circuit breakers. While this number seems huge at first glance, it's better to have extra capacity than not enough. The bathroom outlets and garage/outdoor outlets must be protected by a **ground fault interrupting (GFI) device.** These devices are expensive but they're supposed to prevent you from getting fried if the radio drops into the bathtub or you get caught standing in a puddle when your grinder shorts out. Some GFI's replace the circuit breaker and protect the whole circuit; others replace the outlet receptacle and protect just that outlet. Check what's available with the clerk at the electrical store. By the way, make friends with that clerk; he or she may be an important information source for you.

Once the circuits are worked out, decide how to get the power into the house and to the panel. This is called the **service entrance**. Figure 7-8 shows two examples. Also check for a display at your electrical store.

Basically the electrical design process involves deciding where to put electrical devices in the house, connecting them with circuits, then deciding on the type of service entrance. Here are some general design considerations:

- Locate the service entrance and panel close to the large appliances. Those large wires are expensive and it's wise to keep them short.

- If you design too many electrical devices into the house, you may have to use a 400-amp service entrance. (See NEC Chapter 9, Sect. B and Sec. 220 for details.) By using a gas or oil furnace or a gas water heater, the power requirement can be trimmed back to a regular 200-amp service, but chimneys for the combustion by-products (smoke and carbon monoxide) will be necessary.

- Most wires will be routed through the house by drilling 3/4-in holes in the wall studs. Larger wires (No. 4 or larger) will need to be protected by conduit. Conduit are pipes installed much like plumbing that protect wires from damage (when you drill into a wall for example.) Plan ahead because they require a 2½-in to 3-in hole wherever they go.

Fig. 7-8 *Two typical service entrances.*

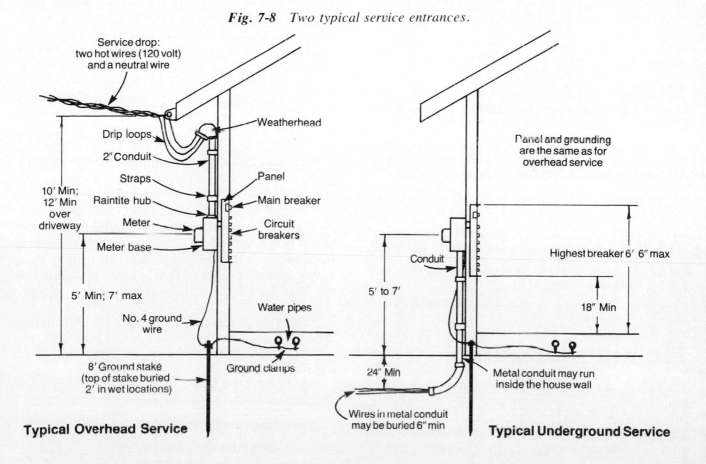

Service drop: two hot wires (120 volt) and a neutral wire

Drip loops
2" Conduit
Straps
Raintite hub
Meter
Meter base

Weatherhead
Panel
Main breaker
Circuit breakers

10' Min; 12' Min over driveway

5' Min; 7' max

No. 4 ground wire

Water pipes

8' Ground stake (top of stake buried 2' in wet locations)

Ground clamps

Typical Overhead Service

Panel and grounding are the same as for overhead service

Conduit

Highest breaker 6' 6" max

18" Min

5' to 7'

24" Min

Metal conduit may run inside the house wall

Wires in metal conduit may be buried 6" min

Typical Underground Service

Electricity used to baffle me. One reason it took me two months to install my electrical system was that I sort of designed the thing with a book in one hand and a wire in the other. If there were better books available it wouldn't have been so tough. I nearly panicked when I discovered that I needed a 400-amp service entrance. My God! I had planned to just copy the display at the electrical store. At that point I even checked with an electrician to see how much it would cost for him to wire the panel but leave the rest to me. That's a reasonable way to go if you can find a willing electrician. I couldn't even afford that expense, so it was back to the books and eventually down to the power company, who were kind enough to give me a diagram of how it should be done.

So I really can't tell you it was easy. In fact, it was downright difficult. As with the house design and the plumbing, I had to go from total ignorance to a working knowledge of the beast. That's the same way my friends Steve and Carl did it—cold turkey. I suppose even Wrench had to learn wiring at some point in time; and my father took the same trail 30 years ago. There are moments of pain and moments of glory in discovering any new territory. The fact remains, I wired my own house. And if I can do it, there's hope for everyone!

For More Electrical Information

National Electric Code. National Fire Protection Association, 470 Atlantic Ave., Boston, Massachusetts 02210 (updated frequently).

Planning Your Home Lighting. U.S. Department of Agriculture Home and Garden Bulletin 138, 1968.

Richter, H.P. *Wiring Simplified.* St. Paul: Park Publishing Co., 1975.

Sunset Editors, ed. *Basic Home Wiring Illustrated.* Menlo Park, Calif.: Sunset-Lane Publishing Co., 1977.

Time-Life Books. *Basic Wiring.* Alexandria, Va.: Time-Life Books, 1976.

Heating Design

I suppose you're one of those hard-driving, demanding types who'll want not only a beautiful, functional, plumbed and wired house, but a warm house too. Preferably a warm house that doesn't cost a fortune to keep warm. OK then, here's how to do it. You'll be relieved to know that I've got only a few pearls of wisdom to offer on this subject. Then we'll get a permit for our sample house and start building.

There are probably more choices to make on the heating system than on any subject so far. There are heat pumps, solar heating, wood burning systems, radiant heat schemes, forced-air furnaces, and so on into the night. To try to cover them in this short space would be folly. This discussion is limited to some general principles.

Principle 1: Design for energy sources that will be available 100 years from now. Everybody has a different crystal ball. My crystal ball says that oil and natural gas will be inceasingly scarce in years to come. It also says that trees, rain, rivers, wind, and sunlight will be available in the year 2085, despite the best efforts of developers and industrial polluters. Therefore, I have designed the sample house with forced-air electric heat and provisions for future wood heat and solar heat systems. Who am I to argue with my own crystal ball?

Your situation and frame of mind may lead you to a different approach. I live in a relatively mild climate where it rains far too much, making electricity cheap. (2½¢ per kilowatt hour.) Electric heat appears to be the best bet for clean, safe, convenient, and available energy. I also have a couple of acres on which I plan to harvest firewood indefinitely. And I like to chop wood, to cleanse my lungs and warm my body. My elaborate solar schemes are still being refined so I get no help from that corner yet.

Principle 2: Design the house to use as little energy as possible. The extreme is an underground house with 2 ft of insulation in the doors and one window, like a buried thermos bottle pointed south. Certainly energy-efficient but not to my taste. On the other hand, there's the house with five separate modules, floor-to-ceiling glass and a dozen skylights. That's more my style but not very sensible.

Achieving a balance of liveability and energy efficiency is the challenge. Working with the model of your house in hand, apply these following three rules:

1. Minimize the heat loss area

2. Maximize insulation

3. Distribute the heat effectively

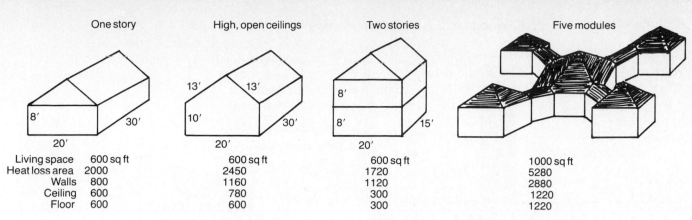

One story		High, open ceilings		Two stories		Five modules	
Living space	600 sq ft		600 sq ft		600 sq ft		1000 sq ft
Heat loss area	2000		2450		1720		5280
Walls	800		1160		1120		2880
Ceiling	600		780		300		1220
Floor	600		600		300		1220

Fig. 7-9 *Comparison of structures for heat-loss surface area.*

Minimize the Heat-Loss Area. The more surface area that separates a heated interior space from an outdoor space, the greater the heat loss. (See Fig. 7-9.) An important factor here is the ability to heat only the spaces that need to be heated. If the kitchen, dining, and bathroom areas can be partitioned from the rest of the house, only this part of the house needs to be heated in the morning as everyone gets up and departs for work or school. Right now I'm writing in a room insulated from the rest of the house and heated by a metal fireplace. Five or six chunks of wood keep this room warm all day. The rest of the house is kept at 60°F most of the time.

Maximize Insulation. The object of insulation is to slow the speed at which the heat leaves the house. Often the largest heat loss is through cracks around windows and doors, but that's a matter of weatherstripping and not really related to the design of the house. From the design standpoint, our knowledge of insulation materials and R values will make the most difference. The higher the R value, the greater the resistance to heat transfer. For example, an R-10 wall will allow twice as much heat loss as an R-20 wall. Table 7-4 and Fig. 7-10 show the insulating qualities of some common materials.

TABLE 7-4 The Insulating Value of Common Building Materials

Material	R Value for 1-inch thickness	Common thickness (inches)	R Value
Aluminum	0.0007	⅛	0.0001
Steel	0.0032	⅛	0.004
Brick or stone	0.11	4	0.44
Concrete	0.11	8	0.88
Concrete blocks	—	8	0.50
Glass	—	single pane	0.88
		sealed double pane	1.64
Dead air space	1.0*	3½	1.0
Gypsum wallboard	—	½	0.32
Tar paper	—	15 # felt	0.06
Plywood	1.25	½	0.6
		¾	0.9
Wood	1.3	¾	1.0
		1½	2.0
		3½	4.5
Solid-core door	1.2	1¾	2.0
Insulating board	2.9	½	1.5
Expanded polystyrene (styrofoam)	3.3	¾	2.5
Fiberglass insulation	3.3	3½	11.0
		5½	19.0

* *The R value does not increase for spaces greater than ¾ in.*

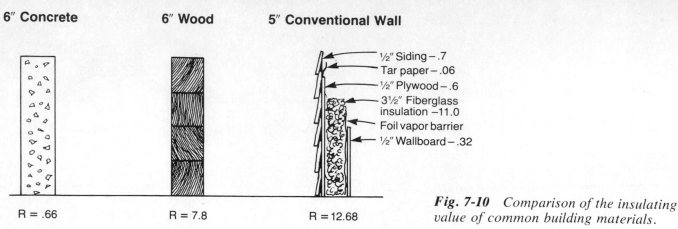

6″ Concrete	6″ Wood	5″ Conventional Wall

½″ Siding – .7
Tar paper – .06
½″ Plywood – .6
3½″ Fiberglass insulation – 11.0
Foil vapor barrier
½″ Wallboard – .32

R = .66   R = 7.8   R = 12.68

Fig. 7-10 *Comparison of the insulating value of common building materials.*

With an *R* value of 0.0001, it's apparent why aluminum windows are not good for cold climates. They transfer heat rapidly and tend to sweat from condensation. And it may come as a surprise to you that a lot more heat is lost through a solid 6-in wood wall than through a conventionally insulated 5-in wall. Heavy or dense materials tend to be poor insulators while light, fluffy materials tend to be good. However, heavy materials such as logs, concrete, and water are good for storing heat, which moderates any change in temperature near them. Shiny materials, such as the foil covering on fiberglass insulation, tend to reflect radiant heat. Hence, the foil is installed on the side of the wall facing the living space.

Conventional house walls built with 2 × 4 studs and 3½ in of fiberglass insulation are roughly R-11 walls. Walls built with 2 × 6 studs and 5½ in of insulation are rated at R-19. Ceilings with 5½ in of insulation are also rated at R-19 and can be increased to R-38 by installing two layers of 5½-in insulation. Table 7-5 indicates how much insulation is needed for different climates.

Distribute the Heat Effectively. In the old one-room log cabin, the fireplace produced the heat for the whole house, though the far corners were chilly and it was hot near the ceiling. In today's larger, extended house, the central fireplace arrangement can have some real problems. Many good minds have been working on these problems though, and there are several systems on the market that are effective in distributing heat from a fireplace. One could write a book about fireplace systems alone.

In Wrench's latest house he's assembled an amazing heating/cooling system that would take several books to describe. A computer network of sensors decides whether to heat a particular room with wood heat, solar heat, wind-generated electricity or heat pump heat at any time of the day or night. I just wandered around the maze of pipes, ducts, valves, and thermostats nodding my head and trying to act like I understood parts of it.

Most owner-builders will opt for a simple, well-proven heat distribution system at least for starters. It's got to be affordable, big enough to keep the whole house warm, and flexible enough to convert to another energy source if necessary. For me, that worked out to be a forced-air ductwork system that can be supplied with heat from nearly any type of central furnace.

Designing a Forced-Air System

Decide on the Air Flow. Basically there are two systems: downflow and upflow (see Fig. 7-11).

Efficient distribution of the warm air is accomplished with ductwork. Round ducts allow better

TABLE 7–5 Quick and Easy Insulation Guide

Average winter temperature	Wall insulation	Floor insulation	Ceiling insulation	Windows
Less than 30°F	R-19	R-22	R-38	Double and storm
30°F to 40°F	R-19	R-22	R-33	Double
40°F to 45°F	R-19	R-19	R-30	Double
45°F to 50°F	R-11	R-11	R-22	Double
50°F or more	R-11	R-11	R-19	Double

Fig. 7-11 *Forced-air heating systems.*

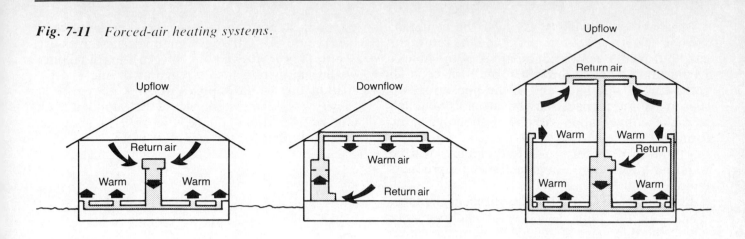

flow than rectangular ducts. By putting the duct-work in the crawl space, it avoids structural modifications to the house and allows the warm air to flow upwardly aided by convection (meaning, hot air rises). If the floor is a concrete slab, the duct-work can be placed in the attic and warm air blown down into the house. In either case the ducts should be wrapped with insulation. The air that is returned to the furnace (called the return air), is best picked up at a high location. This is especially true in a lofted or two-story house where the warmest air ends up near the ceiling and should be recycled back to the lower areas.

Determine the Size and Location of the Furnace. The size of the furnace depends on the heat loss of the house. Usually the power or gas company will take your house design, calculate the expected heat loss, and recommend a furnace size. You can also do it yourself based on their instructions. A 60,000 Btu* furnace would be ample for our sample house in a moderate climate, and large enough to provide for a future addition.

---

* A Btu is a British thermal unit. It's the amount of heat required to raise temperature of one pound of water one degree Fahrenheit.

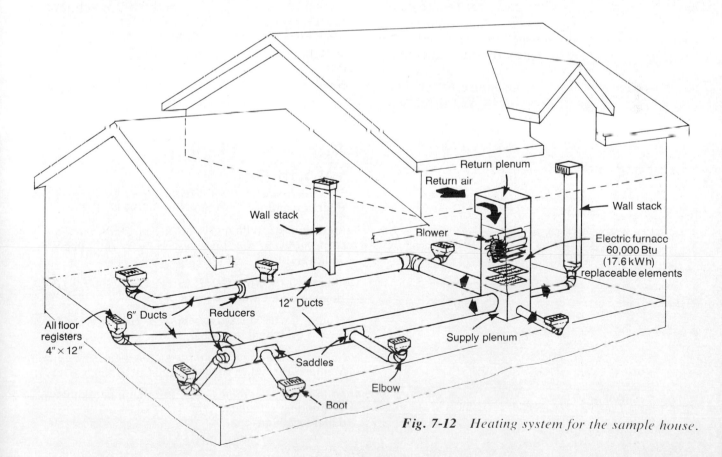

Fig. 7-12 *Heating system for the sample house.*

Furnaces are normally put near the center of the house so the ducts are short and the return air can be drawn evenly from all areas of the house. In the sample house, however, the furnace is located toward the back end of the house for two reasons. First, furnace blowers are noisy enough to make conversation difficult. Secondly, if the furnace is fueled with wood, oil, coal, or gas at any time in the future, a flue (chimney) will be required for the combustion by-products (smoke).

Remember to allow for the furnace plenums, which are the large sheet-metal enclosures to which the ducts are attached. Plenums are often purchased with the furnace, so a custom plenum needn't be made. In the sample house a floor opening is required below the furnace to accommodate the plenum. Since I've selected an electric furnace, it could be located in the crawl space but I prefer to have the furnace where it can be readily serviced.

Design the Ductwork. The trick here is to get the right amount of heat to each part of the house. While it's possible to engineer the system by calculating the heat loss of each room and matching heat register and duct sizes to each room, most professional installers use a simpler method. The small rooms (such as the bathroom, bedroom, laundry room, and so on) each get one register. The larger rooms (such as the living room, dining room, and kitchen) each get two registers. The registers are located at the extremities and colder parts of the rooms, such as under windows. This helps to keep all parts of the house at the same temperatures. Typically 4-in × 12-in registers

with adjustable flaps (to regulate the amount of heat) are used. Other common register sizes are 2½ × 14, 4 × 10, and 4 × 14. An 8-in round duct would supply two 4 × 12 registers or one 4 × 14. A 10-in duct would supply three 4 × 12 registers. Single 4 × 12 registers would be supplied with 6-in or 7-in round ducts. Figure 7-12 shows a ductwork plan for the sample house.

Solar Systems

I installed three portable solar collectors in and around my house. These solar collectors are right on the leading edge of energy technology. If there is any sunshine on or near the house they automatically take up position. My color experimentation is with black, white, and calico collectors. One significant problem remains to be solved: How the heck do you get any usable energy back out of three fat cats after they have absorbed all that sunshine?

Seriously now, I have to refer you to some other sources on solar heating. Despite the appeal of putting that sunshine to work, it took all my available energy just to build a conventionally heated house. I do know this much: Passive systems are generally a lot cheaper than active solar systems. And if you want to heat with the sun, you've got to keep the southern exposure clear of trees. That makes the south side a good place to put the drainfield or the garden.

For More Heating Information

How to Install Central Heating and Cooling. Sears, Roebuck and Co., Catalog No. 42-9998, 1972.

McGuinness, William, et al. *Mechanical and Electrical Equipment for Buildings.* New York: John Wiley and Sons, 1980.

Mazria, Edward. *The Passive Solar Energy Book.* Emmaus, Pa.: Rodale Press, 1979.

Mother Earth News. *Handbook of Homemade Power.* New York: Bantam Books, 1974.

Sunset Editors, ed. *Insulation and Weatherstripping.* Menlo Park, Calif.: Sunset-Lane Publishing Co., 1978.

Uniform Mechanical Code. International Conference of Building Officials, Whittier, California 90601. Updated frequently.

CHAPTER 8

Permits, Schedules, and Estimates

CHAPTER 8

Permits, Schedules, and Estimates

The house design is shaping up. Now is a good time to check it over carefully. We need to decide if this house is right for us *before* getting too heavily commited on the project.

Review the Design

With the model of the house in front of you, try answering these questions:

Does it look right? Do you like it? Are the proportions attractive? Remember that trees and surroundings will enhance the appearance. You can't see all the details like siding, color, roof texture, and so on so you have to use a good dose of imagination at this point.

Can you realistically afford this house? Don't start work on a $40,000 house if you can only afford a $15,000 house. The part of this chapter on estimating describes a way to get a close idea of how much the house will cost.

Can you build this house? You'll probably need to read the rest of this book to answer this one. Make sure the design makes sense for a first-time builder working with limited equipment. Steel or concrete beams, for example, will require heavy equipment to install.

Can you live with the schedule? Do you recall the photos in Chapter 2 showing the different owner-built houses and the time required to build them? Review them again and read the "Scheduling" section of this chapter. Decide if your design is a 1-year or a 5-year project, and what could be done to speed the construction process.

Where could you compromise on this design? Is it a dream house that could be scaled down? Could the design be arranged so you could build part of the house and move in, leaving other parts for the future? This approach has made the difference between success and failure for many a bewildered builder. (See Fig. 8-1.)

How long will this house last? Is the foundation adequate? Sneak a peek under some beautiful old

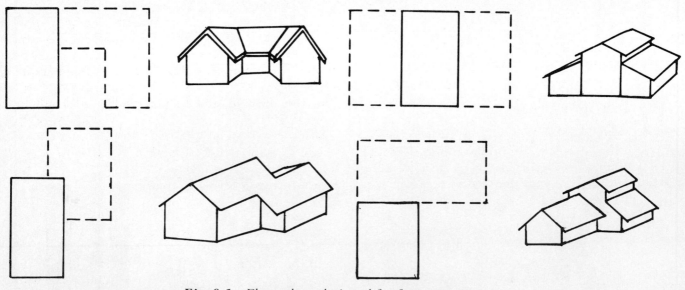

Fig. 8-1 *Floor plans designed for future expansion.*

houses that are starting to sag and you'll see some sad stories. Does the roof design shed water or collect water? Are the floor joists and rafters well ventilated? Is the structure strong enough to withstand Aunt Katharine backing into it with her old DeSoto?

What will it be like to live in this house? Make yourself small and walk up to the house model. Knock on the door. Are you covered by a porch roof or standing in the rain? Can the person inside see you before he or she opens the door? Does the winter wind blow in the door when it's opened? It's a pleasant entry I see. Park your muddy boots and raincoats in that handy closet. The light switches are right by the door. Good planning. Take that soggy bag of groceries into the kitchen, set it on the counter next to the refrigerator and put the ice cream away before it melts. Ahh, fish for dinner. Unfortunately there's no exhaust fan over the range so we'll have fish smell for breakfast too. Go sit with the family in the living room and find out what went on today. The sunset is beautiful. Luckily you made that window so you can see the sunset from the couch, without standing up. But brrrr, it's freezing in here! Who the heck designed that heating system? All the heat goes upstairs and there's no way to keep the living room warm without it getting 85°F upstairs. You'll have to fix that someday.

And so forth. The idea is to mentally road test the house before it's built. Once the design is well formulated and meets all your requirements, the next step is to apply for a building permit.

A Day and a Night at City Hall

This is possibly the most glorious day of my life. All my biorhythms are at their peak. I'm wearing a new pair of slacks, my house plans are tucked in my attaché case and I'm brimming with confidence. Under the facade of conservative threads I'm wearing my "What? Me Worry?" T-shirt. I stride triumphantly into the County Administration Building. Today's the day I get my building permit.

The directory is explicit: "Building Permits, Room 875." Feeling good, I take the stairs up to the eighth floor. My thumping heart tells me I need the exercise. Hm, Room 875, Wedding Li-

censes. The man behind the desk says "I'm sorry sir, the Building Permits office is now in Room 306."

Back down in Room 306 I take a number and stand waiting while six other people transact their business. When my turn finally comes it's another "I'm sorry sir, but first you have to get Legal Lot Verification from the Assessor's Office. They're in Room 654." Back up the stairs, another wait and pretty soon I've got the necessary form in my hands. I hurry back to Room 306 and take another number. I'm beginning to have some doubts about the validity of biorhythms. This time I'm still two customers back when the clock strikes 3 o'clock and the people behind the counter vanish into thin air. For awhile I stand there trying to figure out if this is a silent fire drill or something. Nope. Coffee break time. Then I remember that my parking meter was only good until 2:30 and rush out of the room.

Back again, I finally get to talk to a fat, gruff guy who doesn't look at all pleased to see me. "A do-it-your-selfer, eh?" he growls past the cigar chomped between his teeth. "Let's see your drawings." Producing my plans, I silently expect him to be impressed with their neatness and completeness. "We gotta have kelvinated blueprints," he demands, not even bothering with the perfunctory "I'm sorry sir." After asserting myself to the best of my ability, he growls something about possibly being able to make my drawings into kelvinated blueprints in Room 1005.

Unfortunately the process is long and tedious, though the clerk is real nice for a change. She shows me how to run a large, imposing machine which jams continually spewing huge curling sheets of paper onto the floor. It becomes apparent that I'll never finish in time to get my permit today, but at least I'll be one step closer. I've nearly filled the paper recycling bin single-handedly by the time I'm done. It's late and everyone in the office has gone home except the one helpful lady. Her name is Lucy and she's probably the only person in the county who understands the frustrations of dealing with county regulations. She says Mr. Fat-and-Gruff used to work for an eviction company before he got the job in the Building Permits office.

Lucy needs help blueprinting some skyscraper drawings and I agree to do it. It's sort of a funny feeling to be working on the top floor of the county building at night. The sunset over the mountains and the city lights make a rosy glow outside. The place isn't nearly so forbidding after everyone has gone home. Lucy is pleasant company and I feel some of my muscle tension relaxing. When we're done she gets a twinkle in her eye and produces a bottle of Southern Comfort out of a co-workers desk. We get to talking about all kinds of things. Some of it is scientific and intellectual; for example, we decide that Mr. Fat-and-Gruff might interest anthropologists searching for the Missing Link.

My memory is kind of fuzzy for a space here, but the next thing I'm aware of is the sound of a door opening and military-sounding footsteps at the far end of the room. Where the heck am I? Damn if I'm not curled up in the paper recycling bins with nothing on but my skivvies and counter-culture T-shirt! My new slacks hang over the edge of the bin. I just keep low and quiet as the footsteps approach, covering myself with a big kelvinated blueprint. My tension climbs as the steps walk right up to the bin and I hear the keys in my pants jungle as he picks them up. My heart and my lungs have stopped. Suddenly I get jabbed in the rear with a yardstick. "Yiiii!!" I scream, erupting from the paper bin. "You miserable son of a bureaucrat! It's my house and I'll build it anyway I please!"

The element of surprise has the morning security guard in full retreat. He makes a hasty dash for the door, holding up my slacks to shield himself from attack. I'm in hot pursuit threatening him with a number of dispenser from the counter. "I'll show you what you can do with your county regulations!" In his panic he trips over a waste basket and clangs his head on the corner of a file cabinet. He's out like a dead fish. God, I hope I didn't kill him. He isn't moving. Oh well, I'll just prop him up here behind a desk and it may be weeks before anyone notices any difference between him and the other bureaucrats.

Flick! A light flashes on.

"Frank! Wake up! Wake up!" It's Deb. I'm out in the dining room with a toaster in one hand and sheepishly trying to prop up Bernard (the big stuffed dog) in a chair at the table.

"Frank, you've got to stop worrying about that building permit. You scared me half to death. Jumping up in the middle of the night screaming obscenities and chasing around the place. You've got to relax. It won't be that bad."

My dreams always seem very realistic while I'm having them but they're usually not too accurate with the facts. It turned out that getting a building permit was somewhat easier than my dream had forewarned. There were lines to wait in, number dispensers on the counters, but no one was as bad as Mr. Fat-and-Gruff (nor as nice as Lucy). I asked one guy if I needed kelvinated blueprints. He gave me a puzzled look and said

no, pencil drawings on 8½ in × 11 in graph paper was fine. "What the heck is a kelvinated blueprint?" he asked. Obviously his dream programmer was different than mine.

Try to understand the permit process of your county (or city) early. Call up the building department and ask for requirements for getting a building permit. These days one out of five homes are built by the owner, so most counties have a standard information sheet explaining what drawings and paperwork are required. They may have a summary of the building code requirements. You're bound to discover some things you hadn't planned on, like minimum setbacks from property lines, special foundation requirements or a 3-week delay while the plans are being "processed." The sooner you know the requirements, the better off you'll be.

Every county/city has different requirements but Fig. 8-2 shows some of the drawings that are commonly needed.

Building Codes

In response to tragedies involving structural, fire, and sanitation problems in houses, building codes were developed. For a long time every town and borough in America developed its own building codes, creating the utmost confusion for builders and manufacturers. Now most building officials have gotten together and published "uniform codes." These codes are updated every few years. The primary codes in use are the *Uniform Building Code* (UBC), *Uniform Plumbing Code* (UPC), *National Electric Code* (NEC), and *Uniform Mechanical Code* (UMC).

The Uniform Building Code covers nearly everything related to the structure and livability of the house. The Uniform Plumbing Code covers plumbing, and the National Electric Code covers the electrical, and very thoroughly I might add. The Uniform Mechanical Code covers the heating, ventilation, and cooling systems. The primary problem for novices is sifting through the five million rules to find the hundred or so rules that you really need to know. The *Dwelling Construction Under the Uniform Building Code* book referenced at the end of Chapter 6 is a helpful summary.

Most code rules have very good reasons behind them. For example, a septic tank installed in poorly percolating soil will stink something fierce, making your yard a less than pleasant place for picnics. Building a house on fill dirt can

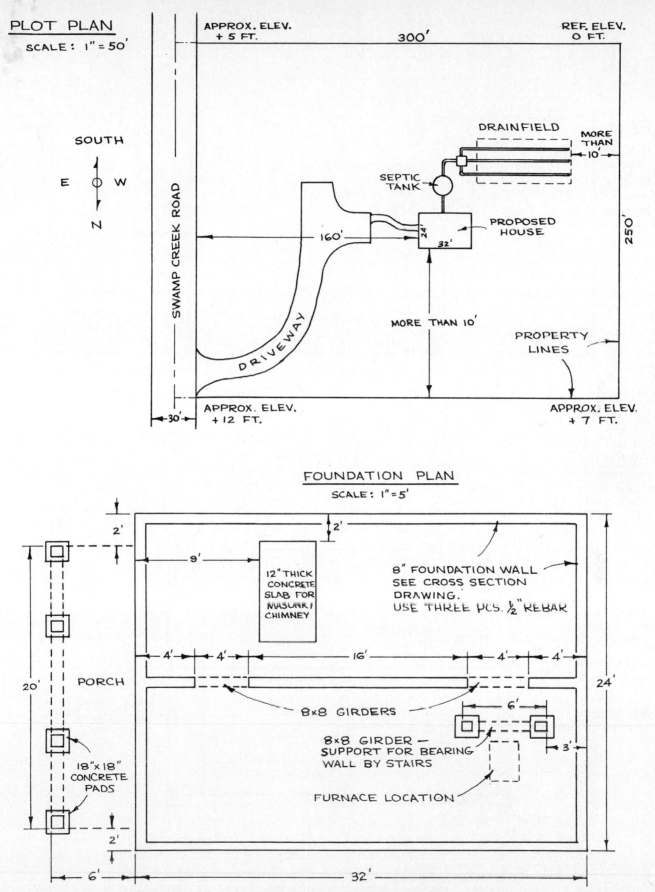

PLOT PLAN

SCALE: 1" = 50'

APPROX. ELEV. + 5 FT.

REF. ELEV. 0 FT.

300'

SOUTH

E ⊙ W

N

SWAMP CREEK ROAD

DRIVEWAY

DRAINFIELD

MORE THAN 10'

SEPTIC TANK

160'

24'

32'

PROPOSED HOUSE

250'

MORE THAN 10'

PROPERTY LINES

30'

APPROX. ELEV. + 12 FT.

APPROX. ELEV. + 7 FT.

FOUNDATION PLAN

SCALE: 1" = 5'

2'

9'

2'

12" THICK CONCRETE SLAB FOR MASONRY CHIMNEY

8" FOUNDATION WALL SEE CROSS SECTION DRAWING. USE THREE PCS. ½" REBAR

20'

4' 4' 16' 4' 4'

24'

PORCH

8×8 GIRDERS

6'

18"×18" CONCRETE PADS

8×8 GIRDER — SUPPORT FOR BEARING WALL BY STAIRS

3'

FURNACE LOCATION

2'

6' 32'

Fig. 8-2 Drawings commonly required for a building permit.

FLOOR PLAN

FIRST FLOOR
768 SQ. FT.

SECOND FLOOR
480 SQ. FT.

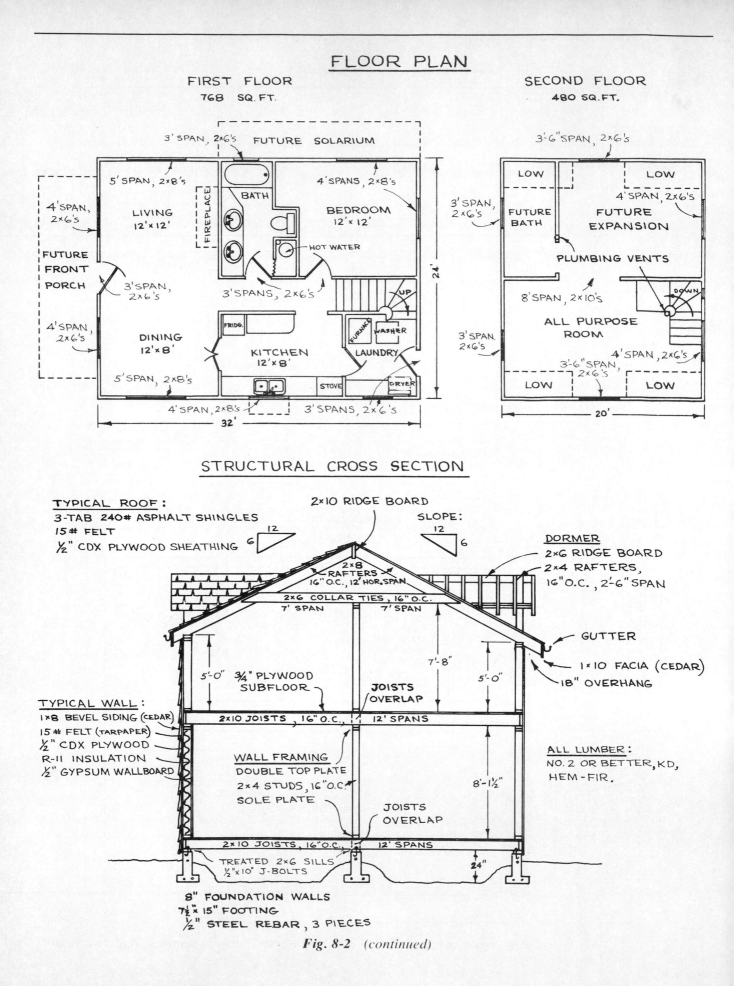

FUTURE SOLARIUM

3' SPAN, 2×6's

5' SPAN, 2×8's

4' SPANS, 2×8's

4' SPAN, 2×6's

BATH

FIREPLACE

LIVING 12'×12'

BEDROOM 12'×12'

FUTURE FRONT PORCH

HOT WATER

3' SPAN, 2×6's

3' SPANS, 2×6's

4' SPAN, 2×6's

DINING 12'×8'

FRIDG.

KITCHEN 12'×8'

UP

FURNACE

WASHER

LAUNDRY

5' SPAN, 2×8's

STOVE

DRYER

4' SPAN, 2×8's

3' SPANS, 2×6's

32'

24'

3'-6" SPAN, 2×6's

LOW

LOW

4' SPAN, 2×6's

FUTURE BATH

FUTURE EXPANSION

3' SPAN, 2×6's

PLUMBING VENTS

8' SPAN, 2×10's

DOWN

ALL PURPOSE ROOM

3' SPAN, 2×6's

4' SPAN, 2×6's

3'-6" SPAN, 2×6's

LOW

LOW

20'

STRUCTURAL CROSS SECTION

TYPICAL ROOF:
3-TAB 240# ASPHALT SHINGLES
15# FELT
½" CDX PLYWOOD SHEATHING

2×10 RIDGE BOARD

SLOPE:

12 / 6

12 / 6

DORMER
2×6 RIDGE BOARD
2×4 RAFTERS, 16" O.C., 2'-6" SPAN

2×8 RAFTERS, 16" O.C., 12' HOR. SPAN

2×6 COLLAR TIES, 16" O.C.
7' SPAN   7' SPAN

7'-8"

GUTTER

1×10 FACIA (CEDAR)

18" OVERHANG

5'-0"

¾" PLYWOOD SUBFLOOR

JOISTS OVERLAP

5'-0"

TYPICAL WALL:
1×8 BEVEL SIDING (CEDAR)
15# FELT (TARPAPER)
½" CDX PLYWOOD
R-11 INSULATION
½" GYPSUM WALLBOARD

2×10 JOISTS, 16" O.C.   12' SPANS

WALL FRAMING
DOUBLE TOP PLATE
2×4 STUDS, 16" O.C.
SOLE PLATE

JOISTS OVERLAP

ALL LUMBER:
NO. 2 OR BETTER, KD, HEM-FIR.

8'-1½"

2×10 JOISTS, 16" O.C.   12' SPANS

TREATED 2×6 SILLS
½"×10" J-BOLTS

24"

8" FOUNDATION WALLS
7½"× 15" FOOTING
½" STEEL REBAR, 3 PIECES

Fig. 8-2 *(continued)*

be like burying a sack of paper money in the woods: both become useless in a period of a few years. Lack of vents in the foundation will seriously shorten the life of the floor joists. Fireplaces without proper chimney clearances from combustibles are a leading cause of home fires and deaths. Putting too many wires in an electrical box increases the chances of short circuits. Every code rule has a reason behind it, though some of them can be quite puzzling.

We'll talk about specific code requirements throughout the book as each subject is discussed to give you some exposure to typical code requirements. Since code rules vary by locality and are frequently updated, **it's imperative that you check with your building department to learn the local rules.** They may not be the same as the requirements discussed in this book. Also, it's a good idea to get approval of your plans for building, septic system, plumbing, electrical, and heating *before* starting each step.

Scheduling

Scheduling is perhaps the most difficult task of all. I was terrible at it. I would tell Deb that I was going over to the house to do a 2-hour task. Returning 4 hours later, I'd have the task only half completed and have discovered a couple more "2-hour" tasks that I hadn't planned on. Such is the life of a amateur builder. Some things go much easier than expected and other things get totally out of hand.

Though individual tasks are difficult to schedule, in general, you can look at how long it took other rank beginners to build a house and get a pretty accurate idea of how long it will take you to build the same house. Here's the system I use:

Type 1 House. Flat site, one story, rectangular shape, conventional floor and wall construction, prefab truss roof, aluminum windows, texturized plywood siding, 3-tab asphalt roofing, one owner-plumbed bathroom, regular 200-amp electrician-assisted wiring job, baseboard heaters, wallboard contracted out, prefab counters and cupboards, and minimal finishing. Plan on 1.5 hours per square foot for Type 1 houses. For example, a 1,000-sq ft house would require 1,500 hours.

Type 2 House. Sloped site, L-shaped floor plan with a porch, conventional floor and wall construction, prefab truss roof with skylights, mostly wood frame windows, bevel siding, 3-tab asphalt roofing, 1½ owner-plumbed bathrooms, regular 200-amp owner-wired electrical service, forced-air furnace heating system, owner-installed wallboard, prefab counters and cupboards, two or three special features (stairs, bay window, and so on), and medium finishing. Plan on 2 hours per square foot for Type 2 houses. For example, a 1,500 sq ft house would take 3,000 hours.

Type 3 House. Lots of site preparation for driveway and house, complex floor plan and foundation, conventional floor and wall construction but three stories in places, complex custom-framed roof structure, owner-built windows and doors, fancy shingled siding, hand-split shake roofing, two owner-plumbed bathrooms, 400-amp owner-wired electrical service, heat pump system, owner-installed wallboard, owner-crafted counters and cupboards, several special features (greenhouse, sunporch, dormers, and so on), and nicely finished with hardwood floors and attention to detail. Plan on 3 hours per square foot for Type 3 houses. For example, a 2,000 sq ft house would take 6,000 hours. If the hubby and spouse argue a lot, add ½ hour per square foot.

Our sample house is roughly a Type 2 house, slightly more difficult in some ways and slightly easier in other ways. A lot depends on the selection of the site, type of windows, plumbing, heating system, number of special features and degree of finishing. My ballpark estimate for the 1,250 ft sample house would be 2,500 hours, for a beginner. Then to bring the house up to "Tour of Homes" standards, probably another 500 hours of detailing.

To convert the total hours required into a calendar schedule, plan on no more than 50 hours per week if you'll be at it full time (40 hours is more likely) and no more than 20 hours per week if you have another full time job. For example, the 2,500-hour sample house would take 125 weeks (2½ years) at 20 hours per week and about a year and 3 months at 40 hours per week. With a part-time helper you could knock a couple of months off the schedule. Table 8-1 shows approximately how much of the time each stage would require.

Several other factors can affect the time required. The experience of the builder and the number of folks working on the house are the main consideration. Don't overestimate the power of inexperienced help though. After estimating the total hours based on square footage and converting that to weeks required, multiply by the following experience and help factors: Inexperienced person working alone: 1.0; extra energetic person with some building experience: 0.8; inexperienced person with part-time helper: 0.9; two inexperienced persons: 0.8; two people

TABLE 8-1 Time Estimate for the Sample House

Hours	Task
300	*Preparation for Building.* Driveway, water and power hookup, build sawhorses, build outhouse or shed, and so on.
60	*Excavation and layout.* Hire bulldozer, set up batter boards.
200	*Foundation.* Build footing forms, pour footings, rent wall forms, pour walls, cleanup and curing, seal concrete, drainage system, backfill.
100	*Build the first floor.* Work carefully and make sure it's square!
160	*Frame and sheath walls, frame second floor.*
120	*Frame and sheath roof.* Half of roof built with trusses, half is custom-framed, dormers not included in this.
100	*Apply roofing.* Including valleys, gutters, flashing, and facia boards.
240	*Windows, doors, and siding.* Mostly preframed windows, two custom-arched windows, prehung doors, bevel siding.
320	*Special features.* Stairs, 100 hr; dormers, 80 hr; bay window, 50 hr; porches, 90 hr.
100	*Septic system.* Dug with backhoe; gravel and dirt placed with wheelbarrow.
200	*Plumbing.* One bathroom, owner-plumbed.
230	*Wiring.* Standard 200-amp service, owner-wired.
100	*Heating.* Electric forced-air furnace, owner-installed.
80	*Insulation.* Fiberglass blankets.
200	*Wallboard, tape, putty, and paint.* Owner-installed.
40	*Counters and cupboards.* Prefab basics only, owner-installed.
250	*Miscellaneous.* Closets, finish trim, shelves, railings, finish floors, and time spent on various problem areas.
TOTAL: 300	Hours to prepare for building
2,500	Hours to build house

with some experience: 0.6; professional specialized crew averaging three people: 0.3. Other factors to consider are weather, financial situation, and health of the builder.

Estimating the Dollar Cost

Armed with a house design and some of the decisions made in scheduling the house (for example, what to subcontract), you can now put a price tag on this dream. Start out by making a rough guess based on the house photos in Chapter 2. Write it down and see how close you come to it with a item-by-item accounting.

The big task here is to compile a bill of materials listing every item that will go into the house. It's drudgery but there's no way around it. Then price each item and add up the damages.

Clear off the table by the phone, place the house model and drawings in front of you, and start listing the materials in sequence the way the house will be built. Plan on a few more materials than seem absolutely necessary. This gives a safety margin for forgotten items and for small design improvements that get made as you are building. Space doesn't permit a stud-by-stud, rafter-by-rafter count here, but Fig. 8-3 and Tables 8-2, 8-3, and 8-4 show the main components of an estimate for the sample house in 1983 West Coast prices.

To get prices for these materials, you have several options. One is to take the list to a building

TABLE 8-2 Dollar Estimate for Sample House (1250 sq ft)

Cost	Item
$ 1,500	*Preparation for building*
	$1,000 Water hookup (fee, digging, pipe)
	$ 100 Temporary power
	$ 100 Driveway (culverts and gravel at turnaround)
	$ 100 Building permit
	$ 200 Miscellaneous
$ 200	*Excavation and layout*
	$ 100 Bulldozer (4 hours)
	$ 100 Batter boards, baling wire, miscellaneous
$ 1,000	*Foundation*
	$ 50 Nails and nonreusable lumber (2 × 8 footing forms will be used as rafters)
	$ 200 Rental of wall forms
	$ 450 Concrete (9 cu yd)
	$ 150 Rebar, J-bolts, form ties
	$ 150 Mudsills, vents, drain pipe, sealer, and so on
$ 1,300	*Floors*
	$ 750 1,600 linear ft of 2 × 10 (includes 100 ea 12 ft)
	$ 500 40 sheets ¾-in plywood underlayment
	$ 50 Miscellaneous
$ 1,000	*Walls*
	$ 450 3,000 linear ft of 2 × 4 (includes 200 lengths at 8 ft)
	$ 60 Headers (100 ft of 2 × 6, 60 ft of 2 × 8, 16 ft of 2 × 10)
	$ 400 50 sheets ½-in CDX plywood sheathing
	$ 90 Miscellaneous
$ 1,200	*Roof framing*
	$ 300 Trusses (9 common, one gable end, 6 in 12 slope, 24-ft span, 18-in overhang)
	$ 350 Custom-framed roof (750 ft of 2 × 8 rafters including 32 16-ft lengths, 24 ft of 2 × 10 ridge, 300 ft of 2 × 6 ceiling joists, 100 ft of 2 × 4 dormer)
	$ 320 40 sheets ½-in CDX plywood sheathing
	$ 230 Scaffolding, gable vents, miscellaneous
$ 1,500	*Miscellaneous structural*
	$ 100 Stairs (add $200 for oak treads and railings)
	$1,000 Porches (front and back)
	$ 100 Kitchen bay window
	$ 300 Miscellaneous
$ 300	*Nails* 150 lb 16d sinkers, 50 lb 8d bright box, 100 lb 8d hdg box, 100 lb 6d hdg box, 50 lb 1½-in roofing, 100 lb 7d siding, 50 lb wallboard, assorted others
$ 800	*Roofing*
	$ 420 1200 sq ft 3-tab asphalt shingles
	$ 200 Tar paper, flashing, gutters, downspouts
	$ 100 1 × 10 facia board (cedar, 200 linear ft)
	$ 80 Miscellaneous
$ 1,400	*Windows and doors*
	$ 800 Windows (total of 16, eight new and eight salvage, double-pane wherever possible)
	$ 250 Doors (two solid-core exterior, five interior, half of them salvage)
	$ 300 Hardware, stops, moldings, weatherstripping
	$ 50 Miscellaneous
$ 1,600	*Siding and exterior trim*
	$1,300 3,200 linear ft 1 × 8 cedar bevel siding
	$ 100 400 linear ft 1 × 4 cedar trim
	$ 100 Stain (two colors, 10 –15 gal)
	$ 100 Tar paper, drip cap flashing, brushes, and rollers

(continued)

TABLE 8-2 (*Continued*)

Cost	Item
$ 800	*Septic system*
	$ 350 1,000-gal septic tank
	$ 150 Backhoe work (6 hours)
	$ 150 Drain pipe, gravel, grade boards, and so on
	$ 150 Miscellaneous
$ 1,500	*Plumbing*
	$ 250 Copper supply piping
	$ 200 ABS plastic DWV piping
	$1,000 Fixtures and appliances (half new, half salvage)
	$ 50 Miscellaneous
$ 1,500	*Electrical*
	$ 200 Service entrance, panel, breakers
	$ 500 Wiring (about 1,500 ft, mostly "12/2 with ground," includes boxes, switches, outlets, and so on)
	$ 800 Light fixtures (modest) and used appliances (dryer, stove, refrigerater)
$ 1,000	*Heating*
	$ 400 Electric forced-air furnace (60 kBtu, 18 kW)
	$ 500 Ductwork
	$ 100 Miscellaneous
$ 600	*Insulation*
	Fiberglass blankets, foil lined:
	$ 350 R-11 floor and walls (2,300 sq ft)
	$ 200 R-19 ceiling (800 sq ft)
$ 800	*Wallboard and paint*
	$ 500 150 sheets ½-in 4 × 8 gypsum wallboard
	$ 200 Putty (25 gal), tape, and paint (15 gal)
	$ 100 Miscellaneous
$ 1,000	*Counters and cupboards*
	Prefab, kitchen and bathroom
$ 2,000	*Miscellaneous*
	Just enough finishing to make it liveable, but not enough to impress anyone
$21,000	*Total for the basic house*

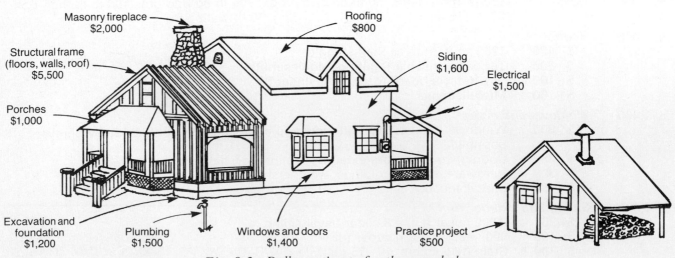

Fig. 8-3 *Dollar estimate for the sample house.*

Masonry fireplace
$2,000

Roofing
$800

Structural frame
(floors, walls, roof)
$5,500

Siding
$1,600

Electrical
$1,500

Porches
$1,000

Excavation and
foundation
$1,200

Plumbing
$1,500

Windows and doors
$1,400

Practice project
$500

TABLE 8-3 Basic Tools

Cost	Purpose
$ 124	*General carpentry:* hammer ($10), handsaw ($12), 16-in tape measure ($7), 100-ft tape measure ($15), combination square ($8), carpenter's square ($12), leather apron ($23), level ($15), plumb bob ($3), T-bevel ($7), cat's paw ($7), chalkline ($5).
$ 119	*Site preparation:* shovel ($10), pick ($12), axe ($12), bow saw ($10), lopping shears ($15), garden hose ($10), wheelbarrow ($50).
$ 475	*Power tools:* skill saw ($35), radial saw ($250), small drill ($20), large ¾-hp drill ($70), saber saw ($20), extension cord ($25), blades and bits ($50), safety glasses ($5).
$ 80	*Foundation:* square-nose shovel ($12), water level ($7), crowbar ($10), sledge ($15), lineman's pliers ($8), hack saw ($5), tin snips ($8), stapler ($15).
$ 48	*Windows and doors:* back saw and wood miter box ($20), wood chisel ($6), hand plane ($11), screwdrivers ($5), portable work light ($6).
$ 52	*Plumbing:* tubing cutter ($6), file ($3), propane torch ($15), two adjustable wrenches ($10), channel lock pliers ($8), two pipe wrenches ($10).
$ 12	*Electrical:* cable ripper ($1), wire stripper ($6), long nose pliers ($3), circuit tester ($2). Already listed: lineman's pliers, ¾-hp drill and bits, screwdrivers.
$ 65	*Wallboard and finish:* utility knife ($3), rasp ($5), wallboard saw ($4), floor lever ($9), metal straightedge ($5), putty knives ($15), trowel ($6), pole sander ($12), long-handled paint roller ($5), nail set ($2).
$ 90	*Helpful extras:* 6-ft step ladder ($40), shop vacuum ($50).
$1,065	*Total for basic tools*

TABLE 8-4 Optional Extravagances

Cost	Item
$ 2,000	Masonry chimney
$ 4,000	Carpets throughout
$ 1,500	Quality light fixtures
$ 1,500	New clothes washer, dryer, and stove
$ 2,500	New bathtub, fancy sinks, and faucets
$ 1,000	Sundeck
$ 4,000	Solarium or greenhouse
$ 6,000	Two-car garage
$22,500	*Total*

materials store. They'll often price material lists for free, though they'll want the list to be very specific. Another way is to get out the Sunday paper and look at the lumber ads. These prices are usually sale prices so recognize that. The big mail order outfits (Sears, Wards, JC Penneys,

and so on) have catalogs that are good sources for prices on tools, appliances, plumbing, electrical, heating, cupboards, and hardware. A few calls on the phone will yield prices for services like plumbing, drywall, bulldozing, and the like, though you won't be able to get any firm quotes until the contractor can actually look at the job he's estimating. After burning out a few calculator batteries, you'll be looking at the price of the most expensive purchase you'll ever make. Add 10 percent for inflation and find a soft place to sit while you think about it.

Looking through the knothole of the amateur owner-builder, it's imperative to develop a realistic schedule and cost estimate for the house design. Now is the time when modifications are easily made, later it will be more difficult. Remember too that while the price tag and hour requirement may make your eyes bulge at the start, it's amazing how steady work and a steady monthly investment can make such a project possible.

Don't compare yourself with professional builders. They can typically build a 1,500-sq ft house in 3 to 6 months, but you're in for a shock if you think you can do it that fast. Speed is their bag and believe me, you can't begin to compete in that realm unless you subcontract most of the work. Where you can beat the professional builder inside out is on price. Besides avoiding the cost of interest, by providing all of the labor you'll cut the dollar cost of the house in half. A professional builder could no more produce our sample house for $21,000 than he could produce a goose that lays golden doorknobs. As the houses shown in Chapter 2 demonstrate, a hard-working amateur has a huge bag of tricks available to keep the costs down.

Getting the Right Perspective

In the last five chapters we've moved from a vague dream about a home to a solid plan of how the house will be built, what systems are necessary, how long it will take to build, and the price tag for dreaming such dreams. If you've read this in two or three sittings, you're probably feeling somewhat overwhelmed by all the decisions to be made. That's one reason why house designs are best developed over a period of several months. It's one thing to blithely sketch a pretty house; it's another to actually plan how that house will be built so all the structural, plumbing, electrical, heating, living, aesthetic, financial, and schedule arrangements fit together. Time consuming, yes,

but the time spent on designing a house is repaid tenfold during construction and repaid a hundred-fold to the occupants of the house.

I hasten to add that the design approach I've described is certainly not the only way to go at it. It's a method that's field tested and it works, but be sure to talk to other owner-builders and see how they did it. Looking back, I feel that the most important points of a design process are that

1. The house ends up being about the right size for your pocketbook and schedule

2. The structure is solid and fairly simple

3. Enough time is spent gathering information that there are no unpleasant surprises during the construction

If I had it to do over again I'd make a concerted effort during the design stage to locate and help some other owner-builders. They can almost always use some menial labor while pouring the foundation, erecting walls, roofing, wallboarding, and painting. Just the experience of being around a construction site is very valuable.

Of course all this house-design brainwork is fine and dandy, but it's entirely based on the assumption that you can locate the business end of a hammer and realize your ideas in concrete and wood. The time has come, dear reader, to park the T-square, white shirt, and house design books and to slip into your overalls.

Part Three

BUILDING

CHAPTER 9

Acquiring Wood-Butcher Skills

There are at least a couple ways to learn to swim. One way is to jump into the deep water at the end of the pier and go for it. Another way is to learn the basics progressively, preferably in the shallow end of a nice warm pool. Skiing is similar. You can take the chairlift to the top of the mountain and point your skis downhill or you can start on a nice gentle slope learning one step at a time. Cooking, reading, baseball, driving, you name it, both approaches to acquiring a skill seem to have their disciples. Whichever way you prefer, you have to get in the water at some point. For building a house, we've arrived at that point.

Do you recall Uncle Barney's arguments against an amateur trying to build a house back in Chapter 1? Barney's theory is to trust each task to an expert, a specialist, rather than to attempt to learn it yourself. That's a good technique if you can afford it and really don't care whether or not you ever learn the skill. However, if your desires exceed your income, you either have to do without the house or figure out how to build it yourself.

Being a believer in a streamlined, progressive step approach to learning, I've set out in this book to take an absolute novice and move him or her along a fast track to being a house builder. What I'm going to do is start you with the most basic of basics and throw you into progressively deeper water until you're nailing up the cupboards in your new house.

This book has a bit of an advantage over some other types of manuals. That is, building a house is a sequential process insofar as you build the foundation first, then the floor, then the walls, and then the roof. Also, once you know how to check a floor to insure it is level and square, you will undoubtedly be able to make the eaves of the roof level or a portion of the roof square. In this way you will build a basic knowledge as we progress, and I will be able to keep the explanations simple and brief. Brevity is also attained by avoiding many topics that are interesting, but not essential. Since there are still a lot of things to remember, I expect that you will need to read the entire book once, then come back to each step when you're ready to use it and read it again.

Practice Projects

Digging

Let's start at the ground floor, or rather, slightly below the ground floor with a step everyone would prefer to skip.

The first hurdle is to get the mind and body accustomed to good old hard, dirty work. Shoveling dirt is a perfect way to achieve this end. It's sweaty, dirty, kills the back, blisters the hands, and, to add insult to injury, could be done by a machine (a backhoe) in about one-tenth of the time it takes to do it by hand. Furthermore, digging in the rain, if you can arrange for this, builds more character per hour than any sport known to man.

Digging a hole for an outhouse is a great first project. It's not a big enough job to be worth hiring a backhoe; it's reasonably necessary unless you don't mind squatting in the bushes or bothering the neighbors every day to use their bathroom; and it serves well as a test case insofar as you must be sure to dig the hole deep enough the first time because it will definitely be a bad job to dig it deeper later on. Besides all that, the outhouse itself provides a good way to practice sawing boards and pounding nails.

In a city or suburban area, an outhouse may be inappropriate or illegal. Check with the Health Department. In such cases, a portable toilet can be rented. At any rate, I'll talk first about an outhouse because it makes such an excellent practice project. The same structure can be built and used as a tool shed.

Select a good location that is at least 100 ft from

Fig. 9-1 *Digging.*

any well, stream, or lake. Avoid sandy and gravelly soils; they allow the effluent to disperse and pollute underground water sources. Also avoid areas where you notice water seeping into the hole as you dig or a rust colored mottling of the soil. Both signs indicate a high water table in the area. Don't take the chance of polluting your neighbors' wells.

For the outhouse design shown later in this chapter, dig a hole 3 ft long, 2½ ft wide, and about 4 ft deep. Assuming normal conditions, this should be sufficient until you get a toilet installed in the house. Dig the sides of the hole straight down (not slanted) and pile the dirt back from the hole so the 3 ft × 4 ft outhouse can sit level over the hole. If you don't want an outhouse, you can dig post holes for a fence, a ditch for the water line, a root cellar, or a hole for the septic tank. (See Fig. 9-1.)

An Apparition from the Past

"Well, well. Look who's gonna build himself a palace!"

"Oh, hi Uncle Barney. This is the last place I expected to see you."

"I was talking with your mom and she said you were gonna start some sorta project today. I thought maybe you could use some experienced help."

"Oh, OK." Good grief, just what I needed. "Today I'm just digging a hole for the outhouse,

Barney. But since you're here, maybe you can give me a hand."

"An outhouse? What the heck for? We should start laying out the foundation. Let's take a look at your house plans."

As soon as I got the plans out of the truck I knew I'd made a grave error. Barney didn't like the plans ("Real carpenters use blueprints.") but he seemed to be determined to lay out the house today and order some concrete anyway. It was also apparent that Barney planned to supervise the job from the director's chair he brought, and I was to follow his directions. I noted Barney was careful not to get dirty, and his right hand was never far from a beer can.

"Here now, Frank. You start stretching the batter board lines and I'll take a good look at these so-called plans."

"I can't start that today Barney. The county hasn't sent me my building permit yet." This seemed like a good occasion to be super legal, and I had to find some way to get rid of the guy.

To make a long story short, after a couple hours of haggling I just picked up my shovel and continued to dig my hole. Barney finally muttered something about it being too crowded in the hole for both of us and left. I was surprised at how eager he was to "help," considering his total condemnation of the project up to that time. Also it was surprising how little he knew about what it took to get a building permit. I suppose I had hurt his feelings, but with any luck at all maybe he would leave me alone for awhile.

Sawhorses

The next step is to try sawing a board to length in the middle of a field. No props now, just you, a handsaw, and any scrap board you can find. As you try this, you'll soon realize why sawhorses were invented. If you can't hold a piece of wood firmly, without bending it, and at about the right height, sawing can be very difficult indeed. You'll notice this right off. Consequently we need some sawhorses and some instruction in sawing before we go any further.

Building sawhorses is a good project for those evenings at home. Acquire the supplies and tools in Fig. 9-2 from a lumber yard or other source and build two sawhorses like the one shown.

Well yes, I know. Those weren't very complete instructions. Suppose you have never sawed a board or hammered a nail before in your life? If that's the case, read the following step by step instructions and also the next two sections on "Carpentry Terminology" and "Basic Skills."

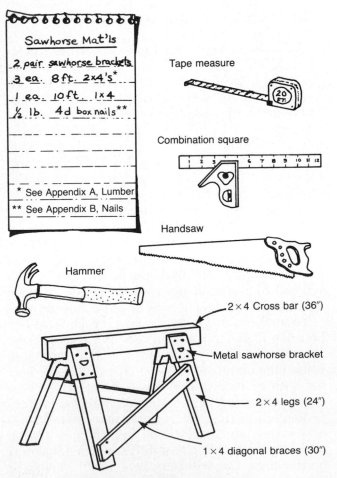

Fig. 9-2 Building sawhorses.

STEP-BY-STEP PROCEDURE

1. For now, use an old chair or a bench as a sawhorse.

2. Mark each board using a combination square before sawing.

3. Saw the legs. I suggest 24-in legs because they will make the sawhorse low enough so you can kneel on the boards as you saw them.

4. Assemble the legs, crossbar, and brackets.

5. Nail the brackets to the legs and crossbar. Try to brace the sawhorse against something while nailing: the floor and your foot, a bench, your legs, or whatever. You need a semi-solid backing to hammer against.

6. If this seemed like a cinch, try beveling the sawhorse legs so it will sit flat on the floor. This takes more skill with the handsaw.

I'll warn you right now that I haven't told you everything. Part of this practice project is learning to use your head to figure things out. As we progress, I will leave more and more to your own ingenuity and will warn you of only the most serious pitfalls.

Carpentry Terminology Illustrated

Figure 9-3 illustrates the following few simple carpentry terms:

Grain: The grain of a piece of wood shows the direction in which the tree grew. When looking at the annual growth rings of a tree, you are looking at the end grain.

Crosscut: A cut across the board (and across the grain.)

Rip cut: A cut along the length of the board (and along the grain.)

Bevel cut: A cut made with the saw blade at an angle to the board.

Plumb: Exactly straight up and down—vertical.

Level: Exactly level—horizontal.

Other carpentry terms can be located in the Index.

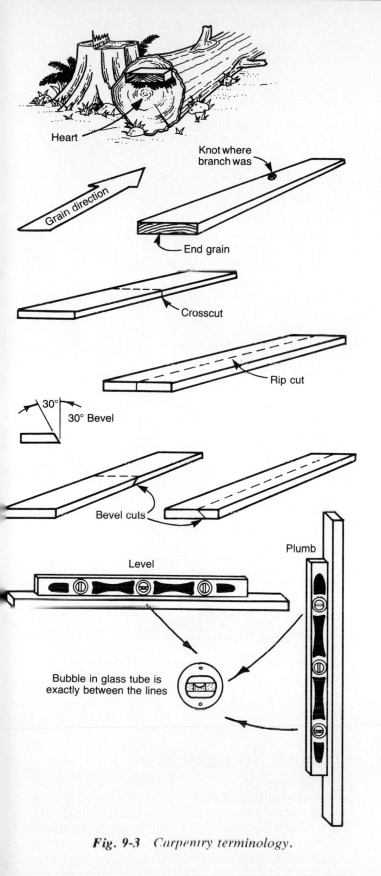

Heart

Knot where branch was

Grain direction

End grain

Crosscut

Rip cut

30°

30° Bevel

Bevel cuts

Plumb

Level

Bubble in glass tube is exactly between the lines

Fig. 9-3 *Carpentry terminology.*

Basic Skills

Sawing with a Handsaw

First mark the board using a pencil and a combination square. (See Fig. 9-4.) Mark both edges and one face of the board to guide the saw cut.

Support the board on a good solid surface. Kneel heavily on the board and hold it firmly with your free hand.

Fig. 9-4 *Sawing with a handsaw.*

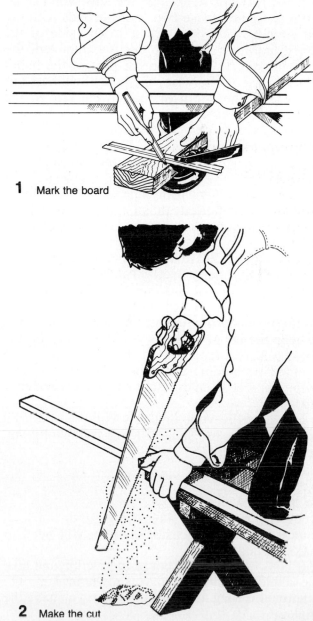

1 Mark the board

2 Make the cut

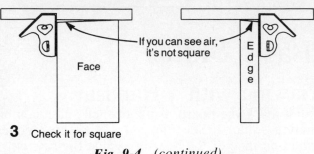

3 Check it for square

Fig. 9-4 *(continued)*

Hold the saw at about a 45-degree angle with the face of the board. Start the cut with a couple of short forward strokes so the saw won't jump around. I guide these two short strokes with my left thumb. Always saw just to the side of the pencil mark, so when the cut is completed, the line will still be visible on the part of the board that you use. Otherwise, the board will be short by at least the width of the sawcut (called the **kerf**).

The saw cuts only on the forward stroke, so apply the pressure then. Use long, steady strokes and don't waste any energy on the return stroke. Be careful not to kink the saw if the saw sticks. If the saw sticks, it is usually due to the board being improperly supported, thus bending and squeezing the saw. Sometimes it may be due to a knot in the board, which calls for short strokes. Other times it can be due to a dull saw, or a saw whose teeth are improperly "set." Take it to a good saw sharpener.

Make sure the saw blade follows both the line on the face and the line on the edge of the board to keep the cut square. As you finish the cut, the corners have a tendency to split off. This can be avoided by holding the falling end of the board with your free hand during the last couple of easy strokes.

Undoubtedly your first few cuts will be a bit ragged and not too square. Keep trying until you get the hang of it.

Using A T-bevel

About 90 percent of all sawing for a house is square cuts. But occasionally you will need to make an angle cut.

If possible, hold the board in position and take the angle on a T-bevel as shown in Fig. 9-5. Tighten the wing nut carefully when you have the angle.

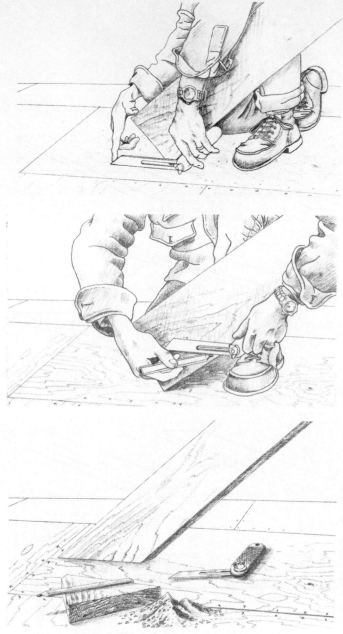

Fig. 9-5 *Using a T-bevel.*

Using a Straightedge

A straightedge is just like it sounds. It's a piece of wood, aluminum, or steel that is used for drawing straight lines. I use a 6-ft piece of angle aluminum that I found in a salvage yard. Some folks use a straight 1 × 6 or a piece of plywood. (A wooden straightedge must be kept dry or it can warp.) The length is a matter of preference and many people prefer eight footers. The sooner you get a straightedge, the better (Fig. 9-6).

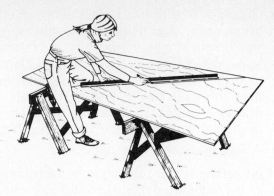

Fig. 9-6 Using a straightedge.

Eyeball for Straightness

Early in your building life you will discover that boards are not created equally straight. At times you'll wonder if there is such a thing as a straight board. How to tell if a board is straight? Give it the old eyeball. Hold one end of the board up to your eye and sight down the board. (See Fig. 9-7.) That previously straight-looking board almost always has a curve to it.

Fig. 9-7 Sighting a board.

The Tight-String-for-Straightness Trick

It's very important that the foundations, floors, walls and roof lines of a house are straight. Figure 9-8 shows an invaluable trick Dad showed me to test a long structure for straightness.

1. Drive an 8d nail partway into each end of the wall, roof, floor, cupboard, or whatever you want to be straight.

2. Tightly stretch a string or piece of twine between the two nails.

3. Insert a small ¾-in-thick block between the string and the wall at each end near the nail. The string will be tight enough to hold the two blocks in place and will be absolutely straight.

4. Now take another ¾-in block in your hand and run it between the wall and the tight string, checking for gaps or narrow spots. Use a hammer or sledge to adjust the wall so it is precisely parallel with the tight string.

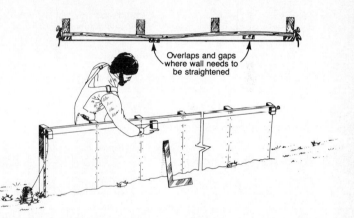

Overlaps and gaps where wall needs to be straightened

Fig. 9-8 Using a tight string to ensure straightness.

Using a Level

A level is used to determine whether a board or structure is level and/or plumb, as was illustrated in Fig. 9-3. Good carpenters use their levels frequently, and before the structure is nailed irreversibly in its final resting place.

Using a Plumb Bob

Occasionally, you will wish to transfer a point on the floor to a point on the ceiling joists, rafters or second floor, or vice versa. Feel free to call on Old Man Gravity and your plumb bob for this. (See Fig. 9-9.)

Typical uses for a plumb bob include the layout of foundations, establishing the location of plumbing holes, chimneys and ridge boards, and checking tall structures for plumb (to be sure the structure is vertical).

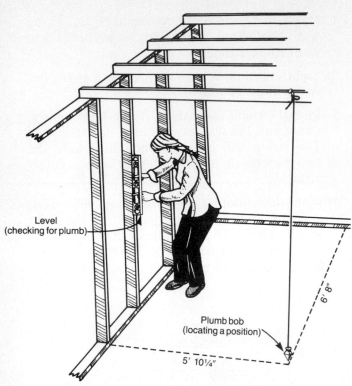

Fig. 9-9 *Using a level and a plumb bob.*

Level
(checking for plumb)

Plumb bob
(locating a position)

6' 8"

5' 10¼"

Nailing

Face nailing is the easiest. The boards are nailed face-to-face. (See Fig. 9-10). Try to avoid any knots in the wood. Just hold the nail with one hand and tap it a couple of times with the hammer to get it started. Move your hand back to a safe retreat, hold the boards, and drive the nail.

Now, for you timid folks, no tap, tap, tap stuff now. Swing that hammer with *authority!* Use mainly your forearm and wrist and a little bit of upper arm and shoulder. Concentrate on the head of the nail and make sure the face of the hammer lands *flat* on the nail head. (See Fig. 9-10.) Wear safety glasses so you can't be seriously wounded by flying nails. Here's how to *avoid splits:*

- Don't nail within 1 in of the end of the board. Preferably further away.

- Avoid nailing two nails into the same grain line.

- Never place two nails close together. Uniform Building Code requires that two nails must be separated by at least half the length of the nail. (UBC Sec. 2510f)

- Don't overdrive the nail. The head of the nail will split the board if you keep hammering on it after it is in.

- Galvanized nails tend to split wood more than bright nails; large nails tend to split more than small nails.

- Some wood splits so easily that you will have to drill a pilot hole (a hole slightly smaller than the nail) before driving the nail. If no drill is available, try blunting the sharp point of the nail with the hammer, then drive it.

- A few splits are to be expected.

End nailing is the primary way of framing the floors and walls. End nailing would be pretty easy if the doggone boards would hold still. Usually holding the board firmly is the most difficult part.

Let's use a wall for example. Wear boots. Steel-toed boots are best. Hold the 2 × 4 with your foot and your hand as you nail them. Your body position is important, so take careful note of the position shown in Fig. 9-10.

You will learn variations of this position as you build. Floor joists usually have to be held with your free hand while you nail. Once you get a few boards nailed together the structure stiffens up and gets heavier, which makes the nailing easier.

Toenailing is definitely an acquired skill. Don't hurry and plan on practicing toenailing the way you would practice shuffling cards or pitching horseshoes.

For example, Fig. 9-10 shows how to toenail a 2 × 4 to a beam. First mark the position of the 2 × 4. Start the first nail at a 45-degree angle to the 2 × 4 with a couple of hammer strokes. Then for practice, try to hold the 2 × 4 in place by putting your foot (or knee, or hand) behind it as you drive the nail home. Alas, you can't hold the 2 × 4. The last few strokes of the hammer, when the hammer hits the board, always knock the 2 × 4 about ¼ or ½ in off the mark. So pull the nail and try again.

This time hold the 2 × 4 about ¼ in to ½ in to the hammer side of the mark. As you drive the last few strokes, hopefully the 2 × 4 will snug right up to the mark. If it goes a ¹⁄₁₆ in too far, try toenailing back from the other side to get the 2 × 4 right on the mark. Once you get one toenail in, the board becomes steadier and this makes it easier to get the other two nails in.

Practice, practice—and if possible watch someone who knows how to toenail. Use 8 penny (8d) bright box nails to avoid splits.

Toenailing can also be used to move boards that are slightly out of place. It is not uncommon to find a floor joist or a stud that, for one reason or another, has slipped ⅛ in or ¼ in out of line when being nailed. A well-placed toenail, nailed with power, can sometimes pull the board into position.

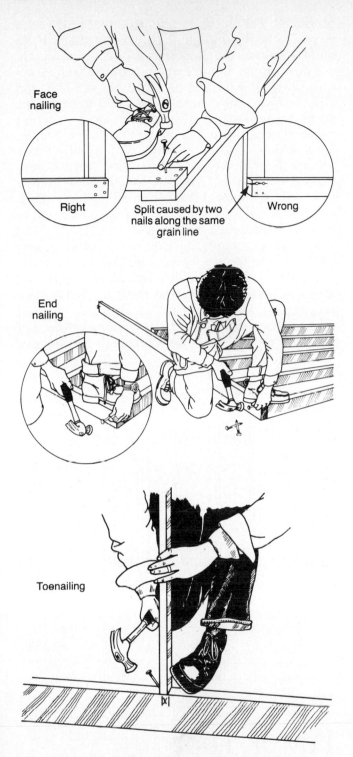

Face nailing

Right   Split caused by two   Wrong
          nails along the same
          grain line

End nailing

Toenailing

Fig. 9-10 *Nailing.*

Fig. 9-11 *Pulling nails.*

Pulling Nails

No self-respecting expert would ever admit to driving a nail in the wrong place. But nonetheless, most experts have a little tool called a cat's paw hidden in their tool box, if not in their apron. (See Fig. 9-11.) And there is only one function for a cat's paw. Need I say more?

Using a Chalk Line

A chalk line is a long string coated with chalk. The line is held taut, then snapped to leave a long straight chalk mark. (See Fig. 9-12.) Chalk lines are frequently used to mark plywood along floors or roofs for trimming.

1. Measure and mark ends of plywood.

2. Shake the canister to get chalk on the line.

3. Stretch the chalk line taut between the marks, tying it to nails if necessary.

4. Snap the chalk line by lifting the line a couple of inches and letting it snap back against the plywood. Be sure to reach as near to the center of the line as possible when snapping it. Also lift the line straight up, not to either side.

Fig. 9-12
Using a chalk line.

Checking for Squareness

"Square" was a nebulous word to me at first. As I built I slowly acquired an appreciation of the importance of making square cuts and square structures. Of course a cut or a cupboard may be rectangular in shape, but it must be "square" when checked with a combination square or a carpenter's square. The use of a combination square was shown in Fig. 9-4. A carpenter's square (also called a framing square or a rafter square) is used in the same way, only on larger things like beams and in window openings. (See Fig. 9-13.)

After sawing a couple posts for my foundation, I soon realized that "pretty square" cuts aren't good enough. To get a really good fit, and a solid structure, those cuts must be perfectly square.

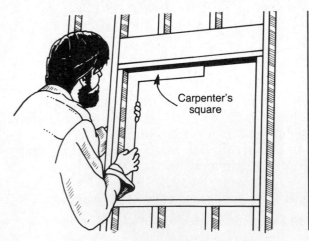

Fig. 9-13 *Checking for squareness.*

Universal Squareness Test

When you get to even bigger structures like foundations, floors, walls, and roofs . . . a minor out-of-square problem can cause you a multitude of problems. The best way to ensure that a structure is perfectly square (perfectly rectangular) is to build it so that the diagonals are of equal length.

For example, if the diagonals of the wall structure shown in Fig. 9-14 measure AC = 142 in and BD = 145 in, they must be adjusted to make the structure square. First, hold one side so it won't move. Toenail side DC to the floor for example. Now which way do we shift it? I usually make myself a little sketch exaggerating the discrepancy as shown in Fig. 9-14. When BD is longer than AC, segment AB must be shifted left. How much to the left? Usually it's an amount equal to half the difference between diagonals AC and BD. In this example the difference is 145 in less 142 in = 3 in, so we would shift AB 1½ in to the left. Then measure the diagonals again, and shift again as necessary. By the third or fourth time you repeat this process, the structure will be square. Then nail a board diagonally or a piece of plywood to the structure so it can't be bumped out of square.

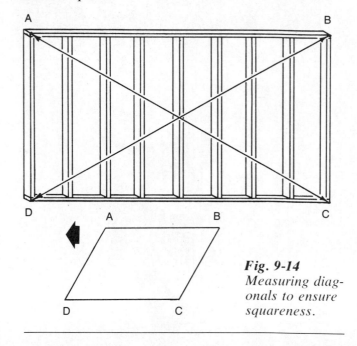

Fig. 9-14
Measuring diagonals to ensure squareness.

Power Tools

If you've experienced the time it takes to work with hand tools, I'm sure that I can interest you in a short discussion of power tools.

Portable Circular Saw (Skill Saw)

Since "portable circular saw" doesn't exactly roll off the tongue very easily, carpenters started calling them by the popular brand name: Skil.

Forty or fifty years later, skill saw is a generic term used for all brands and if you asked a carpenter for his portable circular saw, there's a good chance it'll take him a moment to figure out what you mean. I call it a skill saw for brevity and out of habit.

Whatever the name, it is a tremendously useful saw and an indispensable tool to a builder. It is just the item for cutting plywood and any board that can't be conveniently moved to a stationary saw. Some skill is required to make square cuts, but it's much faster than a handsaw and much cheaper than a radial saw.

The most convenient way to cut a sheet of plywood is to lay it on three 2 × 4's that rest on sawhorses as shown in Fig. 9-15. Measure and mark the cuts, then set the blade of the skill saw so it cuts about ¼ in deeper than the thickness of the plywood. The 2 × 4's support the plywood so it doesn't sag and pinch the saw blade during the cut. If the blade is pinched and you continue trying to saw, you can overload the saw and burn out the motor.

Another common way to burn out a skill saw is by using an extension cord with small gauge wire. Usually you will need a 50- to 100-ft cord to build a house. A No. 16 gauge wire is too small. A No. 14 wire is OK for a 50-ft cord if you don't run the saw too long or overload it. For a 100-ft cord, use No. 12 or No. 10 wire. You can buy a roll of housewire and put an outlet on the end of it as will be shown in Fig. 10-2.

Radial Saw

A radial saw is the most important saw for the amateur builder. (See Fig. 9-16.) The saw travels on an overhead arm. You hold the board stationary against the fence. Buzzzzatt! A fast, perfectly square cutoff every time. This makes for tight joints and a strong house. Also, a radial saw can be easily set up to quickly cut a bunch of studs to the same length.

In addition to square cuts, you can also cut angles, compound angles, bevels and rips with a radial saw. These same cuts can be made with a skill saw, but not as accurately.

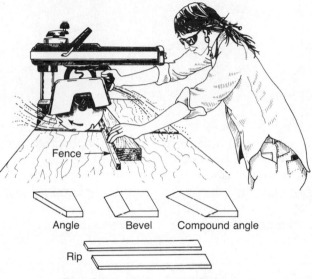

Fig. 9-16   Using a radial saw.

Drill

A small (⅕-hp) drill, shown in Fig. 9-17, is needed for drilling pilot holes, holes to get a saber saw started, and a million small tasks. A larger (¾-hp) drill is needed for the heavy-duty drilling jobs, such as drilling large bolt holes, using hole saws (1-in to 3-in holes), and for the extensive drilling necessary for plumbing and electrical work.

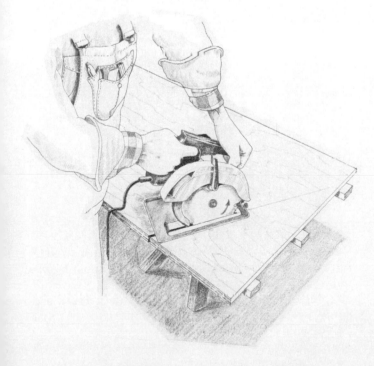

Fig. 9-15   Using a skill saw.

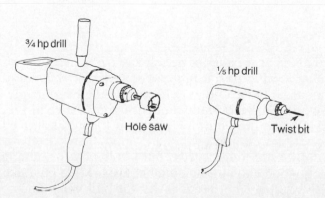

Fig. 9-17   Using a drill.

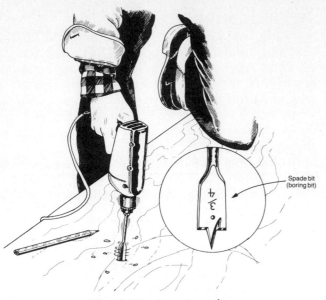

Spade bit (boring bit)

Fig. 9-17 *(continued)*

Happiness is having a good tool. Misery is trying to do a job with the wrong tool. If you've ever tried removing a can lid with a screwdriver instead of a can opener you've got the idea. A dollar invested in a good tool will bring you far more return on your investment in satisfaction and avoided expense than any investment on the market. At least that's what I tell Deb every time the new tool catalog comes in the mail.

Buy good tools and take care of them. Don't leave your power tools or files out in the rain. Don't kink your hand saw by sawing too hard. Don't overload your power tool motors by trying to do a job that is too much for them, or by using a long, small gauge extension cord. Don't chisel through nails with a wood chisel. You may spend as much as $2,000 on tools during this project, but if you buy good tools and take care of them, they will last you a lifetime.

Saber Saw

A saber saw, shown in Fig. 9-18, is used for light duty cuts in plywood and curved cuts. It's light, easy to use, and only takes one hand to operate. For cutting holes in plywood for chimneys and large pipes, it can't be beat.

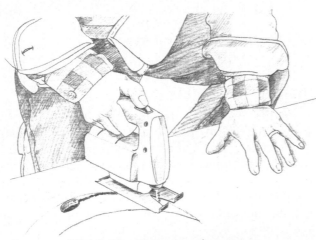

Fig. 9-18 *Using a saber saw.*

Fig. 9-19 *Using a table saw.*

Table Saw

This is also called a bench saw. For me, the primary value of the table saw was in making cupboards and cabinets where I needed to cut plywood very accurately and straight. Notice in Fig. 9-19 that the board or plywood has to be moved past the stationary blade. This is awkward for cut-offs of long boards, but it's handy for ripping plywood and cutting small boards. I like my table saw but it's not a necessity for building a house.

Notes on Using Power Tools

When you buy tools, read the instructions carefully and follow them. The Reader's Digest *Complete Do-It-Yourself Manual* devotes two full chapters to tools and how to use them. Here are some brief notes that apply to the use of most power tools:

• Wear safety glasses.

- Keep the work area where you use power tools clean and clear of obstacles.

- Firmly hold or clamp your piece of work.

- Be sure that saw blades don't strike the wood when you turn them on. Let the blade or bit get up to speed before easing it into the wood.

- Place your hands and fingers so they aren't in line with the saw blade. Keep your knees and feet out from under the workpiece.

- When ripping wood on a radial saw or table saw, keep a push stick handy so you don't need to get your fingers near the blade. Use the anti-kickback mechanisms if the saw has them. If not, don't stand in line with the blade; if the workpiece jams the saw will often throw it right back at you. (For example, 2 × 4's get thrown right through walls this way.)

- On radial saws, make sure there is a bolt at the end of the track to stop the saw.

- You should never have to push or pull a saw very hard to get it to cut. If the saw doesn't move easily through the wood, check to make sure the saw blade is sharp, that the blade guard isn't catching on the wood, or that you haven't mounted the blade on backwards. If the blade starts to smoke, you're overdue for a new blade.

- Let the saw blade come to a dead stop before retrieving the pieces of wood you've cut.

- Keep a first-aid kit in your car or at the site for emergencies. Power tools are fast. They cut through wood like butter. One slip and they cut through fingers and legs like butter too.

Always keep a healthy respect for that cutting edge and work calmly and carefully. Follow the instructions with the tool. About the time you think you are getting pretty proficient with a power tool, you'll get in a rush and it'll bite you. You'll donate enough blood and sweat to this project without donating a finger or two. Please, **BE CAREFUL!**

Now about That Outhouse . . .

Reading has its place, but, given my choice, I'll take someone who has learned by experience.

There isn't a better way to learn to build than by building a small structure. The steps in Fig. 9-20 are for an outhouse, but are nearly identical to the steps taken to build a tool shed or sauna. If you don't want an outhouse, select some other way to practice your wood-butcher skills.

I'm going to throw you into the pool here with some abbreviated instructions knowing full well that you'll have some difficulties. I will give you a couple hints: Make sure the floor and the walls are built square and set the floor on bricks or railroad ties to forestall rotting. I show hand tools because you may not have power at the site yet.

The interior and exterior decoration are up to you. A modified apple crate can serve as a seat. A sack of lime helps to keep the fly problem in check. Sprinkle it in after each use.

Fig. 9-20 *Building an outhouse.*

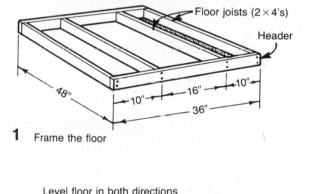

1 Frame the floor

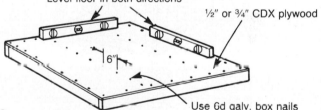

2 Nail plywood to the floor framing

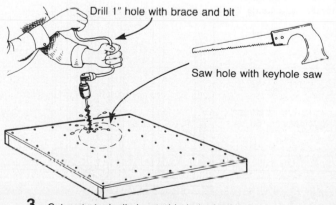

3 Cut a strategically located hole in the floor

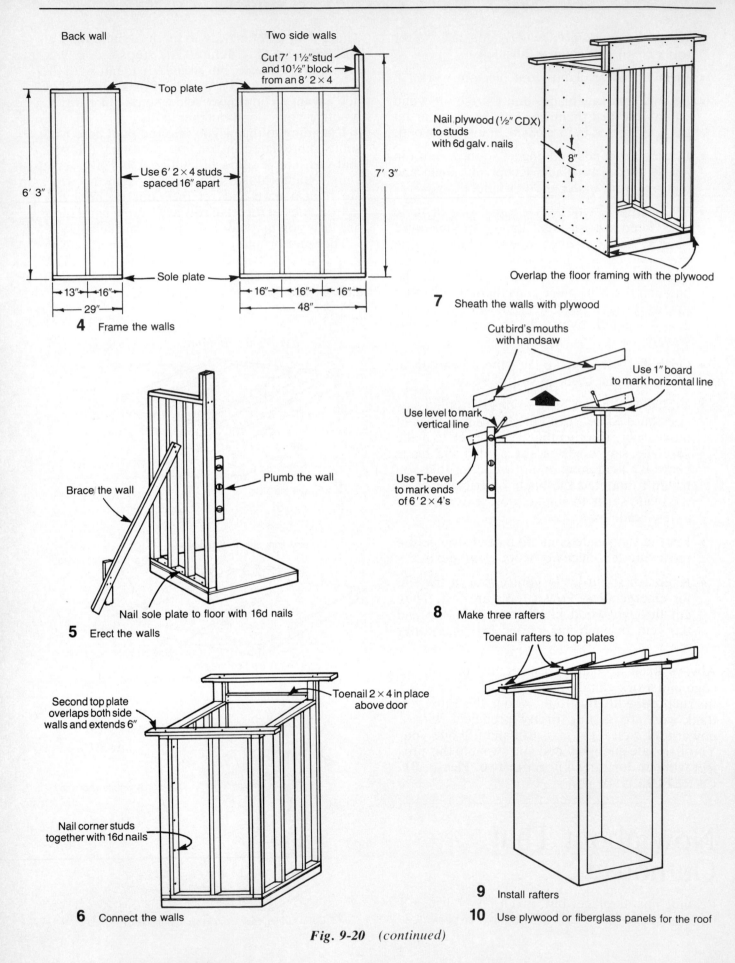

Back wall

Two side walls

Top plate

Cut 7' 1½"stud
and 10½" block
from an 8' 2 × 4

Use 6' 2 × 4 studs
spaced 16" apart

7' 3"

6' 3"

Sole plate

←13"→←16"→
←—29"—→

←16"→←16"→←16"→
←———48"———→

4 Frame the walls

Nail plywood (½" CDX)
to studs
with 6d galv. nails

8"

Overlap the floor framing with the plywood

7 Sheath the walls with plywood

Brace the wall

Plumb the wall

Nail sole plate to floor with 16d nails

5 Erect the walls

Cut bird's mouths
with handsaw

Use 1" board
to mark horizontal line

Use level to mark
vertical line

Use T-bevel
to mark ends
of 6' 2 × 4's

8 Make three rafters

Second top plate
overlaps both side
walls and extends 6"

Toenail 2 × 4 in place
above door

Nail corner studs
together with 16d nails

6 Connect the walls

Toenail rafters to top plates

9 Install rafters

10 Use plywood or fiberglass panels for the roof

Fig. 9-20 *(continued)*

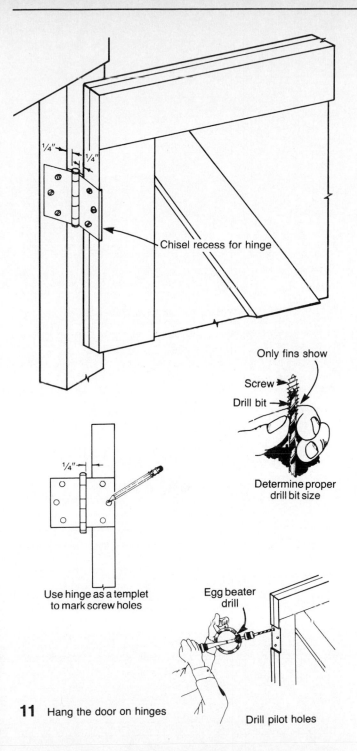

1/4" 1/4"

Chisel recess for hinge

Only fins show

Screw

Drill bit

Determine proper
drill bit size

1/4"

Use hinge as a templet
to mark screw holes

Egg beater
drill

11 Hang the door on hinges

Drill pilot holes

I'm curious about how you worked out certain things. Did you make a material list before you started? Did you think to use a piece of plywood to frame the walls on or was the ground level enough? Did you use a plank between the sawhorses to reach the rafters? Toenailing that 2 × 4 over the door is a bugger, I know. If the 2 × 4 is cut to fit tightly, that frees both hands for toenailing. The door hinges shown are harder to install than strap hinges, but they provide better practice. Mounting the hinges so the door opens and

closes properly is another tough one. I use a yardstick under the door for clearance while I mark the hinge locations. Then I support the door in the open position with blocks of wood while I mount the hinges.

I remember that a lot of things that I now do easily gave me serious problems at first. My novice builder friends had the same sort of problems. Everyone seems to work out their personal ways of doing things and it's all part of the process of learning to build. I fear I've told you so much already that I've taken half the fun out of the project.

There! Done! You've just built your first house. Congratulations. Now you can spend less time concentrating on keeping a tight sphincter and more time concentrating on the house. The material cost may look expensive but I'll bet you gained that much back in experience. If that outhouse is worth 10¢ a use, it just may pay for itself. In fact, if I'd thought to charge Uncle Barney 10¢ for each time he used it I would have made a few bucks on the deal.

Building the basic outhouse/tool shed involves a lot of skills. Sawing, nailing, drilling, floor framing, wall framing, roof framing, sheathing, checking for level and plumb, hanging a door, and more. If you didn't screw up a few things I'd be surprised. They say that experience is recognizing a mistake the second time you make it, or are about to make it. An outhouse or a tool shed is a good place to make those inevitable mistakes.

For More Information

Hylton, William, ed. *Build It Better Yourself*. Emmaus, Pa.: Rodale Press, 1977.

Reader's Digest Association, ed. *Complete Do-It-Yourself Manual*. New York: Random House, 1973.

Wagner, Willis. *Modern Carpentry*. South Holland, Ill.: Goodheart-Willcox Co., 1976.

CHAPTER 10

Prelude to Building

Prelude to Building

prelude \prel′ūd, prē′lood \ n [fr. L *praeludere* to play beforehand, fr. *prae-* before + *ludere* to play] 1: a thing serving as an introduction to a principal event, action, performance, etc.

*   *   *

Let's test the water with our toes before we jump into the next pool. Even if we just inherited $20,000, we can't just buy some boards and nails and start building. There are a bunch of time-consuming tasks that have to be done first. This prelude will give most of us a chance to save some money and ease ourselves slowly into this new occupation. The chore list is shown in Fig. 10-1. You can check off some of the items already.

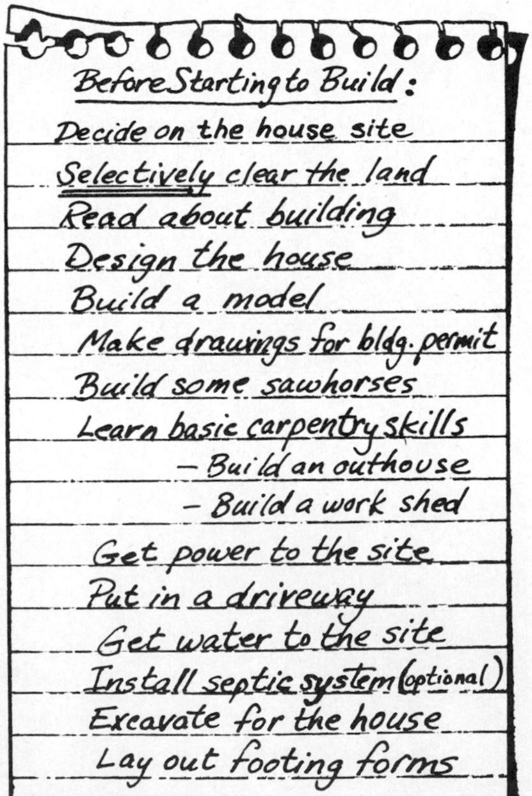

Before Starting to Build:
Decide on the house site
Selectively clear the land
Read about building
Design the house
Build a model
Make drawings for bldg. permit
Build some sawhorses
Learn basic carpentry skills
    – Build an outhouse
    – Build a work shed
Get power to the site
Put in a driveway
Get water to the site
Install septic system (optional)
Excavate for the house
Lay out footing forms

Fig. 10-1 *Chore list.*

The exact order of accomplishing these things will depend on your own inclination and situation. For example, if there are no sewers in the area, a septic system will normally be required. Some houses are located such that the septic drainfield might be destroyed by the bulldozer or cement trucks rolling over it if it is installed too early. But as health regulations become more stringent, it is a good idea to get the septic system installed early. At least have a design for a septic system approved by the Health Department before you start building. The regulations in my district changed twice while I was building the house. The water hook up and power hook up charges skyrocketed and I was glad I took care of them before I started to build. I found it advantageous to accomplish these chores during weekends and evenings while I was still employed at a regulation, upstanding, paycheck-producing job.

Once these tasks are behind you, you'll be in a better position to decide what to do next. Some folks try to save enough money so they can quit their regular jobs and concentrate all their time and energy on building. The house goes up a lot faster this way, but when the income stops it puts a lot of pressure on you financially. The other way is to keep working and bringing home a paycheck and build the house on the weekends and evenings. Financially you are in good shape because you can hardly use $500 worth of materials a month. But it can be mentally and emotionally frustrating because everything goes so slowly. What's more, it takes a lot of fortitude to come home from a full day of work and get out there in the dark and lousy weather to pound a few nails.

During the 3 years it took to build our house, I tried both approaches. I preferred it when I could work full time on the house. The best situation of all was when Deb worked and provided the income and I could just play carpenter all day long. Progress was fast (by my standards) and we had enough money to feed the cats real canned cat-food. The cats loved it, I loved it, and Deb complained only occasionally.

Getting Power to the Site

When you get tired of using a handsaw, call up the power company and get the skinny on getting temporary power to your building site. Find out how much they charge per month for use of their electrical panel and service, and how much per kilowatt-hour. Construction sites usually get charged outrageous rates per kilowatt-hour, but building a house requires only about 1,000 kWh of electricity. It's mainly the monthly charge that kills you. I expected to have temporary power for 6 months and it ended up being a 3-year deal.

When you put in temporary power, put the panel as close to the building site and the work shed as possible. (See Fig. 10-2.) It should also be close to the place where the permanent service will connect to the house. You'll probably need at least one 16-ft pole for overhead service. The installation charge may be less if you install the pole. Dig a hole 3 ft deep for each pole and brace them well. If the power line crosses the driveway, allow at least 15 ft of clearance so the concrete truck can get in. This requires a pole at least 20 ft long.

Check with the power company to compare underground and overhead service installation costs, remembering that underground service will necessitate a 24-in-deep trench to be dug. I went for underground permanent service, even though I had overhead temporary service. It seemed like a mistake when I was digging the 200-ft trench, but now the peace of mind during windstorms and power outages makes it worth it. (I know it can't be my line that's down.) *Make sure that the phone line, driveway light wires, T.V. cable, and so on are in the trench before you cover it up.*

Putting in a Driveway

Normally this is a straightforward task. Sometimes the county or city road department has certain guidelines that must be followed. The road department may even install the drainage culverts, though you will have to buy them. Use

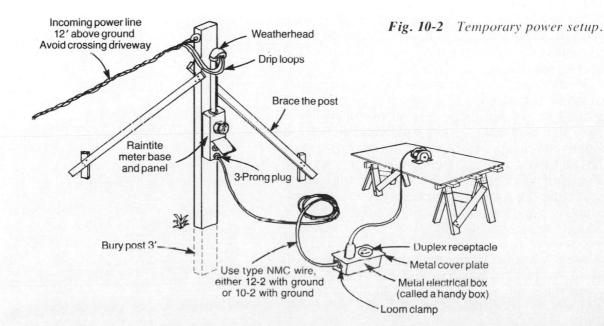

Fig. 10-2   Temporary power setup.

Incoming power line
12' above ground
Avoid crossing driveway

Weatherhead

Drip loops

Brace the post

Raintite
meter base
and panel

3-Prong plug

Bury post 3'

Use type NMC wire,
either 12-2 with ground
or 10-2 with ground

Duplex receptacle

Metal cover plate

Metal electrical box
(called a handy box)

Loom clamp

Getting Water to the Site

Water is not essential to building, but it's very useful. Clearing the property means burning brush and for that you need water on hand. The hose-type levels (see Fig. 10-3) used for getting the excavation and foundation forms level need water, though only a hose full. The concrete foundation should be cured by sprinkling water over it. Small trees and shrubs will need water, not to mention your own hot, thirsty body on those hot summer days. It's a good idea to fill fiberglass septic tanks with water before the rainy season. The rainwater has been known to fill up the hole and float the tank, popping it right out of the ground. Besides all these reasons, if you build your house and then find out that it is impossible to get water for some reason, you'll be up the proverbial creek.

In areas not serviced by a local water district, you'll probably have to drill a well. Check with a local well driller to learn what is needed and how much it costs. Also, the Time-Life book on *Cabins and Cottages* has a good section on do-it-yourself wells and pumps.

When piping the water from the source, be sure to use pipe of adequate size and strength. If the distance to the house is less than 50 ft and not up a steep hill, a ¾-in 100-psi polyethylene pipe should suffice. For longer distances up to 150 ft, a 1-in pipe is normally adequate.

culverts that are 12 in to 24 in in diameter, depending on drainage conditions. Be sure to set them deep enough so that they are covered with about 6 in of fill. Backfill with dirt or "pit run" (unsorted sand and gravel right from the pit). The driveway entrance should be at least 20 ft wide with rocks at each end to hold the backfill. Don't use rocks for backfill; a heavy truck can poke the rocks right through a culvert.

If there are any seriously soft or wet spots along the driveway, have about 6 in of pit run hauled in to build it up. A layer of crushed rock over that will make a very solid driveway. It's a relatively cheap way to avoid stuck trucks. You haven't lived until a full concrete truck with $300 worth of *your* concrete, is stuck in *your* driveway, and the concrete is starting to set up. Occasionally a $300 monument gets poured at the side of a driveway because of this.

It's best to consult a bulldozer driver if the driveway needs dozing or grading. Usually you'll want to get the building site excavated at the same time, since just getting the bulldozer to your property is a substantial cost.

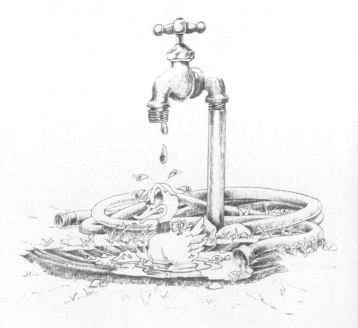

Water lines should be buried about 2 ft deep (42 in in the Mid-West) to avoid freezing. Also think about possible damage from rototillers and post-hole digging. The pipe cannot run within 10 ft of any part of the septic system. Here's another grand opportunity to do some digging. I'll have to confess that I took the easy way out and hired a backhoe with an experienced operator. He dug 300 ft of water line ditch, a septic tank hole, and 250 ft of drainfield trench in 6 hours. I was happy to pay $150 for the task. It took me several days just to refill the trenches after the water pipe and drainfield were installed. God knows how long it would have taken me to dig them.

Logistics

Depending on the distance from your living accommodations to your home site, logistics can be a troublesome problem. Even though I had no theft or vandalism difficulties, I never was comfortable with leaving my tools and some materials on the job site. And I couldn't spare the time to haul everything back and forth. Fortunately there were the remains of a chicken shed on the property which I resurrected into a weatherproof, lockable work shed.

I found that chicken shed to be a life saver. I spent quite a bit fixing it up, but it was worth it. The 12 ft × 20 ft shed was soon stuffed with nails, shovels, picks, salvaged windows, doors, pipes, pencils, raincoats, a wheelbarrow, a lawnmower, rakes, a work bench, a radial saw, buckets, hoses, extension cords, and a few thousand other items. When a downpour started, the shed was my retreat. It was my lunchroom. Heck, it was very possibly my deliverance.

Some people find it best to build a garage with living accommodations, thereby eliminating rent payments, saving commuting time and getting some work space all at once. Another option is to buy or rent a trailer with bunks and a stove, though you still need some work space. Other folks just resign themselves to commuting and work out of the back of a truck. The radial saw stand can be set up under a piece of plywood and the saw motor can be removed to take home each night. While an experienced carpenter can build a house with just a skill saw, that radial saw is a very important piece of equipment to an amateur.

If you do build a work shed, barn, or garage, there are some more logistical considerations. The lumber pile should be close to both the radial saw and the house, so you don't have to carry the boards very far to cut them, or carry them very far to install them. Also, the saw table should be lined up with the door of the shed so long boards can stick out the door while they are being cut to length. You'll need enough room to rip 8-ft boards as well, which requires 10 ft on both sides of the business end of the saw. Usually that can be attained by opening the door.

However the work shed is built, it should provide a dry, secure storage area and a layout and sawing bench. The design and construction is just like a small house except that you may want to put it on pier blocks instead of on a continuous foundation. Dirt floors are OK, but if the shed is heated it will draw a lot of moisture out of the ground. A plastic sheet over the dirt will prevent this. The Agricultural Handbook No. 438 referenced at the end of this chapter shows plans for 43 small structures. A shelter of some type can make the time you spend building your house considerably more pleasant.

Other Preparations

Wheels

You'll need wheels. Wheels that can transport 4 × 8 sheets of plywood, bath tubs, beams, and stoves. Old pickup trucks qualify. So do station wagons, Veedub busses, and cars with roof racks. Some folks solve the problem with an old trailer or by borrowing or renting a truck when necessary. A lot of lumber stores deliver to your building site if your order is large enough. But to take advantage of the salvage yards, garage sales and other second-hand sources, you're going to need wheels, or access to wheels.

There's No Such Thing as a Free Lunch

Approach each project like you are responsible for everything. Don't turn down offers of help, encourage them. But don't depend on free help. Depend only on yourself. For example, you are ready to plumb and you have an offer from an experienced plumber to help you next month. Go ahead and start on the plumbing now and approach it as if you will be doing the entire job. And if things work out that the plumber actually will be available next month, you will probably still be able to use his help and will be doubly appreciative of it. If he gets tied up and can't help you, your house will not be needlessly delayed.

Why this approach? Because the minute you begin to delegate responsibility to others you sort of relax and figure that "Joe will take care of that." Sometimes this approach works very well. But most of the time, Joe is as overly optimistic as you are about the amount of time and effort that a project will take and how much time away from his job and his family that he will be able to spend helping you. And you certainly can't blame Joe if your house isn't the number one thing on his list. My experience is that the energy I spend on learning and accomplishing a task myself is always rewarded with results, but depending on someone else to do a task frequently leads to disappointment.

Now of course you can't always do everything yourself. You can't work on your own teeth for instance; it's too damned awkward and you can't see what you are doing. So when you determine you need help, approach it on a businesslike basis. Hire someone with a good reputation and pay the going rate. If he is a friend, maybe you can repay his services with an equal amount of your services. But if you find yourself making excuses for delays, cost, or construction problems because of the performance of others, remember that there is only one person in charge of building this house.

On Working Alone

Plan on it. You will surprise yourself with what you can do. Don't strain yourself or take chances. Think things out carefully and work safely and methodically. Rely on your brains and ingenuity. You'll find that it takes only one person to build most houses.

Nonetheless there are some jobs that are expedited by having help. Pouring concrete is a three- or four-person job, although one person can do it by pouring small sections one at a time and providing a rebar connection between them. Erecting long walls takes two or three strong people, although a person can use a jack or rope and pulley systems to do it alone. The same goes for large beams and rafters. Putting sheet rock on the ceiling is much easier with two people, although one person can do it with props or by renting a ceiling jack. Stacking lumber and installing roofing go much faster with two people. Chinking log cabins and painting can be speeded up remarkably by inviting a bunch of folks over for a work party.

While it is important to learn that you can do most anything by yourself, it can be a decided advantage to get help in some situations, especially from someone who is also building his or her own home. Besides having a sympathetic soul to commiserate with, you will find that some jobs go more than twice as fast with two people working on them. After working alone, it's a real treat to work with someone else for a few days.

Excavating

First make sure of the location of your property corners and lines. Try not to rely on the location of fences or roads. Find a survey monument or marker to use as a reference to measure from. Be sure to correct magnetic compass readings to true north-south directions. Some states have laws requiring all registered surveys to be available to the public. These can be a big help in locating property lines. If necessary, hire a surveyor. But be certain to build the house on property that you own.

Foundations must sit on flat, undisturbed soil. The excavation scrapes away the looser topsoil and gets down to the more solid, settled soil below. The depth of the excavation depends on the topography of the site, the depth of the frost line, and the house design. No part of the foundation may rest on fill dirt and all stumps and debris must be removed.

Review your foundation design carefully. Figure 8-2 shows both a plot plan and a foundation plan for the sample house that we'll build in this book. Lay out the foundation with stakes to within a few inches. Normally the excavation should allow about 3 ft space around the foundation. This way you'll have room to work on the forms and the concrete truck can still get fairly close. If the foundation is stepped, stake the areas that will be on the same level. Use a water level to determine the amount of slope to the site; land can look deceptively flat to the inexperienced eye.

A water level costs about $7 and attaches to a garden hose as shown in Fig. 10-3. You can also make one with some clear plastic tubing and hose fittings. Fill the hose and tie the tubes to the stakes. Adjust the water level by pouring some water in one of the tubes. Since water seeks it's own level, this gives you a level reference line to measure from. Make sure the hose is not kinked. A $200 transit can be used to accomplish this same task a bit more conveniently.

Now suppose this excavation is for a two-story house, thus requiring the base of the footing to be 18 in below the final grade. The base of the footing must be at least 6 in below the original grade

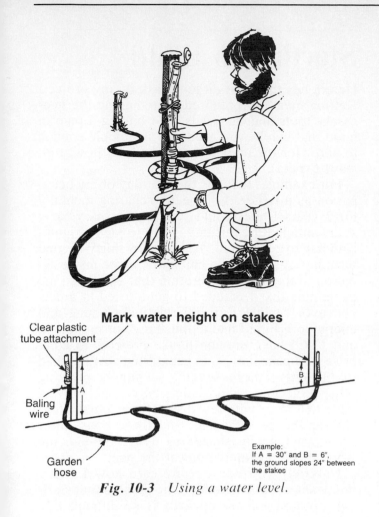

Mark water height on stakes

Clear plastic
tube attachment

Baling
wire

Garden
hose

Example:
If A = 30" and B = 6",
the ground slopes 24" between
the stakes

Fig. 10-3 *Using a water level.*

Notice that there is going to be a flat spot and a big pile of dirt after the bulldozer is gone. Since that big pile of dirt may not enhance our landscaping plan, we might as well use this dirt to change the grade level as shown in option 2. This allows the foundation to be 12 in shorter, which makes the house look lower, requires less excavation, and saves a couple of hundred dollars on the concrete bill. The footing still sits on undisturbed soil and is 18 in below the final grade.

How do you get the dirt all rearranged to the final grade? Well, Deb decided she wanted a flower garden out there and shoveled it all into place. I thought this was an excellent way. Otherwise you might have to hire the dozer driver to come again after the foundation is poured.

Even flat building sites require some excavation in order to get down to undisturbed soil and to get below the frost line. Dirt expands when it freezes and can easily buckle the foundation. One way to avoid the need for a bulldozer is to dig trenches for the footings as shown in Fig. 10-5. Dig carefully with a square-nose shovel and the excavation can be used as the form to pour the concrete into.

Remember that for a one-story house, the final grade next to the foundation wall must be at least 12 in above the base of the footing, which requires some piling of dirt around the foundation wall. In cold climates, deeper footings will be required. Check with your building department. Also, the crawl space must be 18 in as measured from the ground to the bottom of the floor joists.

If you do hire a bulldozer driver, make sure you leave room to maneuver and to pile the dirt around the site. Have the driver pile the dirt so the cement truck will have room to get in, but close enough so it won't be too much trouble to

to be sure it is on undisturbed soil. And suppose the difference between A and B is 24 in, as illustrated. We have two options for the foundation design as shown in Fig. 10-4.

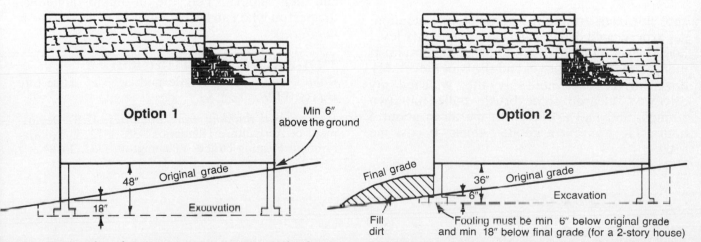

Option 1

Min 6"
above the ground

48"

Original grade

18"

Excavation

Fig. 10-4 *Plan the excavation to minimize the foundation height.*

Option 2

Final grade

36"

Original grade

6"

Excavation

Fill
dirt

Footing must be min 6" below original grade
and min 18" below final grade (for a 2-story house)

Note: In cold climates, the footing may be required to be as much as 42" below final grade to be below the frost line.

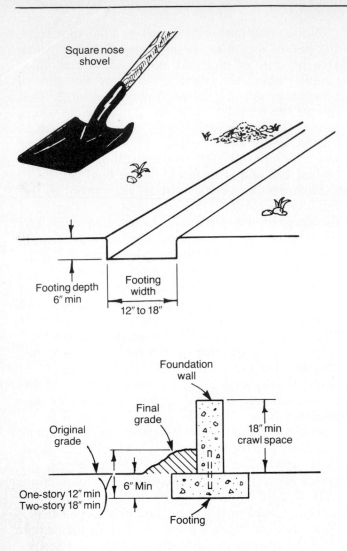

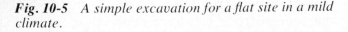

Important:
The footing must be below the frost line. Check local codes.

Fig. 10-5 *A simple excavation for a flat site in a mild climate.*

shovel it back around the completed foundation. An experienced driver can get the site pretty level just by eye. You will have to make minor adjustments later with a shovel and the water level. My dozer driver excavated my site, widened my driveway, cleared some brush, pulled out two stumps, and had a beer with me all in about 4 hours. He was extra good, I think. It cost me $100.

Starting to Build

Herein lies the most difficult task of any project: Getting mentally prepared, marshalling the necessary tools and materials, and having a plan in mind that can be followed, but is flexible enough to allow for the unexpected (which by now, I always expect).

For example, today is the first day of my occupation as a writer. I've been preparing mentally for 2 years I guess, and I've kept a diary for 4 years anticipating this day. I had to pick a significant day to get started. Today is my thirty-second birthday. A year and a half ago I wrote my resignation to the huge corporation that employed me explaining that I planned to get started in a new endeavor "while I'm still relatively young and energetic." First I had to finish my house though, and I still have enough house projects to comprise a 10-page list. So to get started writing I had to blatantly ignore several seemingly pressing other projects, read an article on writing, move a table to a place with few distractions, build a fire, answer the phone twice, and three hours after breakfast I finally started to write. Behold, the words begin to tumble out of the pen.

Naturally you don't really care how I write; you want to build a house. The point is that starting any new major endeavor is so difficult that you must reach into the very depth of your soul and character and squeeze out every available ounce of determination and enthusiasm in order to finally get started.

Your life style will change. Your entire attitude and approach to living must change. The clothes you wear and the way people relate to you changes. That's why you must set a date, having done all the necessary planning and ground work, and then "Poof!" You, the all-thumbs amateur, the person who can barely pound a nail, become a *builder*.

For More Information

Cabins and Cottages. Alexandria, Va.: Time-Life Books, 1979.

Recreational Buildings and Facilities, U.S. Department of Agriculture Handbook 438, 1972. U.S. Government Printing Office, Washington, D.C. 20402.

CHAPTER 11

The Foundation

The Foundation

Poof you're a builder!

Now don't expect any miracles. You feel just the same as the day before you "poofed." You still can't drive a nail very well and you have to measure the boards to tell the difference between a 2 × 6 and a 2 × 8.

Maybe you'll be better prepared than I was. Even after a couple of practice projects I was remarkably clumsy and slow. Knowing that the foundation was crucial to the whole project, I went slowly and carefully but still committed some classic blunders. Blunders that could have been avoided if I'd talked with Dad, but at that point in my life I was dead set against asking for advice. I wanted to confront this beast on my own terms. So I did it my way and played out the four-star comedy that resulted. Since I didn't ask for any advice, the only advice I got was from Uncle Barney. Worse yet, I took it. Yes I know, present the brown helmet award and call me Dummy. It's painful to remember, but I'd better come clean or you may make the same mistakes. Lest I confuse you with my blunders and misguided reasoning, I'll save that part for the end of the chapter and tell you now the right way for a beginner to build a foundation.

This is a long chapter so brace yourself. The instructions are for a standard foundation for a house with a crawl space. Unless you have experienced help, pouring a basement is a risky undertaking. As I mentioned in the design section, you'll be wise to stick to foundations less than 3 or 4 ft tall. I get into a lot of detail in this chapter because it's critical that the foundation turn out right. Be patient, I'll let up on the detail after this chapter.

Layout

The purpose of the layout is to ensure that the foundation forms will be set up squarely and with the correct dimensions. Lines attached to boards (called batter boards) are used to establish the house corners and are set back from the area where you will be driving stakes, sawing, and hammering.

After the bulldozer has excavated for the foundation, you probably won't be able to find hide nor hair of your excavation stakes. So the first job is to lay out the house position again, this time very accurately.

Materials and Tools

You'll need a 50- or 100-ft steel tape measure, a reel of baling wire (200 ft minimum), a pair of lineman's pliers, a sledge hammer, carpenter's square, plumb bob, hammer, nail apron, compass, handsaw, at least four 2 × 2 stakes (12 in long), at least twelve 2 × 4 stakes (4 ft long), at least eight batter boards (6-ft long 2 × 4's work well), and some 8d bright box nails. If you can snag a helper to hold one end of the tape measure, do it.

STEP-BY-STEP PROCEDURE

1. Use the plot plan that was submitted for the building permit (Fig. 8-2) to locate two corners of the house. Figure 11-1 illustrates this step and all subsequent steps. Measure a specific distance from the property lines for the first corner and drive a 2 × 2 stake. (These are often called hubs.) Then lay out the second corner on the north-south or east-west line from the first stake. Use a compass (and remember the magnetic correction) or the shadows cast by the sun to do this. If north-south orientation isn't your bag, just point the house the way you want it to sit for the best view and to get the sun on the sundeck. Drive the second 2 × 2 stake, then put 8d nails in both stakes, which are the precise length of the house apart.

2. Determine the best location for the 2 × 4 stakes and the batter boards. The lines must

intersect over the house corners and the batter boards should be at least 2 ft back from where the foundation forms will go. The lines should be higher than the future foundation walls. Drive the 2 × 4 stakes firmly into the ground with a sledge hammer. Nail on the batter boards. Use your carpenter's square to get the batter boards approximately square with each other. You'll have to decide if it's easier to put the batter boards inside or outside of the excavation.

3. Stretch a line between two batter boards so it is directly over the two 2 × 2 stakes marking the house corners. I found that a light wire like baling wire works best because it won't sag from day to day. Use a plumb bob to get the line (the wire) exactly over the nails in the stakes. This line locates one side of the house. A shallow saw cut in the batter board will help hold the line in place.

4. Measure the width of the house from the two corner stakes and stick a couple of nails in the dirt approximately where the third and fourth house corners will be. The nails must be the length of the house apart (32 ft apart in our example) and the diagonals between the corners must be equal. Keep measuring and moving the nails until you get it right. Refer back to Fig. 9-14 (Universal Squareness Test) if you need to. When you get it square, drive in the last two 2 × 2 stakes and put a nail in them to mark the exact house corners. This is a tedious, trial-and-error process so be patient.

A handy thing to know when trying to set up two lines at right angles to each other is the 3-4-5 triangle rule. When one side measures 3, the other side measures 4, and the diagonal (or hypotenuse) measures 5, with the two shorter sides at right angles to each other. With the sample house, one side measures 24 ft (or 3 × 8 ft), the other side is 32 ft (or 4 × 8 ft), so the diagonal will be 40 ft (or 5 × 8 ft) when the two sides are at right angles, or square to each other. You'll have to get the diagonals equal by trial and error if your house dimensions don't work out to be a 3-4-5 triangle.

5. With all four corners located, stretch the last three wires and use the plumb bob to get the intersections of the wires directly over the nails. The wires can be tightened by bending the nail in the batter board and twisting it with pliers, thus winching the wire tight.

6. After the lines are in place, check for equal diagonals both between the line intersections and the nails. When you are satisfied, make a shallow saw cut in each batter board to hold the lines in place. The layout should now look like the last drawing in Fig. 11-1.

Fig. 11-1 *Laying out the foundation.*

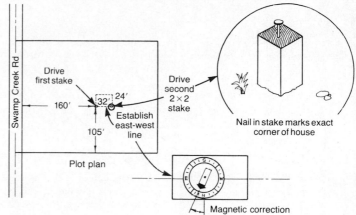
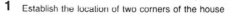
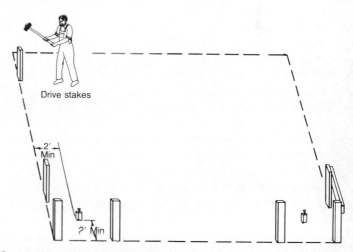

1 Establish the location of two corners of the house

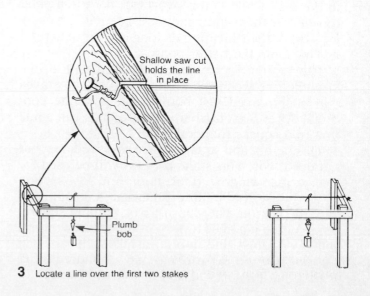

2 Set up the stakes and batter boards

3 Locate a line over the first two stakes

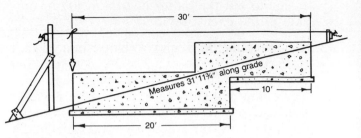

4 Locate the third and fourth corners

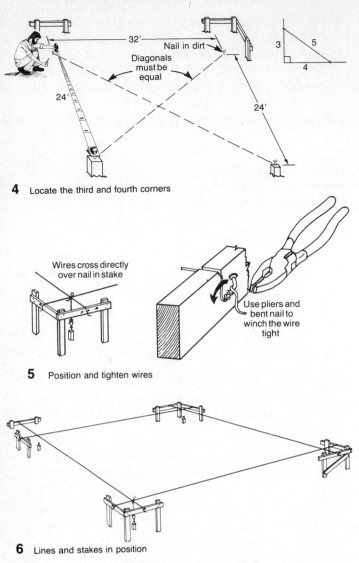

5 Position and tighten wires

6 Lines and stakes in position

Fig. 11-1 *(continued)*

The batter board lines will be used to locate the foundation wall forms accurately. Since they get in the way while you're working, it's often good to remove them until they are needed. The saw cuts mark their location as long as you're careful not to bump the batter boards.

The footing forms will be located with a line stretched between the corner stakes. The stakes will be removed just before you pour the concrete. After that, the batter board lines will guide you to a square foundation.

This layout and squaring process is tedious, but get used to it. This same process will be used for every wall, window, door, floor, and roof section in the house. In building, squareness is king; and pity the person who doesn't worry about squareness. When it comes time to fit the walls together, put in windows and doors, install wallboard and flooring, put on the roofing, and a hundred other tasks, that person will be in a world of hurt!

Layout on Slopes

When the house is located on the side of a hill, often a stepped foundation will be necessary. In such a case, you will probably have better luck measuring along the batter board lines, rather than at ground level. Make sure the lines are level (Fig. 11-2), as you will get a false measurement if you measure along a sloped line.

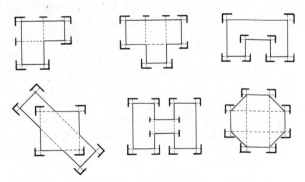

Fig. 11-2 *Take all measurements along level lines.*

Layout of Nonrectangular Houses

Since L-shaped, T-shaped, and H-shaped houses are merely a combination of rectangles (Fig. 11-3), the same batter board system is used. It gets a bit more complicated, but the principle of setting up rectangles with equal diagonals is the same.

Fig. 11-3 *Layout of nonrectangular houses.*

Footing Forms

Now is a good time to review your design carefully, before you cast it in concrete. Is the septic tank downhill from the bathroom? Do you want the underground power to come up through the house wall or on the outside of the wall? Are any minor changes necessary to orient for a view or for sunshine? Make any modifications that you see fit as you go.

The first decision is whether to pour both the footing and the wall at once or pour them separately. Most novices (and contractors) prefer to pour the footing first, let it cure, then pour the walls.

Materials and Tools

The footing forms for our sample house will require:

- About 300 linear ft of 2 × 8 (mostly 16-ft lengths for use later as rafters)
- 70 linear ft of 1 × 2 for crossties
- 120 16-in 1 × 2 stakes (or make your own)
- 25 lb double head nails
- 220 ft of ½-in (No. 4) steel rebar (precut at the lumber yard)
- 12 ea prebent corners
- Another 200 ft of baling wire
- A big roll of twine or heavy string.

Besides the tools needed for the layout, you'll need a square-nose shovel, hand level, crowbar, skill saw, hack saw, and a regular shovel.

STEP-BY-STEP PROCEDURE

1. Level the soil in a 2-ft wide path with a square-nose shovel. (See Fig. 11-4.) The footing must sit on undisturbed dirt, so sweep away loose rocks and dirt. Use a long straight board and a hand level to get the ground level.

2. String a line between the corner stakes to guide the placement of the forms. The outside 2 × 8 (or 2 × 6) will be offset from the line by 3 in to 4 in as shown. The stringline locates the outside of the foundation wall. The stringline locates the outside of the foundation wall.

   Note: Plan ahead for brick facing so either the footing or the wall allows a surface to support the bricks.

3. Set the outside 2 × 8's on edge around the entire footing. Start at a corner by nailing two 2 × 8's together. Then position the 2 × 8's, drive the stakes, and nail the stake to the 2 × 8. Use a level or water level to maintain the top edge of the forms at the same height. A crowbar works well to pry up the forms to level them. Dirt will be shoveled over the gaps left at the bottom.

4. With the outside forms in place around the perimeter, the inside forms are located with spreader sticks. These sticks (15 in for the sample house) hold the 2 × 8's at the proper spacing while the inside forms are staked and nailed. You can check for straightness with a tight string.

5. Place the rebar in the forms, then nail on the crossties every 4 in. Hang the rebar from the crossties with baling wire. The rebar should be a least 1 in away from the forms and the ground. At joints, the rebar should overlap 12 in and be wired together.

   Note: Bending and cutting rebar at the site is possible, but a lot of work. To cut it with a hack saw, saw at least half way through and then bend it (using leverage) until it snaps. Wear safety glasses.

6. Cut or buy a bunch of 18-in to 24-in pieces of rebar to have ready to stick in the wet concrete. These pieces connect the footing to the wall, and must be spaced a minimum 24 in on center and extend 12 in minimum above the footing. (See Fig. 11-8.)

7. If you need to place conduit for underground power, do it now. The incoming end should be 24 in below final grade.

8. Shovel dirt against the forms to fill any gaps and to keep the forms from bulging from the weight of the concrete.

9. Check the forms carefully for proper dimensions and level. Place and recheck the batter board lines to be sure the wires intersect over the corner stakes. Then remove the corner stakes, spreader sticks and anything else inside the forms. Your forms should now look like Fig. 11-5.

10. Usually the building inspector will want to see the forms before the pour is made. Give him a call before you set up a date with the concrete truck.

11. Figure out how much concrete you need using the formula given in Fig. 11-6 and place your order with the concrete company.

    Always order a bit more concrete than you think is necessary. It's a bummer to run short. Normally you'll want the standard "ready-mix." This should be 1 in to 1½ in maximum size aggregate (stones), five or six bags of cement per cubic yard, air-entrained (which makes the cured concrete less likely to crack in freezing weather and may cost extra), maximum slump of 4 in, and a 28-day compressive strength of about 3,000 psi. While you're talking to the concrete company, ask them how long their chutes are that come with the truck (they often can bring extra chutes), how much clearance their trucks need (to get under powerlines), and if there are any extra charges (for minimum quantities or for extra time required at the site for example). Arrange to pour on a day when you can get helpers. Don't pour if the temperature is near freezing; it radically reduces the strength of the concrete and makes it pucker. If your site is such that a truck can't get to it, make arrangements with the company for a truck with a concrete pump and a boom.

Well, that's all I can think of. Looks like we're ready to pour.

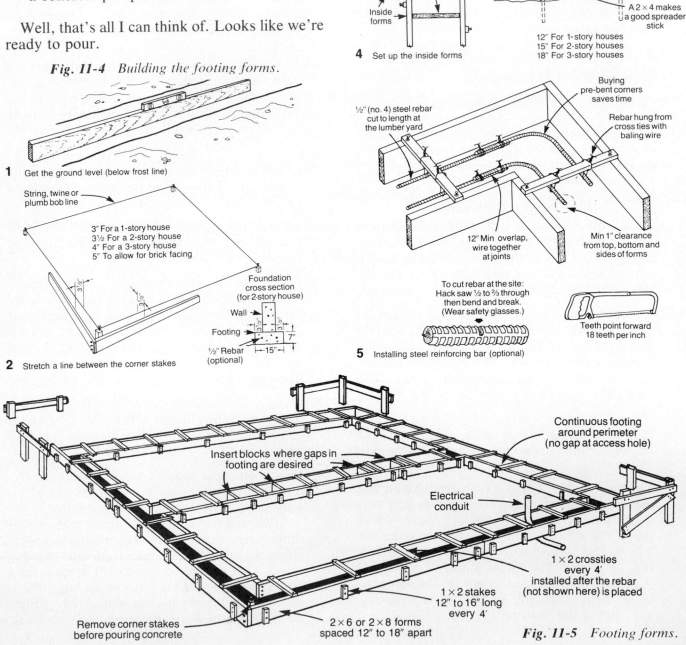

Fig. 11-4 *Building the footing forms.*

3 Set up the outside forms

Use crowbar to get the forms level
Outside forms needn't be cut
Top of stakes are below form top to allow "strike-off" of wet concrete
Toenail forms together
Two stakes at joints
Using double head nails speeds form removal

4 Set up the inside forms

Top view
End view
Outside forms
Level inside and outside forms
Spreader sticks (temporary)
Inside forms
Gaps will be covered with dirt
A 2 × 4 makes a good spreader stick
12" For 1-story houses
15" For 2-story houses
18" For 3-story houses

½" (no. 4) steel rebar cut to length at the lumber yard
Buying pre-bent corners saves time
Rebar hung from cross ties with baling wire
12" Min overlap, wire together at joints
Min 1" clearance from top, bottom and sides of forms
To cut rebar at the site: Hack saw ½ to ⅔ through then bend and break. (Wear safety glasses.)
Teeth point forward 18 teeth per inch

5 Installing steel reinforcing bar (optional)

1 Get the ground level (below frost line)

2 Stretch a line between the corner stakes

String, twine or plumb bob line
3" For a 1-story house
3½ For a 2-story house
4" For a 3-story house
5" To allow for brick facing
Foundation cross section (for 2-story house)
Wall
Footing
½" Rebar (optional)
15"
7"

Remove corner stakes before pouring concrete
2 × 6 or 2 × 8 forms spaced 12" to 18" apart
Insert blocks where gaps in footing are desired
Electrical conduit
Continuous footing around perimeter (no gap at access hole)
1 × 2 stakes 12" to 16" long every 4'
1 × 2 crossties every 4' installed after the rebar (not shown here) is placed

Fig. 11-5 *Footing forms.*

Footing Plan

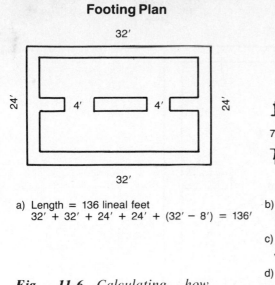

a) Length = 136 lineal feet
32' + 32' + 24' + 24' + (32' − 8') = 136'

Cross Section

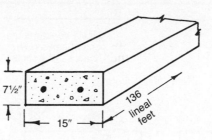

b) Cross-sectional area = .78 sq ft
7½" × 15" = 112.5 sq in ÷ 144 = .78 sq ft

c) Volume = 4.0 cu yd
.78 sq ft × 136 lin ft = 106.1 cu ft ÷ 27 = 4.0 cu yd

d) Add 5% to be sure you have enough

Order 4¼ cubic yards

Fig. 11-6 Calculating how much concrete is needed.

1 square foot = 144 square inches

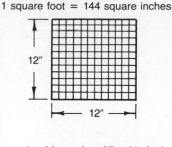

1 cubic yard = 27 cubic feet

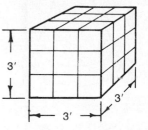

Pouring the Footings

There's no easy way to pour concrete. Plan on hustling your buns all day and hope that nothing goes wrong. Be ready ahead of time, and don't order the concrete until you're ready. Make sure the truck can get into and out of the the driveway. The truck is *huge*. It needs 12 ft to 16 ft of vertical clearance, 10 ft of width, and the turning radius is approximately 2 miles, I swear. The driver is usually in a hurry so you should have one or two of your many assistants on hand to help things go smoothly. By the way, provide them with some old gloves; concrete really dries out hands and even causes open sores on some people.

Materials and Tools

Wheelbarrow, two shovels, extra stakes, three 6-ft 1 × 6's for paddles, a 24 in × 30 in piece of plywood for a splashboard, gloves, seventy-five 20-in pieces of rebar, garden hose, 4-mil plastic to cover the concrete, and a float and trowel for any flat surfaces.

General Procedure

As the truck is coming up the driveway, someone should use a garden hose to wet down the dirt in the forms. This keeps the concrete from drying out too fast. Direct the truck driver as he backs up to the forms. The chute swings in an arc and you have to figure out where to park the truck so you can get the "mud" into the forms. Don't get the truck too close to the excavation. The 40,000 to 60,000 lb can cave in the dirt. Position the truck and add or remove sections of the chute; the trick is to get the forms filled with concrete with as little work and excitement as possible. Figure 11-7 gives you some possible problems to anticipate.

The driver gives you a skeptical look as he assembles the chute and climbs up to his controls on the back of the truck. (He also has controls in the truck cab if needed.) He pulls a lever and some concrete presently plops out of the drum and slowly creeps down the chute. "How's this?" he asks.

"Fino," you say, yelling over the noise of the truck. "Just give me the regular mix."

He controls the flow and wetness of the concrete and adjusts the angle of the chute. You swing the chute over to the forms and make "let 'er rip," "that's enough," "chute up," "chute down," and "hold it" motions with your hands as the concrete pours out. The helpers frantically paddle and push the concrete down the forms and complain about how heavy and unmanageable the stuff is. It doesn't take long to discover that it takes more than one set of hands to carry off this performance.

Concrete flows, but very stiffly and slowly. By using a shovel, a 1 × 6, or 1 × 4 like you would a canoe paddle, you can eventually get the stuff where you want it. Frequently "rod" the concrete by moving your paddle up and down, thus getting it to flow into the corners and around the rebar. Once the forms are full, take a short board and "strike off" the concrete level with the top of

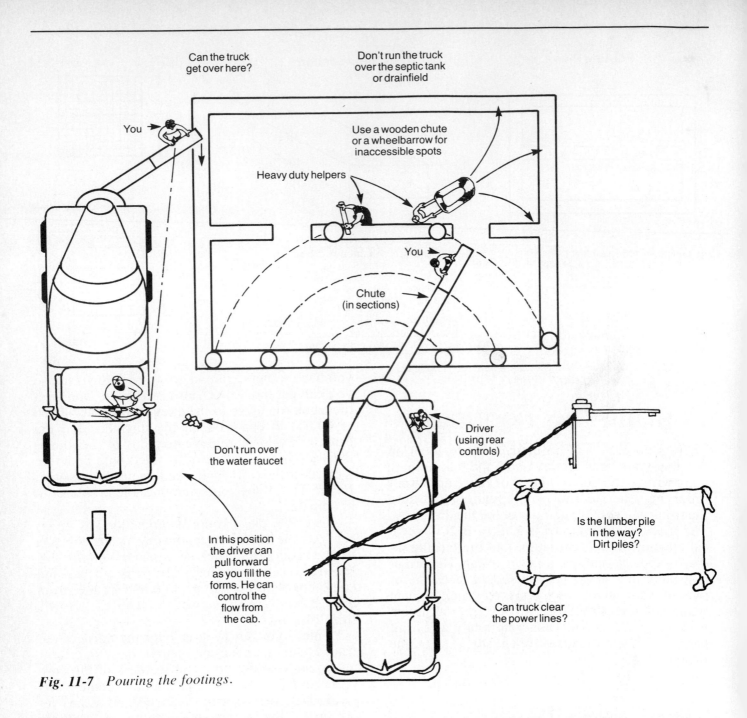

Fig. 11-7 *Pouring the footings.*

the forms. (See Fig. 11-8.) The crossties get in the way every 4 ft, naturally. Leave the top surface rough so it makes a bond with the wall you'll be pouring.

While the concrete is wet, put in the short vertical pieces of rebar which will connect the footing to the wall. Space them no more than 24 in O.C. and leave at least 12 in sticking out above the footing.

With a smidgin of luck, the driver will tell you if you're doing anything drastically wrong. In fact, my driver gave some real helpful advice, but that's not the sort of thing you should count on. With another smidgin of luck there will be enough

concrete to fill the forms and some left over. Have some extra forms staked out on the ground somewhere so the excess can be made into pier pads, walkway stepping stones or a boat anchor. Have a float and trowel handy to smooth the blocks and some water to wash off the tools and the helpers after its all over. And to keep things going smoothly, it helps to remember your checkbook (to pay the driver) and some beer (for the helpers).

Concrete will set up in half an hour or so on a hot, windy day. On such days it's likely to crack if you didn't wet down the ground before you poured. On cool damp days it may take a few

Fig. 11-8 *Strike-off, wall connection, and smoothing tools.*

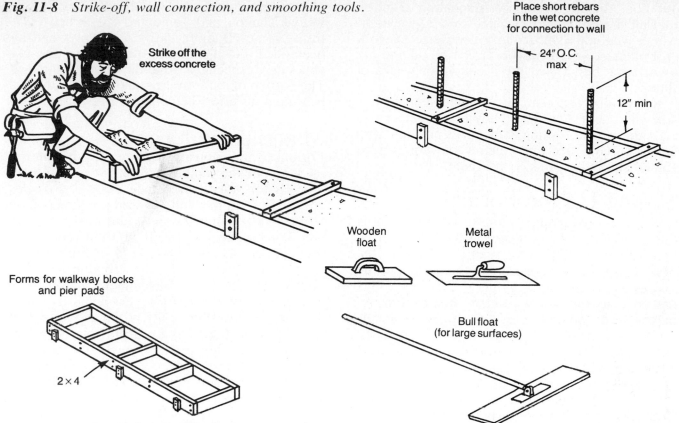

Strike off the
excess concrete

Place short rebars
in the wet concrete
for connection to wall

24" O.C.
max

12" min

Wooden
float

Metal
trowel

Forms for walkway blocks
and pier pads

2 × 4

Bull float
(for large surfaces)

hours to set up. Under water, even longer. The set-up time also depends on the mix. A dry mix is strong, but sets up quickly.

The curing process takes a week. The slower it cures, the stronger it will be. Cover the concrete with plastic right after you're done pouring. If the weather is dry, sprinkle the concrete with water each day and recover it with plastic. After the week is up, (No fudging now! Seven full days per Uniform Building Code Sec. 2605c), strip the forms off with a crowbar and a hammer. Do it carefully so the 2 × 8's can be used as rafters. Scrape the concrete off with the square-nose shovel or a hoe.

Barney

Old Uncle Barney couldn't stay away for long. I was flailing away with my hammer, trying to beat the foundation wall forms into submission, the day he dropped by. By this time I was painfully aware that I must be doing some things the hard way and had started to accept the fact that talking with an experienced person would be a big help. After all, Barney had said he spent 30 years in the trades; he probably knew a lot about foundations.

Barney needed no coaxing. He took the opportunity to expound on all the errors I'd made so far

and to make some recommendations for the next few steps. As much as his personality grated on me, his reasoning sounded good, so I followed his advice. Once again, I'll save these gems for later in order to keep the possibilities for confusion at a minimum.

Foundation Wall Forms

Decision: Rent or Build Your Own?

Forms for foundation walls take more time to construct and to set up than the footing forms. You'll have to decide whether to rent the forms or to build them and reuse the lumber later in the house. I built my forms, but next time I'll try to rent them. Building the forms is good practice for sawing and nailing because the forms are only temporary, but it takes time. Time to build them and time to take them apart and clean them. In addition, the height of my foundation required that all the plywood be ripped to a 32-in width, which was not a particularly good size to work with later. I saved maybe $250 by building my own forms but it cost me a couple of weeks of time.

Rental forms are usually available in only a few sizes, like 2 or 4 ft high. They bolt together and sit right on the footing, braced in position with 2 × 4's. They are fairly heavy, so you'll need some help and a truck to get them.

Setting Up the Forms

The first and overriding thing to keep in mind is that concrete is heavy, heavy! Like 150 lb per cu ft heavy. Like if the wall is three feet high, the pressure of the wet concrete at the bottom of the wall is 450 lb per sq ft trying to burst the forms and ooze out onto the ground. So don't scrimp on materials or braces here. (Many professionals don't brace the forms, but amateurs will have better results if they do.) Forms for walls less than 3 ft high are typically made of ¾-in plywood on a 2 × 4 frame as shown in Fig. 11-9.

The forms are held together with form ties. There are many types of form ties. The type I like are metal rods that are inserted through holes drilled in the plywood. Wedge clamps (also called "shoes") slip on the ends of the form ties so the wall cannot bulge. Form ties are located between each stud and spaced vertically every 2 ft. They are available for several different wall thicknesses. You buy the form ties and rent the wedge clamps normally.

Sections of wall forms will be bolted together. (See Fig. 11-12.) Some folks just nail the sections together, but I'm a scaredy-cat; I bolt them.

The forms are removed by tapping off the wedge clamps and removing the braces. The forms come off more easily if the plywood has been oiled, but this makes the plywood less useable later, especially if the oil has an odor to it. The portion of the form tie sticking out of the wall is broken off after the concrete has cured.

Materials and Tools

For our sample house, you'll need: 2-ft-high rental forms or the lumber and nails to build them, 2 × 4's for stakes and braces, 1 × 2 crossties, about 150 form ties and 300 wedge clamps, about 60 ea ⅜-in × 4-in bolts, nuts, and washers, a drill and ⅜-in spade bit, 160 ft of ½-in rebar, 6 ea prebent corners, 14 ea 6-in × 12-in vents, 160 lin ft of ½-in × ¾-in level strips, conduit and short sections of 3-in pipe, a 4-in diameter × 8-in coffee can, 20 lin ft of 1 × 10 for access holes and girder supports, and probably a few things I've forgotten. Bring all the tools from previous steps, including a hand level, water level, skill saw, and plumb bob. A radial saw will help if you're building your own forms. You can get 32 J-bolts now; you'll need them when you pour.

STEP-BY-STEP PROCEDURE

1. Build the wall forms. (If you rent forms, skip to step 2.) Start by making a short stud wall as shown in Fig. 11-10. The forms should be 2 to 8 in taller than the wall will be to allow room for the concrete as it's poured. Mark the plates, frame the form as shown and nail them together using two 16d nails at each joint. Eight-foot sections are normal. Nail the plywood to the studs with 6d nails spaced about 12 in apart. Since you have to take these

Fig. 11-9 *Foundation wall forms (walls 3 ft or shorter).*

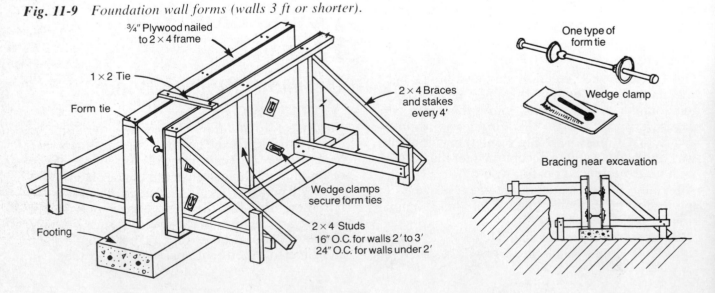

¾″ Plywood nailed to 2 × 4 frame

1 × 2 Tie

Form tie

2 × 4 Braces and stakes every 4′

Wedge clamps secure form ties

2 × 4 Studs
16″ O.C. for walls 2′ to 3′
24″ O.C. for walls under 2′

Footing

One type of form tie

Wedge clamp

Bracing near excavation

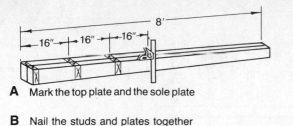

A Mark the top plate and the sole plate

B Nail the studs and plates together

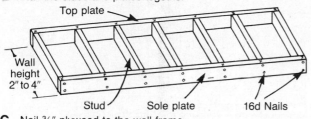

Top plate

Wall height 2" to 4"

Stud   Sole plate   16d Nails

C Nail ¾" plywood to the wall frame

Fig. 11-10 Building the wall forms.

forms apart later, use bright nails (rather than galvanized). They're easier to pull.

2. Set each section of the outer forms on the footing (Fig. 11-11). Start at a corner and use a plumb bob hung from the batter board lines

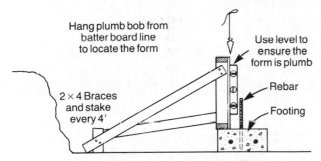

Hang plumb bob from batter board line to locate the form

Use level to ensure the form is plumb

Rebar

2 × 4 Braces and stake every 4'

Footing

Fig. 11-11 Set up the outside forms.

to locate the forms. Drive stakes and position the braces to secure the forms. The forms must be plumb (exactly vertical).

3. Bolt the sections of forms together and brace them every 4 ft (Fig. 11-12). Continue until all the outside forms are set up.

4. The forms should now look like Fig. 11-13. Recheck them for plumb and make sure they are directly under the batter board lines. Then use a tight string to be sure they are lined up straight. (See Fig. 9-8, "Basic Skills.")

5. Mark on the forms where the top of the concrete wall will be. Start in one corner and measure up from the footing (18 in for the sample house). Fasten one end of the water level (connected to a garden hose) in that corner and take the other end and proceed around the forms, carefully marking the top of the soon-to-be foundation wall as shown in Fig. 11-14. This takes some care—adjust the water level carefully, don't let the stationary end slip, and look out for kinks in the hose. With a little luck, and if there are no bubbles in the hose, the marks will match when you get back to the original corner.

6. Nail the level strips to the outer forms. Level strips should be approximately ½ in × ¾ in and can be ripped from 1 × 2 or 1 × 4 if they're not available at the lumber store. Align the top edge of the strips with the marks on the forms.

Fig. 11-12 Bolt the outside wall forms together.

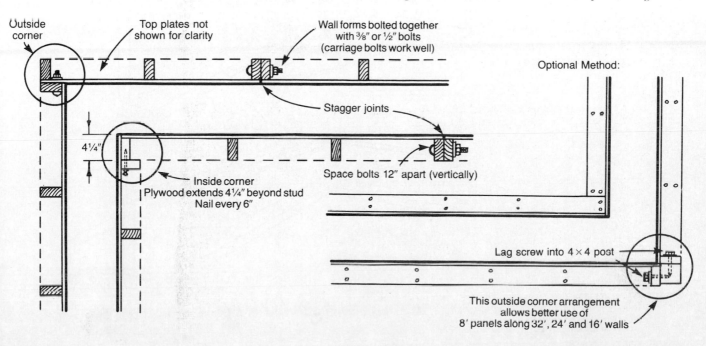

Outside corner

Top plates not shown for clarity

Wall forms bolted together with ⅜" or ½" bolts (carriage bolts work well)

Optional Method:

Stagger joints

4¼"

Inside corner
Plywood extends 4¼" beyond stud
Nail every 6"

Space bolts 12" apart (vertically)

Lag screw into 4 × 4 post

This outside corner arrangement allows better use of 8' panels along 32', 24' and 16' walls

Top View

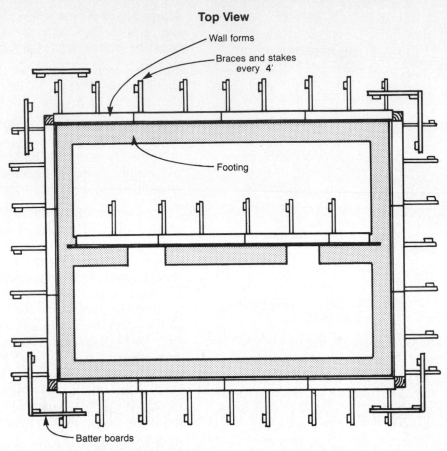

Wall forms

Braces and stakes
every 4'

Footing

Batter boards

Fig. 11-13   Outside wall forms erected and straightened.

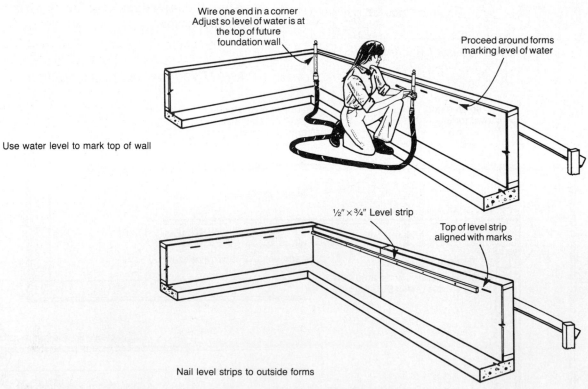

Wire one end in a corner
Adjust so level of water is at
the top of future
foundation wall

Proceed around forms
marking level of water

Use water level to mark top of wall

½" × ¾" Level strip

Top of level strip
aligned with marks

Nail level strips to outside forms

Fig. 11-14   Installing the levels strips.

Digression: Here's where I got my first lesson about using Uncle Barney's advice. His suggestion was to align the bottom edge of the strips with the marks, reasoning that this way there would be no ½-in × ¾-in recess in the final concrete wall. Partway through pouring the wall I realized that it was darn near impossible to strike off the concrete to the bottom edge of the strip. To run a float along the top edge of the strip would have been much quicker and more accurate. The way I did it took so long that the concrete was starting to set up and I still had to place the J-bolts, pay the driver, direct him to turn around, float and trowel the walkway steps, and possibly a dozen other things. Two weeks later, as I chipped off the excess concrete with a cold chisel, I realized that the siding would have neatly covered the recess anyway. So much for that smart idea. I wanted to mention this to Barney but he'd left for Vegas on vacation.

7. Now set up the inside forms. By building a few 4-ft sections you can stagger the joints as shown in Fig. 11-12. A way to handle the inside corners is shown in the same drawing. *Don't brace or bolt the inside forms yet.*

8. Locate and install the form ties as shown in Fig. 11-15. They should be about 2 in to 4 in below the lower edge of the vents. Normally 10 in down from the top of the wall and 6 in up from the footing are good locations. For walls less than 20 in tall, one row of form ties is enough. Place them between each pair of studs and 6 in up from the footing. The top of the forms will be held together by 1 × 2 crossties.

Drill holes for the form ties in the outer forms and insert them. Then drill holes in the inner forms and slip the form ties through as the inner form is tilted into position. An extra pair of hands helps here. Secure the form ties by tapping on the wedge clamps.

For the sample house, we'll use one row of form ties 6 in up from the footing. The piece of rebar will be wired to the form ties.

9. When the inner forms are set up and tied to the outer forms, place the rebar in the forms. Use baling wire to secure it to the form ties, or hang it from the crossties as we did for the footings. Don't nail on the crossties until after the rebar is placed; then nail the 1 × 2 crossties every 3 or 4 ft.

10. Stake and brace the inner forms and bolt them together. Braces are spaced every 4 ft. When you can't take a step anywhere without tripping over a brace, you have enough. Try bumping the forms with the sledge; they shouldn't budge.

11. Locate and install the foundations accessories. This includes the vents, access holes, girder supports, conduit, and pipes. First locate everything with a marking pen, including the J-bolt locations.

    a. Vents must be 1.5 sq ft for every 25 lin ft of foundation wall that is exposed to the outdoors. (UBC Sec. 2517c6) There must be a vent within 3 ft of each corner. Vents are usually placed flush with the top of the foundation.

Important: A vent has no strength to support a floor joist so you must locate the vents between the floor joist positions. That means you have to lay out the floor joists now with a marking pen. Typically vents measure 6 in × 12 in and can be purchased (plastic ones) or built with cedar 1 × 10 (ripped 8 in wide) and screen stapled on. Just nail the vents between the forms and notch out the level strip where necessary (see Fig. 11-16).

    b. Access holes and girder supports are also made from cedar 1 × 10 boards

Fig. 11-15 *Set up the inside forms using form ties.*

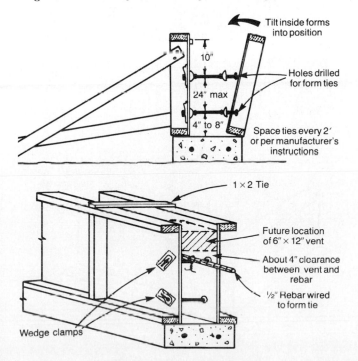

Tilt inside forms into position

10"

Holes drilled for form ties

24" max

4" to 8"

Space ties every 2' or per manufacturer's instructions

1 × 2 Tie

Future location of 6" × 12" vent

About 4" clearance between vent and rebar

½" Rebar wired to form tie

Wedge clamps

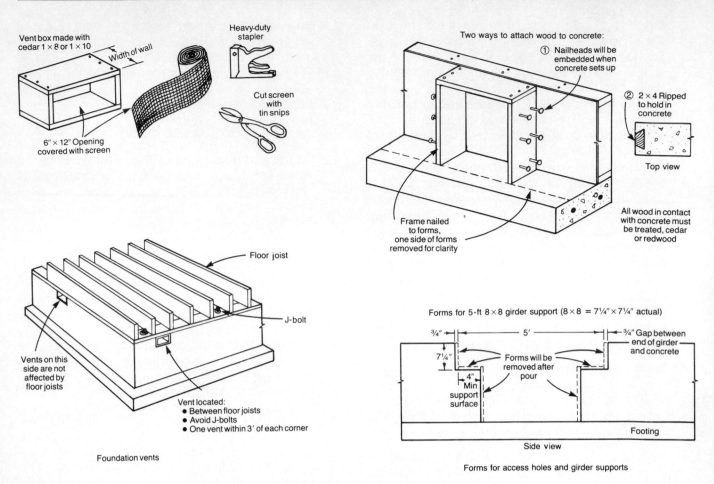

Fig. 11-16   *Vents, access holes, and girder supports.*

(ripped 8 in wide). Just locate the boards so the concrete can't flow where you want the opening. Nail the boards to the forms. The Uniform Building Code Sec. 2517c2 requires a minimum unobstructed opening of 18 in × 24 in for access to the crawl space. Figure 11-16 shows how to permanently mount the boards to the concrete so hinges for an access hole door can be installed. In the case of girder supports, the boards will be removed so the girder will rest on solid concrete (with a vapor barrier in between). Measure carefully.

c. J-bolts (also called anchor bolts) are stuck into the wet concrete and are used to attach the mudsills. (See Fig. 11-17.) That's because it's rather difficult to nail into concrete. J-bolts must be ½ in diameter and 10 in long. Locate them every 6 ft and within 12 in of each corner (UBC Sec. 2907e). Mark the locations clearly and be careful not to spatter the marks with concrete when you pour. Of

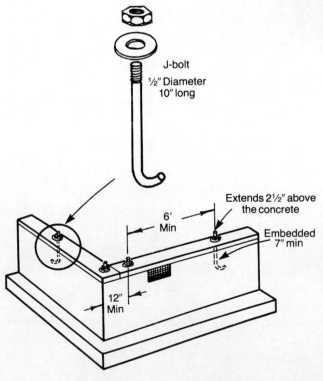

Fig. 11-17   *J-bolt placement.*

Fig. 11-18 *Forms for the foundation walls.*

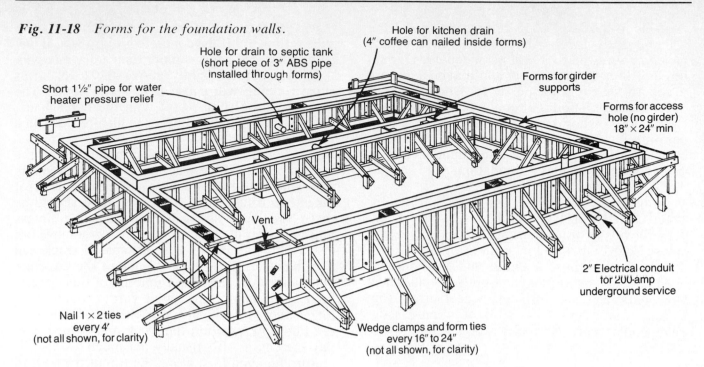

Hole for kitchen drain
(4" coffee can nailed inside forms)

Hole for drain to septic tank
(short piece of 3" ABS pipe
installed through forms)

Short 1½" pipe for water
heater pressure relief

Forms for girder
supports

Forms for access
hole (no girder)
18" × 24" min

Vent

2" Electrical conduit
for 200-amp
underground service

Nail 1 × 2 ties
every 4'
(not all shown, for clarity)

Wedge clamps and form ties
every 16" to 24"
(not all shown, for clarity)

course, J-bolts have to miss all the vents, access holes, and floor joist locations.

d. Plumbing and electrical holes are easy to provide for now and a real pain to do later on. For the sample house we'll provide four holes. For the main sewer drain we'll put a short section of 3-in ABS pipe right through the forms. Similarly, a 1½-in pipe will be placed for a drain in case the water heater pressure-relief valve is ever activated. The hole in the center wall will be a 4-in or 5-in diameter coffee can nailed into the forms. This allows a loose fit for our sloping 3-in main drain. Finally, the electrical conduit is already in place and sticking out of the footing. Since the water supply line comes in under the footing, no hole is needed for it.

Check your electrical and plumbing designs to determine your needs, remembering that your local plumbing and electrical codes may require different arrangements than I've shown here.

12. Make a final check of the forms for squareness, proper dimensions, straightness, plumb walls, and level strips. (See Fig. 11-18.) Make sure all the accessories are in place and the walls are firmly braced. If the concrete is being pumped to the site, be aware that it may be soupy enough to squeeze out of a ½-in gap. When you're ready, call up the inspector and see if he wants to check the forms before you pour.

By now you'll begin to understand why so many older houses weren't built on continuous foundations. The forms are a lot of work and require a lot of material. It seems that a small house could be framed in the time it takes to build a foundation. That's a fact, but the value of a continuous foundation can be reaffirmed by looking at some of the beautiful old houses with priceless craftsmanship that are now sagging or leaning due to inadequate foundations. You can also look at it this way: you're actually serving an abbreviated apprenticeship before building your house, and the good foundation will be an added bonus.

Pouring the Foundation Walls

This is the big day. You can't possibly be overprepared because there are scads of things to think about. Have the forms for the chimney foundation, pier pads, stairs, and walkway steps ready if you need them. If the truck can't get within 20 ft of the rear wall, you may need to build a chute with 2 × 10's and support it with *heavy duty* sawhorses. (There can be as much as 1,000 lb of concrete in a 15-ft chute.) A wheelbarrow isn't much help on walls unless you have a good way to get it high enough to dump into the forms.

You'll need to make a couple of hand floats that just fit into the forms so you can easily run the float along the level strip getting the top of the wall perfectly level. Any board with a handle nailed or screwed to it will be adequate. Line up three or four trusty helpers and maybe a camera to record the event for posterity. Don't forget to order the right amount of concrete and give good instructions on how to get to your place. The temperature is well above freezing, right? OK then, clear the decks and man the cannons—we're gonna pour a first-class foundation.

Materials and Tools

Two shovels, extra stakes and braces, a plywood splash board (about 32 in × 48 in), gloves, three 6-ft 1 × 4 paddles, a chute and sawhorses if needed, two hand floats, J-bolts, a float and trowel for any flat surfaces, wheelbarrow (optional), water and hose, roll of 4 mil-plastic, checkbook, beer, and lucky rabbit's foot.

General Procedure

Having already poured the footings, there should be less surprises this time. Back the truck up to the forms. Get the chute where you want it and hold a shovel or the plywood splash board to guide the mud into the forms. The paddlers are in for a workout. There's more concrete to move this time but the trick is to pour it so you don't have to move it very far.

Rod the concrete by moving the 1 × 4 up and down. This fills the mud around the rebar, into the corners and under the vents. When the concrete no longer settles as you rod, that's enough. *Be sure to rod very thoroughly around the vents and girder supports,* otherwise there will be an unpleasant surprise waiting for you when you strip the forms next week. Figure 11-19 shows the voids that can occur with inadequate rodding.

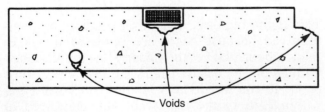

Fig. 11-19 *Voids resulting from inadequate rodding.*

As much as possible, try to fill the forms by working from one end continuously to the other end. If the forms start to bulge, quit pouring immediately and brace that spot some more. Also stop rodding in that location. Then finish pouring

that part later after giving it a half hour or so to start setting up. If the concrete is extra wet, it will tend to bulge the forms more than a dry mix will.

Fill the forms slightly above the level strips since rodding will cause it to settle some. The excess will be scraped off with the hand floats, but low spots will be a bugger to fill after the truck leaves.

Once you start pouring, continue until the forms are full. There will be a visible joint in any spot where an hour elapses between pours. If it takes two truckloads, the second load should be there less than an hour after the first one, unless you can stop at a girder support or somewhere else where there won't be a joint. A truck can haul between 6 and 9 cu yd. Often the concrete company allows a specific amount of time for the pour, like 8 minutes per cubic yard, and if it takes longer there may be an extra charge. If you need the time to reposition the truck and chutes, just pay the charge. It's usually not much.

Run the hand float along the top of the wall to get it flat and level. The cement paste floats to the top as you work and makes a smooth surface. Poke down the stones.

Wiggle in the J-bolts so they are implanted well. They should extend about 2½ in above the top of the wall to allow for the mudsill, sill sealer, washer, and nut. Remember that the edge of the mudsill will be flush with the outer edge of the wall and the J-bolts must not interfere with the floor header. Hence, for 2 × 6 mudsills, the J-bolt should be back about 3 in from the outer edge of the wall. Double check that the bolts are spaced no more than 6 ft apart and within 12 in of every corner. Also place any post supports or pins in the pier pads. Watch to see that the J-bolts aren't sinking into the mud.

After satisfying yourself that everything is in order, wash off your tools, cover the forms with plastic for curing and find a place to rest your aching body. Pouring the foundation is probably the largest single job in building a house and the most physically demanding. That's probably the reason it gives you such a great sense of satisfaction when it's done (Fig. 11-20).

A Four-Star Comedy

Normally I'd be too embarrassed to fess up to this before God and all the people. Over the past couple of years though, I've heard enough stories

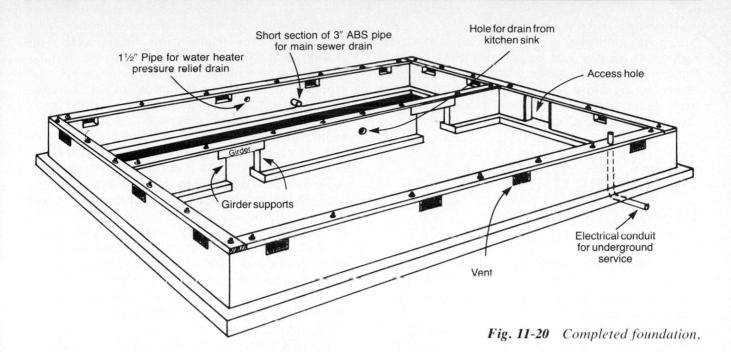

Labels on figure:
1½″ Pipe for water heater pressure relief drain

Short section of 3″ ABS pipe for main sewer drain

Hole for drain from kitchen sink

Access hole

Girder

Girder supports

Electrical conduit for underground service

Vent

Fig. 11-20   *Completed foundation.*

to come to the conclusion that I'm not the only one with the brain of a small tadpole. Nearly all beginners have their problems with foundations. Unfortunately, nobody advertises their mistakes, so beginning builders march through history making the same blunders.

To set the stage, let me explain the mistakes I made long before the big day when we poured the concrete. Mistake number one was believing that there was just one best way to pour a foundation. There are hundreds of good ways to do it, and this chapter describes only one of them. Mistake number two was thinking that Barney knew the single best way; but there I go trying to pin it on him. He gave me free advice and I decided to take it. I would have been more cautious if I'd known his ''30 years in the trades'' consisted of selling fencing and portable toilets to construction outfits. He hadn't even been out to a site for the last 10 years and his previous experience amounted to being a sightseer on big *professional* jobs. By listening to him I concluded that everybody poured the footing and the wall at the same time, that shiplap and baling wire was standard for building forms, bracing was optional, and you didn't have a hair on your chest if you couldn't pour 25 yd singlehandedly. Dear me, dear me, the dangers of getting all your information from one source!

The big day dawned with me frantically trying to wire the rebar in place before the truck showed up. I didn't finish before I was brought to my feet by the sound of huge limbs snapping off trees as the truck tried to get up the driveway. Then there wasn't room for him to turn around so he backed out the driveway, turned around on the highway,

then backed all the way in to the site. I was off to my typical start.

The first truckload went OK, though my single helper (who shall remain unnamed by mutual agreement) and I were busier than we'd been in a long time. We were relieved when the driver left to get the second load; we thought we'd get the top floated and the J-bolts placed before he got back. Dream on, kids. The guy must have been a frustrated A. J. Foyt. We'd just got the top semi-floated (which was a problem due to the incorrect level strips) and started looking for the J-bolts when who comes rolling back up the driveway?

The second load was a first-rate circus. The driver added lots of water to the mix to get it to flow down the long (35-ft) chute to the rear wall. That made the concrete easy to paddle, but when it was about 10 in from the top of the forms we heard this funny ''twang'' noise, followed by ''twangg, twangg, twanggg,'' and the forms started to bulge. Oh my God! The wires were breaking. Panic, shouting, and scurrying for braces ensued. We were out of braces and stakes naturally. It was amazing the driver could keep a straight face as he told us what to do.

By waiting awhile for the concrete to start setting up, I avoided having a paved crawl space but I ran up a pretty good bill with the concrete company. The day continued in the fashion in which it had started: I jumped down off the forms onto a board with a nail in it, the forms got so slopped up with concrete that we couldn't find the J-bolt marks, we managed to pour some concrete down the electrical conduit, and some other trivial stuff that I won't bother you with.

I had lots of time to reflect on my mistakes during the next two weeks as I chipped off the high spots on the foundation with a cold chisel. My diary has some choice, but unprintable, comments during those "foundation days." I had a good, level, strong foundation in the end, but I sure did it the hard way.

Finishing the Foundation

The forms may be removed after seven days of curing. During hot weather, sprinkle the concrete with water occasionally and recover it with plastic. Use this week to put in the foundation drains and to plan the floor.

Removing the forms is straightforward. A crowbar and a cat's paw are helpful tools. Remove all the nails so you don't step on them, and scrape the concrete off the plywood immediately, before it can harden any more. Voids and low spots can be filled with a cement called topping mix though those spots will have little strength. Snap off the protruding ends of the form ties.

Coat the outside of the foundation wall with emulsified asphalt wherever the wall is in contact with the soil. This black, gooey stuff can be painted on with an old broom or a big paint brush. The emulsified asphalt waterproofs the wall and without it the concrete will transmit moisture into the crawl space.

Foundation drains are used to carry water away from the foundation. Rainwater and groundwater that accumulate next to the foundation will eventually get into the crawl space and promote rot. The foundation drains are installed level around the foundation and then slope away from the house to a downhill location or to a sump as shown in Fig. 11-21. The sump is a hole filled with gravel and works like a drainfield, holding the water until it percolates into the soil.

Downspout drains should be installed now also. These should be 4-in plastic drainpipes (not perforated) that lead the large amounts of rainwater that the roof collects away from the house. Although the downspout drains often lead to the same place as the foundation drains, they shouldn't be connected because they might end up flooding your crawl space with downspout water. Neither drain should be connected to the septic drainfield.

Backfill around the foundation after it's coated and the drains are installed. By shoveling some dirt over the inside footings as well, the entire footing will be below the frostline even when the house is not heated. Plant some grass seed immediately. Until the grass grows, this area will be a muddy mess every time it rains.

Cut and place the girders and posts. Be sure all cuts are absolutely square. Put a moisture barrier (for example, an asphalt shingle or some 4-mil plastic) between any wood and concrete and allow a ½-in air gap between the concrete and the ends of the girders as shown in Fig. 11-22 (UBC Sec. 2517c5).

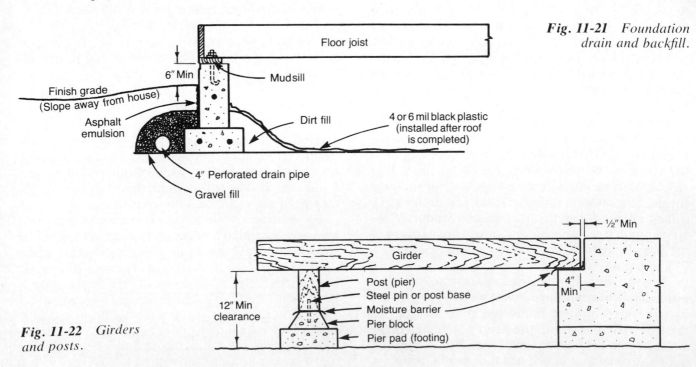

Fig. 11-21 *Foundation drain and backfill.*

Floor joist

6" Min

Finish grade (Slope away from house)

Mudsill

Asphalt emulsion

Dirt fill

4 or 6 mil black plastic (installed after roof is completed)

4" Perforated drain pipe

Gravel fill

½" Min

Girder

Post (pier)
Steel pin or post base
Moisture barrier
Pier block
Pier pad (footing)

4" Min

12" Min clearance

Fig. 11-22 *Girders and posts.*

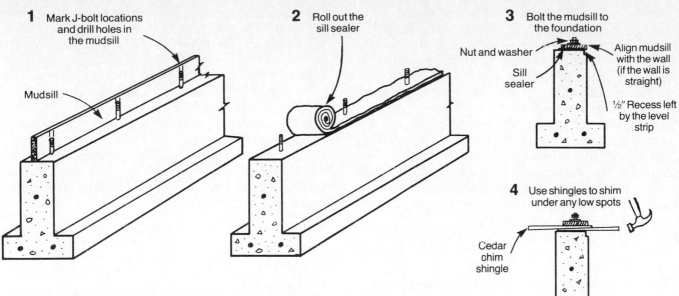

1 Mark J-bolt locations and drill holes in the mudsill

Mudsill

2 Roll out the sill sealer

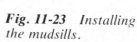

3 Bolt the mudsill to the foundation

Nut and washer

Sill sealer

Align mudsill with the wall (if the wall is straight)

½" Recess left by the level strip

4 Use shingles to shim under any low spots

Cedar shim shingle

Install the mudsills and sill sealer. Mudsills are pressure-treated 2 × 6's (or occasionally 2 × 4's) that will not rot even when subjected to moisture. (See Fig. 11-23.) Lay them out carefully to form a straight, square, perfect rectangle, preferably the exact size of the intended floor. By snapping a chalkline from corner to corner on the top of the foundation wall you can be sure the mudsills are straight, even though the foundation may have some (perish the thought) minor bulges.

The sill sealer looks like a roll of fiberglass insulation that is about an inch thick. It is laid out between the mudsill and the concrete and seals any gaps between them, thus denying vicious and unruly insects access to the crawl space. The sill sealer will squeeze down to about ⅛ in when the mudsills are bolted on. Also install metal termite

Fig. 11-23 *Installing the mudsills.*

shields if you're building in termite country. Make sure the mudsills are installed perfectly level. If there are low spots, shim them up with shim shingles.

Make gravel for foundation drains. An easy, cheap way to make a small amount of gravel, for foundation drains for example, is by shoveling dirt onto an inclined screen as shown in Fig. 11-24. Get some ½-in or ¾-in screen (also called wire cloth) from the lumber or hardware store and nail it to a 2 × 4 frame. The gravel rolls down the screen and the sifted dirt goes on through, ready for the garden or flower bed.

Fig. 11-24 *Making your own gravel.*

Whew!

Talk about a marathon chapter! You see that building a foundation is more complicated than you might expect. Basically there are just two steps: (1) build the forms, and (2) pour in the concrete. However, since the foundation is the toughest part of building a house, and since it's the first major task, I broke it down into a lot of small steps. Like any huge task, no one step is very hard. From now on the instructions will be more concise. I figure that if you can pour the foundation, the rest of the house won't give you many problems. You now know how to take a big job, break it into small steps, and just keep at it until it's done.

Of course some of the small steps can be unexpected and formidable. For example, the truck that delivered Wrench's second load of concrete broke down right there in the driveway. The big drum screeched to a halt and there was no way to get the mix out of the drum. The driver called the office and got no answer. So here's Wrench, halfway through a pour with concrete setting up both in his forms and in the truck. Even with his experience, this looked like a real stumper. With characteristic fearlessness and aplomb, Wrench got out his tool box, disassembled the drum drive mechanism right there on the spot, and 15 minutes later the drum was turning again. I was impressed. Wrench thought it was great sport. "It's not every day you can get an opportunity to take a $100,000 cement truck apart and see what makes it tick."

Personally, I'm not that enamoured by opportunities to work on mechanical contrivances. My tool box contains only a large hammer and a rag. I can't always fix things, but I can always get them to the point where there's no question as to whether they will ever work again or not. Let this be fair warning to any cement trucks that break down in my driveway.

CHAPTER 12
Building the Floor

CHAPTER 12

Building the Floor

I don't know if you're like me, but if you are, there are some things about building a house that are in your favor. What I mean is that, well, I'm not too good looking or lucky or clever and I very seldom get big pleasant surprises. I can rest assured that I'm not going to get a great job modeling or in public relations on account of my looks. No one in my family owns a big successful business where I might suddenly get an important, well-paid job. And the chances of me inheriting any large sum of money are remote at best. That sort of stuff just never happens to me, if you know what I mean. The only way I ever succeed at something is to work harder at it than everyone else, stick to it longer, and consistently think things out better (in my opinion, of course). And then rewards have ranged from moderate to nonexistent. I know. . . . grump, moan, and complain.

But when you're used to that, building a house is a very rewarding experience. It's a lot of hard work and a test of your imagination and ingenuity, but you are amply rewarded for your efforts. There are no giant steps forward (like a salesman gets when he makes a big sale) and no giant steps backward we hope. Just a day of work resulting in a day's progress toward a house. Or a lazy, restful day resulting in not very much progress. I like it that way. I've given up on winning the Irish Sweepstakes. All I want is a fair return on my efforts and an opportunity to work as hard as I can. Building your house is like that. You're taxed on the things you buy but you're not taxed on your labor. You put in 2,000 to 4,000 hours of labor and in return you get a nice house and no mortgage. If there is a day you want to do something other than building, you do it. Maybe just lie in the sun, read a book, or go sailing. That happened to me several days. No work and no progress. . . . as expected. If you try to take short cuts or be lucky ("Oh, this floor looks

pretty square and I don't have time to check it."), you'll find out soon enough that taking the easy way out is not a very good idea.

Floor Framing

Framing the floor is mostly fun and exciting. Finally the house itself is being built. There are several ways to build floors but this discussion concentrates on the most common method.

STEP-BY-STEP PROCEDURE

1. First mark the corners of the house on the mudsills. If the mudsills are on perfectly, the corner of the sills will be the corners of the house and the floor. Measure the diagonals between the corners to be sure the floor will be square. (See Fig. 12-1.)

    Snap a chalk line along the mudsills between the corners. If the mudsills are perfectly straight, the edge of the mudsills will be the edge of the floor. If the mudsills are a bit out of line, don't worry about them. But build to the chalk line rather than to crooked sills.

2. Pick out your two straightest planks and use them as the first header and the first floor joist. Nail the header and the first joist together at a corner with three 16d nails. Then align them with the mudsills, or the chalk lines on the mudsills. Carefully toenail these headers to the mudsill with 8d box nails, keeping them in alignment. Once you get a couple of toenails in the board it becomes steadier and easier to put the other toenails in.

3. Place enough headers to take care of the first half of the floor. Each additional header is

126 / Practical Housebuilding

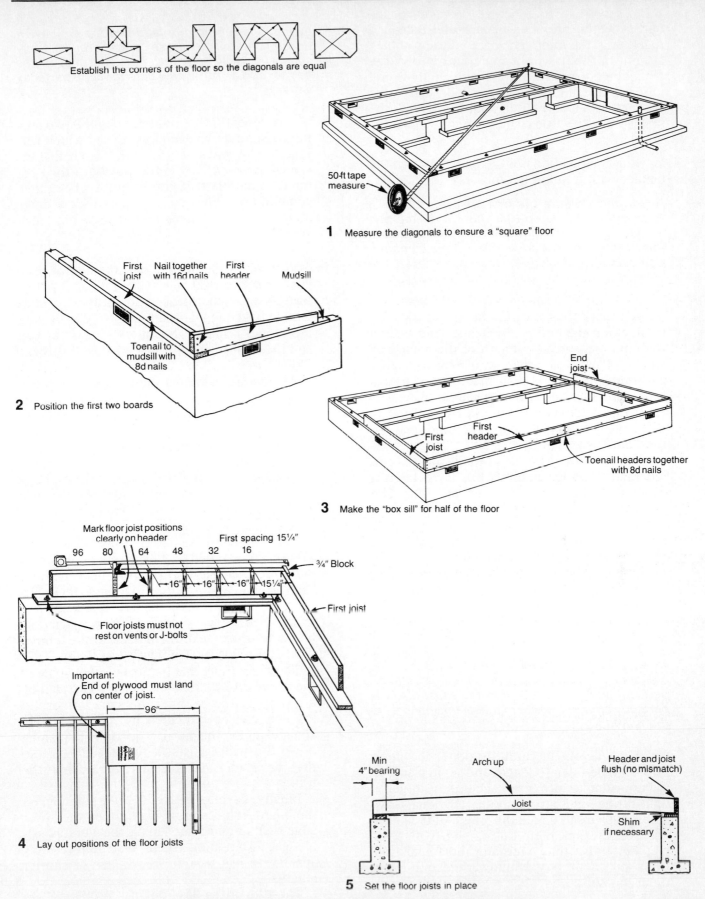

Establish the corners of the floor so the diagonals are equal

1 Measure the diagonals to ensure a "square" floor

50-ft tape measure

First joist Nail together with 16d nails First header Mudsill

Toenail to mudsill with 8d nails

2 Position the first two boards

End joist

First joist First header

Toenail headers together with 8d nails

3 Make the "box sill" for half of the floor

Mark floor joist positions clearly on header

First spacing 15¼"

96 80 64 48 32 16

¾" Block

First joist

16" 16" 16" 15¼"

Floor joists must not rest on vents or J-bolts

Important:
End of plywood must land on center of joist.

96"

4 Lay out positions of the floor joists

Min 4" bearing Arch up Header and joist flush (no mismatch)

Joist

Shim if necessary

5 Set the floor joists in place

Fig. 12-1 *Building the box sill and placing the floor joists.*

butted to the previous header and toenailed together at the joint. They are a little wobbley at first, but they will become far more sturdy when the floor joists are in place. Nail the end joist in place at the end of the floor. Check the corners with a carpenter's square. This is the first half of the "box sill," which defines the floor.

4. Lay out the positions of the floor joists on the header with a long tape measure.

Important: Be sure the first spacing is 15¼ in (rather than 16 in) so that the end of the plywood subfloor will land in the middle of a floor joist. Mark the header clearly using a combination square. (See Fig. 12-1, step 4—showing a view of the first corner from inside the foundation.)

As you mark the header, it helps to tack a ¾-in block on the end of the header and stretch the tape measure the length of the header as shown. This avoids the cumulative error that can build up if you measure just 16 in at a time, and it ensures that the first spacing is 15¼ in. After you've marked the header, mark the far mudsill where the other ends of the floor joists will be placed.

I can remember this 15¼-in business seeming really strange at first, but once I put a piece of plywood on, it all became clear. The end of the plywood absolutely must fall at the center of the joist. The natural tendency of 99 percent of novices is to lay everything out on 16-in spacing and then wonder what to do when the plywood end lands on the edge, rather than the center, of a joist. By the time you have laid out the floors, the walls, and the rafters, this spacing technique will become second nature to you.

Note: Whenever framing anything, keep the plywood in mind. Plywood is manufactured perfectly square, so the framing must also be perfectly square. Frame to make efficient use of the plywood.

5. Once the layout is carefully done, the floor joists can be sawn to length and set in place. It is important to saw very accurately to get a tight-fitting and strong floor. A radial saw helps. Be sure to check the length of the joists and don't whack them off too short. Also remember those that overhang the foundation for a porch or bay window, and don't cut them too short.

Note: A time-saving alternative here is to set the joists in place, then saw off the ends that

overlap the center foundation with a skill saw. It doesn't matter if those ends are perfectly square anyway; but the ends that nail to the header and all other saw cuts must be perfectly square.

Lay several joists out between the two supports. Sight down each joist to see which way it arches. Always put the arch (also called the *crown* or *crook*) up, so the weight of the floor tends to flatten it. If you forget to check this, some of the arches will be up and some down and the floor will be uneven. (See Fig. 12-1, step 5.)

6. Nail through the header into the end of each joist with three 16d nails. (See Fig. 12-2.) Then toenail the joist to the mudsill at both ends with two 8d nails. Use a tight string to be sure the header remains absolutely straight as you nail the floor joists to it. (See Fig. 9-8, "Basic Skills.")

As you are installing the floor joists, check frequently to be sure that when measuring from the end of the floor, 48 in, 96 in, 144 in, and 192 in measures to the *center* of a joist. This ensures that the end of each piece of plywood will land on a joist. Also measure the diagonals frequently to be sure everything is square. A floor that is out of square can cause you really serious problems later on.

A couple of things can happen that never get mentioned in the books. Like a few of the joists are so arched or warped that there is no way you can use them in the floor without there being a big bump. Discard them for now. You will probably be able to use them in short pieces for blocking or for window headers. If not, take them back to the lumber yard and tell them you can't use them.

The other thing that happens is that some of the so-called 2 × 10's are 9¼ in wide, as they are supposed to be, but others are 9 in or 9½ in. This can be remedied by placing a shim shingle under the narrow joists and planing down the wider joists at the header end, so that the header and the joists are flush at the top.

Finally, despite the No. 2 grade stamped on the board, it has a great big knot on the lower edge right at midspan. This is the worst possible place for a knot. Again, discard it for now and try to use it somewhere else in shorter lengths.

These modern-day quality problems are one reason that it's smart to order more lumber than the precise number of boards

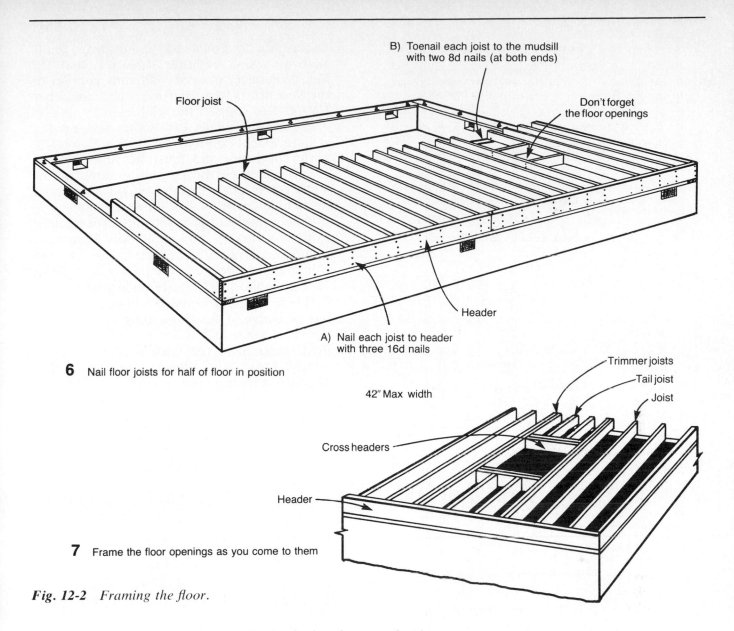

B) Toenail each joist to the mudsill with two 8d nails (at both ends)

Floor joist

Don't forget the floor openings

Header

A) Nail each joist to header with three 16d nails

6 Nail floor joists for half of floor in position

42″ Max width

Trimmer joists

Tail joist

Joist

Cross headers

Header

7 Frame the floor openings as you come to them

Fig. 12-2 *Framing the floor.*

needed. A few extra planks will also be handy to use as scaffolding when you're putting on the roofing and siding.

7. Frame the floor openings for chimneys, furnace plenums, stairways, and the like as you come to them. Figure 12-2 shows a floor opening and Fig. 12-3 illustrates the procedure for framing a floor opening.

Keep your wits about you when building floor openings. Mark the cross header positions (both sets) before nailing them, so you don't end up with an opening 3 in too short after the double cross headers are added in step 4. Make sure everything fits snugly.

When face nailing the double trimmer joists together, stagger the nails as shown. Also, 16d nails will poke through unless they are nailed at an angle, and those points will give you a heck of a scratch.

8. Floor overhangs for porches, bay windows, and so on can be simple or difficult depending on the direction the floor joists run. The simpler case, shown in Fig. 12-4, is when the overhang can be an extension of the floor joists.

When the overhang is perpendicular to the direction of the floor joists, pay attention to the sequence in which each board is nailed. Most joints must be nailed together *before* the doubler is installed. If you get two boards doubled, then realize that a 16d nail isn't long enough to attach the cross joists, you will have to use some joist hangers to solve the problem. (See Fig. 6-16.) Toenailing is not adequate for floor joists.

No matter how the overhang is framed, the joists must extend inward (into the house floor) twice as far as they overhang. (See Fig. 6-14.)

Fig. 12-3 *Framing floor openings.*

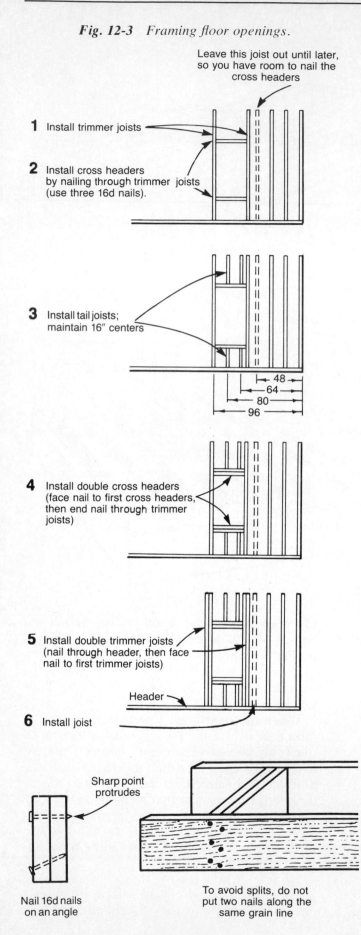

Leave this joist out until later, so you have room to nail the cross headers

1 Install trimmer joists

2 Install cross headers by nailing through trimmer joists (use three 16d nails).

3 Install tail joists; maintain 16″ centers

48
64
80
96

4 Install double cross headers (face nail to first cross headers, then end nail through trimmer joists)

5 Install double trimmer joists (nail through header, then face nail to first trimmer joists)

Header

6 Install joist

Sharp point protrudes

Nail 16d nails on an angle

To avoid splits, do not put two nails along the same grain line

9. Double the floor joists to support heavy loads such as walls, bathtubs, libraries, and the like. Where an interior wall runs parallel to the floor joists, the floor must be framed per Fig. 12-5 to support the wall.

For bearing walls (walls that support the structure) the double joists function as a girder or beam and must be supported by posts or another bearing wall. For nonbearing walls (partitions), either doubled floor joists or blocking every 2 ft will be adequate.

Where nonbearing walls run perpendicular to the floor joists, no special floor framing is required. However, bearing walls require solid blocking between the joists and a girder or wall supporting directly underneath.

Where the floor supports a big cast-iron tub full of water, a floor-to-ceiling collection of books, a water bed, a hot tub, or any similar load, make sure the floor framing is extra sturdy. Two typical ways to frame a floor that supports a bathtub are shown in Fig. 12-6.

10. Bridging or blocking is used to prevent the floor joists from shifting or wobbling (a lateral stability concern). The Uniform Building Code Sec. 2506g requires floor bridging or blocking a minimum of every 8 ft. (See Fig. 12-7.) Solid blocking is stronger than bridging and is easier for a novice to install. Solid blocking is required under all bearing walls, posts, and stairway stringers.

I like to block the joists every 4 ft so the edge of the plywood will have a solid blocking. This isn't necessary if the subflooring is ¾-in tongue-and-groove (T&G) plywood or if

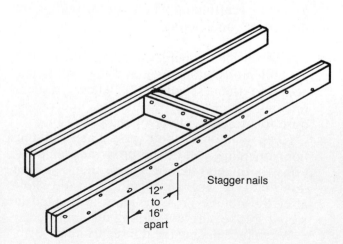

Stagger nails

12″ to 16″ apart

Simple Overhang by Extending Floor Joists

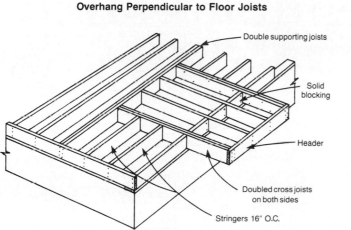

- Extended joist
- Doubled joists at each end
- Headers
- Solid blocking

Where Walls Run Parallel to Floor Joists

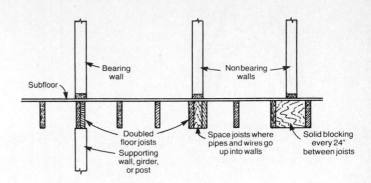

- Bearing wall
- Nonbearing walls
- Subfloor
- Doubled floor joists
- Supporting wall, girder, or post
- Space joists where pipes and wires go up into walls
- Solid blocking every 24" between joists

Overhang Perpendicular to Floor Joists

- Double supporting joists
- Solid blocking
- Header
- Doubled cross joists on both sides
- Stringers 16" O.C.

Fig. 12-4 *Framing floor overhangs.*

Where Walls Run Perpendicular to Floor Joists

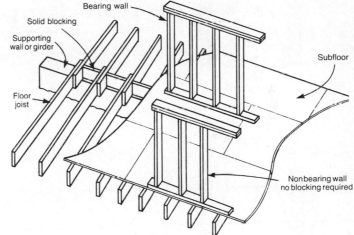

- Bearing wall
- Solid blocking
- Supporting wall or girder
- Floor joist
- Subfloor
- Nonbearing wall no blocking required

Fig. 12-5 *Floor framing to support walls.*

Where Tub is Parallel to Joists

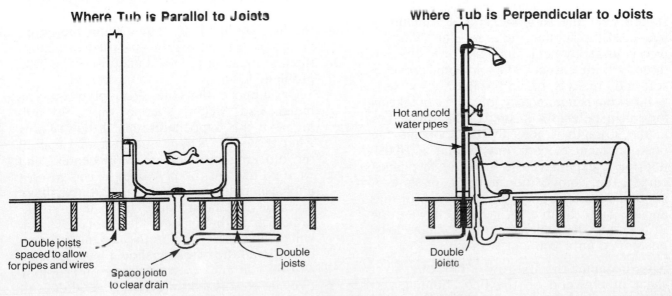

- Double joists spaced to allow for pipes and wires
- Space joists to clear drain
- Double joists

Where Tub is Perpendicular to Joists

- Hot and cold water pipes
- Double joists

Fig. 12-6 *Typical floor framing under bathtubs.*

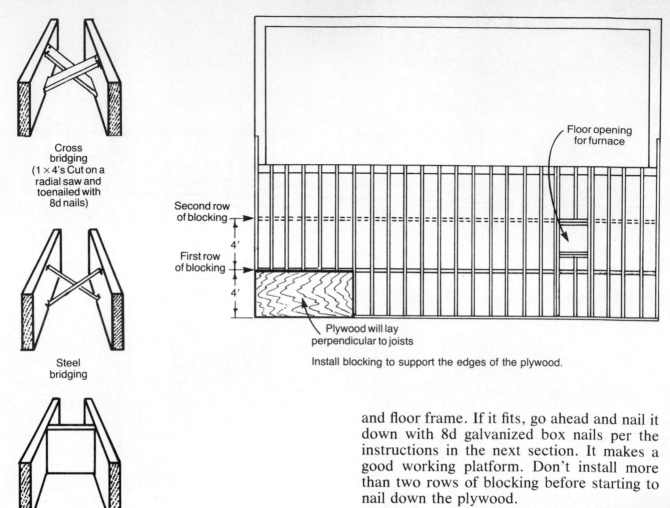

Cross
bridging
(1 × 4's Cut on a
radial saw and
toenailed with
8d nails)

Steel
bridging

Solid
blocking

Second row
of blocking

First row
of blocking

4′

4′

Plywood will lay
perpendicular to joists

Floor opening
for furnace

Install blocking to support the edges of the plywood.

Fig. 12-7 *Bridging and blocking.*

a second layer of wood or chipboard is installed over the subfloor. But it's definitely needed if the subflooring is regular ¾-in plywood underlayment and will be covered directly with carpet. The blocking prevents soft and creaky spots in the floor.

Blocking is not an easy job, and can take as long as installing the joists. It's advantage lies in the strength it adds to the floor and the postponement of bankruptcy. That is, finish flooring is expensive, be it oak or fir planks, carpet, tile, composition flooring, or slate. A good solid subfloor can be covered with rugs and be quite livable for years, allowing the $1,000 to $5,000 expense for finish floors to be postponed until you can afford it.

11. After installing the first row of blocking, lay a sheet of plywood on the floor framing and make sure the edges line up with the blocking and floor frame. If it fits, go ahead and nail it down with 8d galvanized box nails per the instructions in the next section. It makes a good working platform. Don't install more than two rows of blocking before starting to nail down the plywood.

Here is some advice on avoiding mistakes. The first time I do something, I always try to work all the way through a stage of the process on a small scale, to see if there's anything I haven't thought of before doing it full scale. Floor framing is a good example. Suppose that second joist was spaced 16 in from the first, rather than 15¼ in, and the entire floor was framed before you tried to nail on the first piece of plywood and discovered the problem.

Or suppose that the T&G plywood you bought had "Scant" stamped on it. No fooling now, this happened to me. I didn't know what the heck "scant" meant until I had all the blocking installed on 48-in centers and went to nail on the plywood. It was a most unpleasant surprise to learn that this fluke plywood had been milled from 48-in stock, rather than the proper 48½-in T&G stock which allows ½ in for the tongue. Consequently, all the blocking should have gone on 47½-in centers . . .#@*!! and other expletives deleted! Luckily T&G plywood doesn't really require blocking. But as a result of this

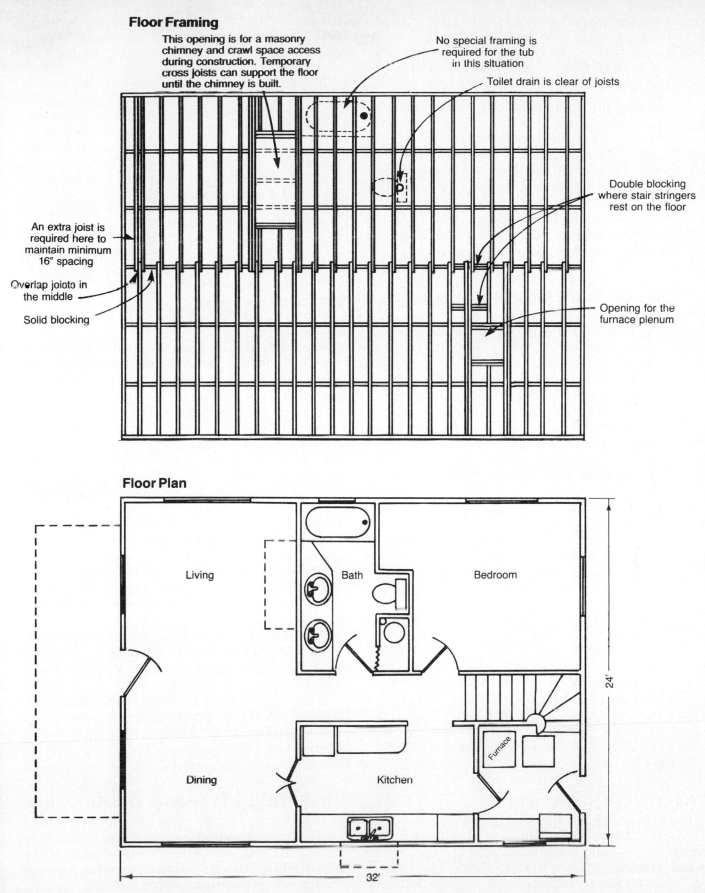

Floor Framing

This opening is for a masonry chimney and crawl space access during construction. Temporary cross joists can support the floor until the chimney is built.

No special framing is required for the tub in this situation

Toilet drain is clear of joists

An extra joist is required here to maintain minimum 16″ spacing

Overlap joists in the middle

Solid blocking

Double blocking where stair stringers rest on the floor

Opening for the furnace plenum

Floor Plan

Living

Bath

Bedroom

Dining

Kitchen

Furnace

32′

24′

Fig. 12-8 *Making sure the floor framing matches your floor plan.*

and other experiences I found it best to go as far as possible with each process so I didn't spend two weeks on step A, only to learn during step C that somehow step A was all screwed up.

12. Framing the second half of the floor is much the same as the first half. Since the joists overlap at the center support, the spacing is slightly different and may require an extra joist. Install solid blocking the length of the overlapped joists. In the case of the sample house, the center blocking serves three purposes: it blocks the overlaps, provides support for a bearing wall, and is backing for the third row of plywood. (See Fig. 12-8.)

13. Take time to make the following checks on the floor framing before you nail on the subflooring:
    a. Is the floor perfectly square (diagonals equal)?
    b. Are the edges of the floor straight? Use a tight string.
    c. Are all the openings for stairways, chimneys, furnace ducts, dumb waiters, and so on correctly located and square?
    d. Are the overhangs correctly located?
    e. Are the joists spaced so the ends of the plywood will land on the center of a joist?
    f. Is the first row of blocking in place?
    g. Are all walls, bathtubs, stairway stringers, water beds, and so on adequately supported?
    h. Are bathtub and toilet locations placed so the drains aren't right over a floor joist?

OK then, let's nail on the subfloor.

Subflooring

Nailing subflooring to a properly framed floor can be a fast and satisfying process, though it too has its pitfalls.

Material Selection
The first decision is the selection of a subflooring material. The most commonly used material is 3/4-in plywood underlayment because of its strength and versatility. The plywood should be marked "Underlayment—interior with exterior glue." Most other grades of plywood have voids under-

neath the first ply and should not be used for flooring. The exterior glue is necessary to withstand the exposure to the weather during construction.

If you choose plywood, the next decision is whether to use tongue-and-groove (T&G) or regular plywood. The T&G concept, be it for plywood or boards, is to lock pieces of material together at the edge joint. Some finish-flooring boards are "end-matched T&G," which means that both the ends and the hedges lock together. Figure 12-9 shows an example.

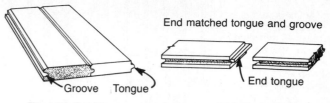
End matched tongue and groove

Groove   Tongue

End tongue

Fig. 12-9 *Tongue-and-groove (T&G) boards.*

The advantage to T&G plywood is that blocking the floor framing at the plywood edges isn't absolutely necessary. Many builders install carpets directly over T&G subflooring and save a lot of time by not blocking the floor every 4 ft. My experience was that by the time the lumber truck dumped the plywood in the yard and I had used it for foundation forms and had wrestled it into place with various tricky techniques, a lot of the tongues got smashed and grooves got dented. Consequently it was difficult and time-consuming to install. Since I'd already blocked the floor for strength reasons, it was a total waste of time for me to bother with the T&G plywood. Next time I'll either block or use T&G plywood, but not both. (Even with T&G plywood, blocking is required every 8 ft for adequate lateral stability.)

In a lot of older houses, 1 × 8 boards were used for subflooring and were nailed diagonally across the floor joists. Plywood is faster to apply, has no cracks through which the wind can blow, and is stronger than 1 × 8's.

If the floor was framed using beams rather than floor joists, the common choice for flooring is 2 × 6 T&G decking, which serves both as subflooring and finish flooring. This heavier flooring can be used to span up to 4 ft between supports.

Installing Plywood Subflooring
Plywood must be laid perpendicular to the floor joists. Try to avoid laying any sheets less than 32 in long, as you lose the spanning strength of the plywood with shorter pieces. If you didn't design the floor length as some multiple of 4 ft, you may end up with some short pieces.

STEP-BY-STEP PROCEDURE

1. Start in any corner, preferably with a full sheet of plywood. Make sure it is on squarely. Any misalignment will magnify itself as you move across the floor.

2. Nail the plywood to the joists with 8d galvanized box nails or ringed flooring nails as shown in Fig. 12-10. Space the nails every 6 in along the edges of the plywood and every 8 in along the floor joists. Don't scrimp on nails. This pattern is to avoid squeaky floors.

3. Progress lengthwise down the floor to the end. Space the ends of the plywood 1/16 in apart by tacking a couple of 8d nails at the end of each piece of plywood. This prevents buckling if the plywood soaks up water and swells before you get the roof on.

4. Measure the edges of any openings before they are completely covered with plywood. Then measure and mark the plywood when it is nailed in place and saw the opening with a skill saw and/or a handsaw. Don't saw into the floor joists or into any nails. Set the skill saw blade so it will cut exactly 3/4 in deep.

5. After reaching the end of the first row, go back and make sure the blocking for the second row of plywood is 48 in from the edge of the installed plywood. Start the second row of plywood so that the end joints are as far as possible (preferably 4 ft) from the end joints

of the first row. This technique, illustrated by Fig. 12-11, is called **railroading** and prevents weak spots in the floor by staggering the joints. Use 8d nails to space the edges 1/16 in from the first row of plywood.

6. Continue this process right across the floor to the other side:
   a. Nail in a row of blocking
   b. Nail down a row of plywood
   c. Measure and cut out floor openings as you come to them
   d. Space 1/16 in between all joints

What is the greatest misery in nailing on floors? Well, it could be the smashed fingers. I always sprinkled a bunch of nails on the floor and then nailed away as fast as I could go. After all the toenailing and awkward positions required to block the floor framing, I felt like a regular speed demon when I started nailing on the plywood subfloor. But just as I was getting a good rhythm

Fig. 12-10 Nailing plywood subfloor to the joists.

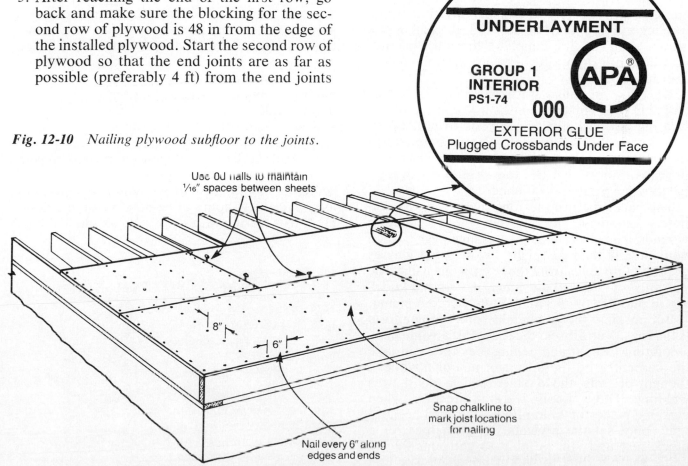

Use 8d nails to maintain 1/16" spaces between sheets

UNDERLAYMENT

GROUP 1 INTERIOR PS1-74   APA®   000

EXTERIOR GLUE
Plugged Crossbands Under Face

8"

6"

Snap chalkline to mark joist locations for nailing

Nail every 6" along edges and ends

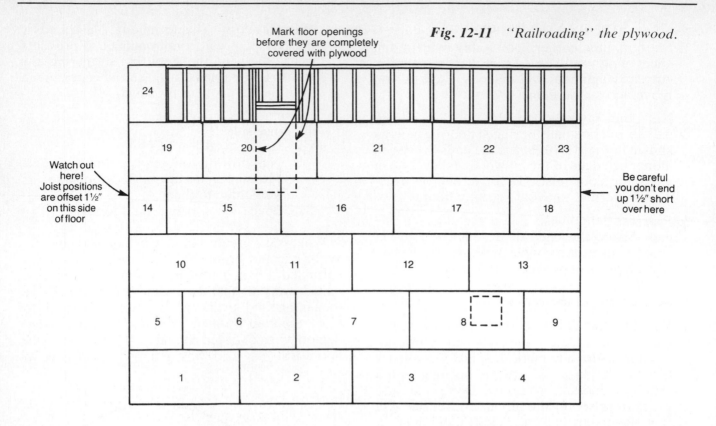

Mark floor openings before they are completely covered with plywood

Fig. 12-11 *"Railroading" the plywood.*

Watch out here! Joist positions are offset 1½" on this side of floor

Be careful you don't end up 1½" short over here

24

19   20   21   22   23

14   15   16   17   18

10   11   12   13

5   6   7   8   9

1   2   3   4

going, I'd get sloppy and smear my thumb and forefinger with one of those "sink the nail in one blow" hits. The first couple of times are painful, but the tenth or twelfth time in the same day is excruciating.

Kneeling on those sprinkled nails is another entry in the misery contest. I even took to wearing old basketball knee pads to protect my bare knees. My knees were always bare except for the first couple weeks after buying a new pair of jeans. Yours will be, too. You'll see what I mean when you start building.

The backache from leaning over all day is definitely another entry. And those puncture wounds from nails that go "pranggInggg" when you don't hit them just right. I wore safety glasses to protect my eyes from those pranging nails. The nuisance of steamed up safety glasses and kneeling in puddles on the rainy days were no picnic.

But possibly the granddaddy of all discomforts is the toes. Yes Moses, the toeses! If anyone finds a solution to bent and aching toes they can take the carpentry business by storm. For a while, I thought it was those steel-toed boots I was wearing. Those boots were a lifesaver when carrying pieces of ¾-in plywood around because you could set the plywood on your toes for a quick rest and still be able to get your fingers under the plywood to pick it up again. The first

day that I changed to wearing tennis shoes to solve the toe-cramp problems, I damn near amputated all ten toes with one of those quick rests. Then I had a heck of a time picking up the plywood to get it off my toes. Ow! Ow! Ow! That's almost as embarrassing as the time. . . . well I don't think I'll put that one in print. Ask me sometime and I'll tell you about it.

Anyway, this bent-toes problem doesn't seem to depend on the kind of shoes you wear or your amateur status. Even my father, who has built scores of floors, had to straighten out his toes after every few hours of helping me on my floor. He was wearing wing tips. Yeah, no kidding,

wing tips. Bent toes seems to be an occupational hazard that nobody gets used to.

It is quite a thrill when the floor is done. Just walking on a nice, solid, flat floor feels good and makes a very satisfying noise. No more tippy-toeing your way over the floor joists. You can lay out all the rooms on the floor and set up apple crates and sawhorses to simulate counters and walls. We spent hours just playing house and even made some design changes. And I was beside myself with joy knowing that I now had a nice level platform out of the mud and dirt to work on.

Now is a great time for a party! Set up a record player and invite some friends over for a dance and a picnic. You deserve a day or two off anyway.

Extraneous Tidbits

Tools

As you go shopping around for tools, I suggest you stick pretty much to basic, time-honored tools, especially at first. Being a nut for tools, I came frightfully close to buying some expensive and not-too-useful tools. For example, I seriously considered buying a transit ($200) to get the excavation and foundation level. I later found that a $7 water level works just fine. I almost bought a motorized mitre saw ($150), but soon found that a $20 back saw and mitre box will cut just about as accurately. I did buy a framing hammer, which I used for a while but found it too heavy to suit me. As you outfit yourself during the few months prior to building, stick with the basics until you are pretty sure of the need for a more specialized tool.

Storing Lumber

After dumping your savings account on the lumber store counter, the lumber store sends a truck to dump a bunch of lumber on your building site. Take care of this lumber as it can get wet, warp, rot, become a home for ants, or simply disappear before your very eyes.

Have some old 4 × 4's (called **cribbing**) to keep the lumber from sitting directly on the ground. Often the lumber store provides these and brings them with the truck. Lumber trucks have rollers on their beds. The standard way of unloading the truck is to:

1. Remove the chains or ropes securing the load

2. Put the truck in reverse

3. Back up fast

4. Slam on the brakes

Fig. 12-12   *Unloading the lumber.*

The momentum of the lumber rolls it right off the truck onto the ground. With a little luck, one end of the lumber will be sitting on the ground and the other end will still be on the truck. (See Fig. 12-12.) Tell the driver to wait for a minute while you toss the 4 × 4's every 4 ft under the lumber pile. (Need I mention not to reach under the precarious pile while doing this?) Then the truck drives off and the lumber lands with a resounding crash. Not exactly a finesse method, but unless you have a fork lift handy, it is the quickest and easiest way to do it. And no, the driver will *not* hang around while you unload it piece by piece.

Be sure to locate the lumber pile on dry, level ground so the pile will not be twisted. A twisted pile will result in twisted wood every time. This is especially true if the lumber sits there for a few months. You may have to use a hydraulic jack or a small car jack to level the 4 × 4's, or just restack the boards. Often restacking will be necessary anyway to get to the boards that you will need first.

Stacking or restacking lumber is immeasurably faster with two people than it is by yourself. Stack the lumber close to the building site. By putting **stickers** (boards or slats of wood laid perpendicularly to the lumber pile) between each few layers of lumber, the boards will be ventilated and can dry out. Even though you have ordered kiln-dried lumber, it still has some moisture in it. If it dries out too fast it may crack and split, so keep the lumber in the shade or covered up.

By stacking the lumber neatly and tightly, there will be less warpage. Loose lumber warps within a few days as it dries out. Also, tightly stacked lumber can be secured by wrapping a chain around it, tightening the chain with a chain tightener, and locking the handle of the chain tightener to the chain with a bicycle lock as shown in Fig. 12-13. This prevents lumber with feet from walking away.

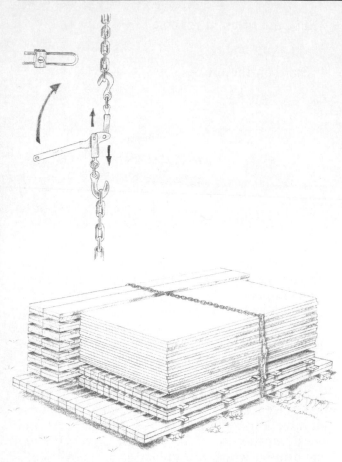

Fig. 12-13 *The lumber pile.*

Finally, cover the lumber with plastic sheeting to keep it dry. Black plastic is better because clear plastic tends to degrade in the sunlight. The plastic tends to get torn on the corners of boards and has to be replaced occasionally.

If you have beams that are going to be visible when the house is finished, take special care to keep them stacked straight, well-ventilated, and in the shade. Otherwise, huge cracks may form if they dry out too quickly. Also avoid buying beams that are part of the heart (center) of the tree. You can determine this by looking at the end of the beam. Usually the wood containing the heart splits and cracks more than other parts of the tree.

Visit Other Construction Sites

Some stellar revelations came while I was looking at other houses being built, like the use of foundation form leveling strips and a good way to build stairs. I really like to look around construction sites, but I always feel a little paranoid that someone will think I am ripping off stuff. I know I wouldn't appreciate someone poking around my place unless I knew them, so I usually try to go when the owners are there so I can meet them. Or I take Deb or Garth with me and try to act like they are the ones interested in it. Meanwhile, I measure the size of the garage door headers, try to figure how the plumbing and electrical panels are hooked up, and see how the roof overhang is framed. You have to get there real quick after you see a house being built, though, or the buggers will finish it in about 2 months and have people living in it! That gets a bit discouraging after you have spent 2 months just to pour the foundation.

Get Used to Slow Progress

Concentrate your energy on doing a good job, rather than on speed. You will be much happier with the final result. I've always figured I could do things about as fast as anyone else, and have enough of the "Type A" time-conscious personality that I hate to waste time. But in building a house, every time I'd try to hurry, I would screw something up. The key to success was to plan as far ahead as possible and then keep going at a steady, careful pace.

I never did learn how to schedule myself. Literally everyone asks you how long it'll be until you move in. I started out saying "Oh, about 6 months," because after listening to Uncle Barney I thought it should take only 3 or 4 months. After 6 months, I was just framing the roof, but I thought I must be nearly finished. When a year had elapsed I was working on the windows and doors, but by then I had returned to a regular job. Between the job and commuting, my pace had slowed radically. Impartial observers would call it a virtual standstill. I continued to estimate a 6-month completion date for another year and finally devised a stock answer to the inevitable question: "I've quit guessing," I'd say. Under pressure, I'd set an estimated completion date 10 years in the future.

Nearly all my owner-builder acquaintances had the same experience. It's one of the character-building aspects of building a home yourself, or venturing into any new territory. The tendency is to compare your speed to that of professional builders, and there's no comparison. They can build a house in 3 months alright, but it takes 30 years to pay for it.

Scheduling and the Learning Curve

Scheduling is a subject you can't fully appreciate until you are familiar with all the steps of building a house and the unusual situations encountered

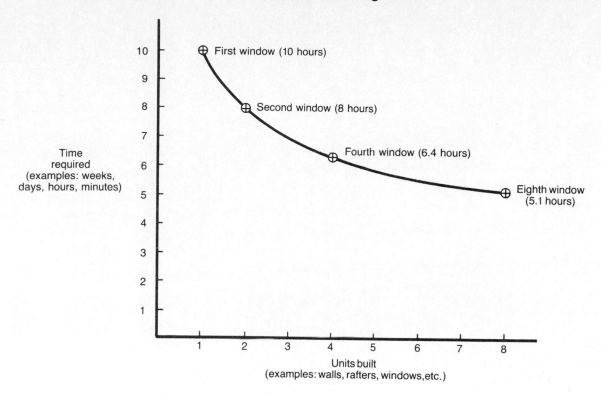

An 80% Learning Curve

First window (10 hours)

Second window (8 hours)

Fourth window (6.4 hours)

Eighth window
(5.1 hours)

Time
required
(examples: weeks,
days, hours, minutes)

Units built
(examples: walls, rafters, windows, etc.)

Fig. 12-14 *Predicting your improvement using a learning curve.*

by the marginally financed do-it-yourselfer. The first time you attack any new task you have to marshal the tools, material, and know-how together and it always takes a small part of forever to finish the task. The second time goes quite a bit faster. You will go through a learning curve where the first wall you build may take 3 days, the second wall 2 days, and the tenth wall about 4 hours. (Fig. 12-14 illustrates this improvement.) It took my friend Steve and me a whole day to put three log rafters in place. By the twenty-fourth rafter, he was putting them in place in 20 minutes by himself, with a rope-and-pulley system powered by his truck.

An Important Approach

There's an important approach to take when beginning any difficult or very large and involved stage of housebuilding. The approach has a lot of names, but it goes something like this: After spending what seems like an appropriate amount of time pondering, sketching, and planning a job, decide what the first step has to be and do it.

From there, figure out the next step, and so on.

While this may not strike you as profound, it's surprising how many people won't attempt a project because they can't work out all the details in their heads. Plumbing a house is a good example. After 2 weeks of planning, sketching, and procrastinating, I finally decided that no matter what, there had to be a main drain running from the kitchen across the house toward the septic tank. So I installed one. Then I made a short branch to the bathroom sinks and a long branch to the laundry room. Before too long the house was plumbed. The biggest step was to quit worrying about it and get started.

The attitude "Well, at least I can do this much" is my way of getting started. A wise philosopher said, "A journey of a hundred miles begins with a single step." Steve says "You know how to eat an elephant, don't you? . . . one bite at a time." An interesting analogy, coming from a vegetarian. But the point is that the one-bite-at-a-time approach works and nearly every owner-builder uses it.

CHAPTER 13

Wall Framing

Wall Framing

"Stand back, stand back! Make way for the walls!" This entry of my diary buzzes with excitement. A strong, level square floor supported by a sturdy foundation sits solidly and majestically on the site. Next to it is a pile of lumber that's ready to become my house. Having earned my Junior Woodchuck status building the foundation and floor, I'm eager to get on with it. "Enough of this groveling down on the ground; I'm ready to build to the sky!"

Framing the Exterior Walls

Since our enthusiasm still outshines our ability, let's be prudent and begin with a wall at the back of the house. Dig out the house plans and quickly review the design considerations for walls in Chapter 5.

Make a sketch of the wall before starting to frame it and determine all of the proper dimensions. (Fig. 13-1 outlines some of the details.) I have some personal preferences here. For the sake of appearance, I like the tops of all windows and doors to be at the same height, 6 ft 10 in. Also, I like the walls to be 8 ft 1 in or 8 ft 1½ in high because it makes the wallboarding task easier later on. (With ½-in to 1-in clearance from the ceiling, the 8-ft wallboard is easier to position and doesn't require so much cutting heightwise.) Use your own personal preferences, but stick to the building code. The Uniform Building Code Sec. 1407 requires a minimum ceiling height of 90 in after the ½-in ceiling wallboard and ¾-in finish flooring are applied.

Figure 13-2 shows dimensions for the west wall of the sample house. I build all exterior walls as bearing walls, even though the side walls of the sample house bear more weight than the end walls. (Notice that, structurally speaking, the house would stand without need for the end walls

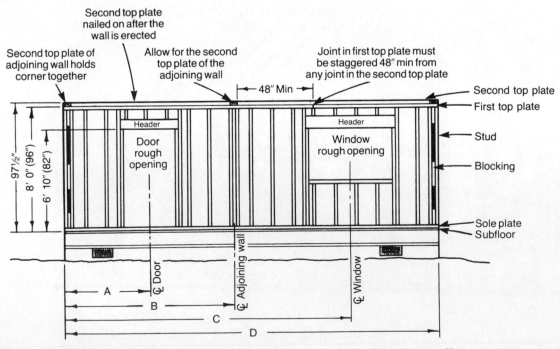

Fig. 13-1 *Wall framing details for a typical bearing wall.*

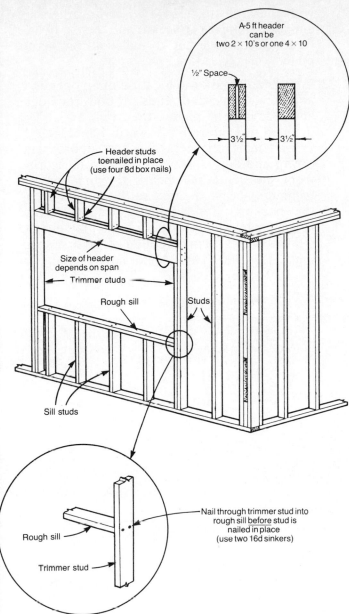

Framing detail of bearing wall

A-5 ft header can be two 2 × 10's or one 4 × 10

½" Space

3½" — 3½"

Header studs toenailed in place (use four 8d box nails)

Size of header depends on span

Trimmer studs

Rough sill

Studs

Sill studs

Rough sill

Trimmer stud

Nail through trimmer stud into rough sill before stud is nailed in place (use two 16d sinkers)

as long as the side walls were well braced for lateral stability.) Nearly the entire weight of the floors, ceilings, and roof is carried by the side walls and the central bearing wall.

The size of the window and door headers is determined from Table D-4, Appendix D. As you decide on the header size for each opening, it helps to determine the worst load condition on the side and central walls and use those size headers for all openings. This way all of the 3-ft to 3-ft 6-in door and window openings in the first floor walls (except in the nonbearing interior walls) are bridged by a double 2 × 6 header; 4 ft to 4 ft 6 in openings are bridged by 2 × 8's; and larger openings up to 5 ft 6 in are bridged by 2 × 10's. That's something I can remember and it frees up my finite brain capacity to think about other things as I build, like the proper size of the openings.

The size of rough openings depends on the type of doors and windows you plan to purchase. It seems like almost all doors and windows have different frame sizes. The guideline here is to frame the rough opening ½ in larger in both height and width than the outside dimensions of the door or window frame. The ½ in necessary in order to shim the frame to a level and plumb position. For example, I plan the window in the west wall to be a double pane measuring 48 in × 36 in and framed with 2 × 6's. This means the unit will measure 51 in × 39 in minimum, and the rough opening will be 51½ in × 39½ in. You have to work out the rough opening size this way for every window and door.

Wherever two walls adjoin, the continuous wall must be framed, as shown in Figs. 13-1 and

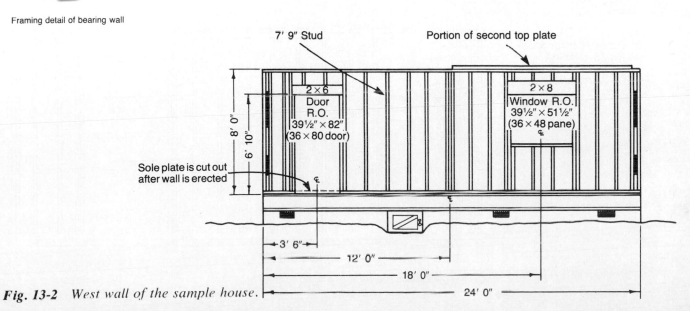

7' 9" Stud

Portion of second top plate

2×6
Door R.O. 39½" × 82" (36 × 80 door)

2×8
Window R.O. 39½" × 51½" (36 × 48 pane)

8' 0"

6' 10"

Sole plate is cut out after wall is erected

3' 6"

12' 0"

18' 0"

24' 0"

Fig. 13-2 *West wall of the sample house.*

13-10, to provide a backing where the walls connect. Without such a backing, there would be nothing to nail the interior wallboard to in the corners of those rooms. If you can keep the wallboard in mind while framing the walls, it will save you a lot of trouble a few months from now.

All right then, let's build the wall shown in Fig. 13-2.

STEP-BY-STEP PROCEDURE

1. Sweep off the floor where the wall will be built. Locate the wall so it will be in its final position when it is tilted up. <u>Select some long straight 2 × 4's for sole and top plates.</u> Saw them to length and sight them for straightness, placing them side by side, on edge, and with the arch up. (See Fig. 13-3.) Align the ends and, if necessary, butt and toenail two or three pieces together to make a continuous plate.

2. <u>Layout</u> and mark the location of all openings, adjoining walls, and studs on both the edges and faces of the plates. (See Fig. 13-3.) The detailed procedure is shown in Fig. 13-4.

First mark the centerlines of doors, windows, and adjoining walls. Then, measuring from the centerline, mark the locations of trimmers and studs next to the openings and the backings for adjoining walls. Finally, spacing as you did for floor joists, lay out the stud locations 16 in on center. (The first spacing is 15¼ in, right?) Make sure the edge of

Fig. 13-3 Framing a wall.

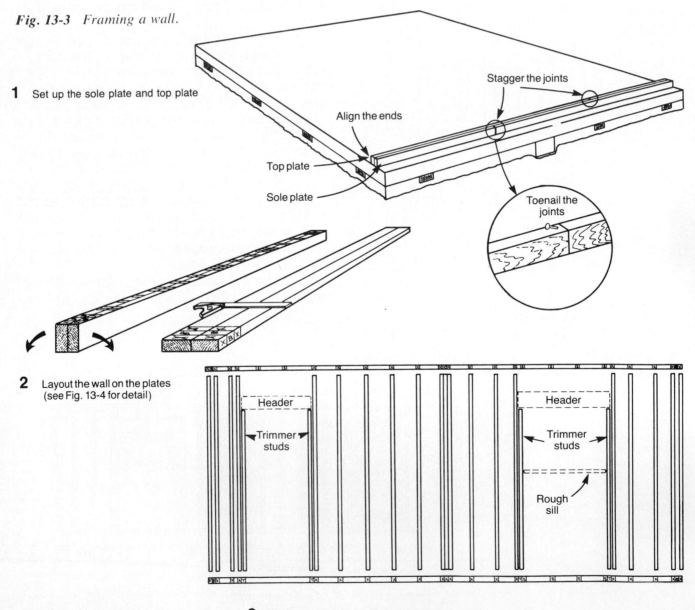

1 Set up the sole plate and top plate

Stagger the joints

Align the ends

Top plate

Sole plate

Toenail the joints

2 Layout the wall on the plates
(see Fig. 13-4 for detail)

Header

Trimmer studs

Header

Trimmer studs

Rough sill

3 Place the studs and trimmer studs

each sheet of plywood sheathing will land on the center of a stud. The X, T, B, and S marks on the plates help to avoid confusion here.

3. Cut all the pieces and set them in place. Put the plates on edge 8 ft apart. Make a count of the X's and cut that number of 93-in studs on the radial saw. (Nail a block to the saw table 93 in from the blade so you don't need to measure every stud.) Similarly, count the T's and cut that many 80½-in trimmers. The blocks can be anywhere from 6 in to 24 in long and the sill and header studs will have to be measured individually for each door and window.

Sight each stud and trimmer as you lay it out between the plates, placing it with the arch up. If some arches are up and some down, the wall may appear wavy after it's wallboarded and painted. Severely arched or warped studs (more than ⅜ in of warping in 8 ft) should be discarded and used in shorter

lengths for blocking. You'll have some problems here if you bought green lumber, rather than *kiln-dried* (KD) lumber. The wall should now look like the lower drawing in Fig. 13-3.

4. Saw the headers and the rough sill. Mark the sill position on the trimmers, measuring down from the top of the trimmer by the height of the rough opening. This is 39½ in for the window in the wall we're building.

5. Nail the wall together. Beginners will find it best to start with the rough openings, and then nail in the studs. The sequence for framing the openings is mildly tricky and is illustrated in Fig. 13-5. All joints get two 16d nails, except the header-to-stud joint, which get three 16d nails.

You'll undoubtedly screw up the sequence occasionally and have to toenail a sill or a header in place. That's OK, but don't make a practice of it.

Fig. 13-4 *Wall layout details.*

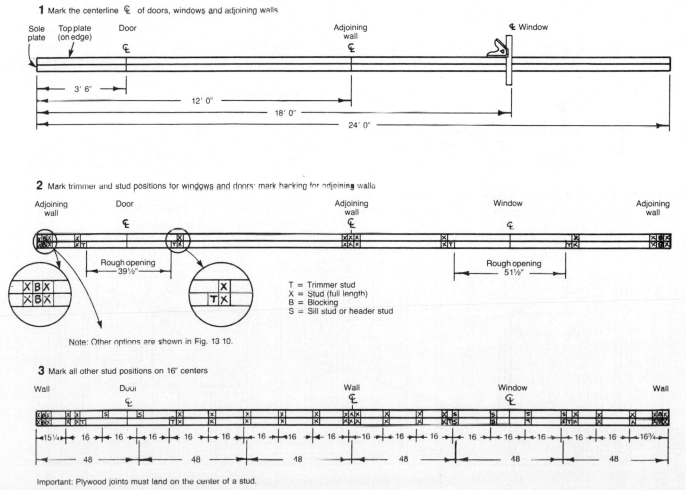

Note: Other options are shown in Fig. 13-10.

Important: Plywood joints must land on the center of a stud.

Fig. 13-5 *Nailing sequence for window openings.*

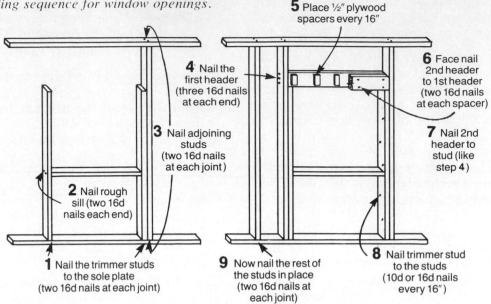

5 Place ½″ plywood spacers every 16″

4 Nail the first header (three 16d nails at each end)

3 Nail adjoining studs (two 16d nails at each joint)

2 Nail rough sill (two 16d nails each end)

6 Face nail 2nd header to 1st header (two 16d nails at each spacer)

7 Nail 2nd header to stud (like step **4**)

1 Nail the trimmer studs to the sole plate (two 16d nails at each joint)

9 Now nail the rest of the studs in place (two 16d nails at each joint)

8 Nail trimmer stud to the studs (10d or 16d nails every 16″)

6. After all the studs, trimmers, headers, sills, and blocking are in place, <u>saw and install the sill studs and header studs.</u> Where the header studs meet the header, toenail with four 8d box nails. All other joints get the standard two 16d sinkers.

The wall is now framed and looks like Fig. 13-2, except that it's lying flat on the floor.

7. <u>Make sure the wall is straight and square.</u> (See Fig. 13-6.)

Toenail the sole plate to the subfloor at both ends with 8d box nails. Use a tight string to straighten the sole plate (per Fig. 9-8, "Basic Skills"). Then toenail the sole plate to the floor every 6 ft and recheck for straightness. A sledge hammer helps to tap the sole plate into position.

With the sole plate straight and stationary, measure the diagonals of the wall and adust it until it's square. Then toenail the top plate to the subfloor with two 8d box nails so nothing can move. Also check the door and window rough openings to be sure they are square. Now we're ready for plywood sheathing.

In most cases I like to nail on the plywood before erecting the wall because it's much easier to apply it now than later. It also ensures a square wall, but it does make the wall heavier when erected. Since I always had to get help when erecting the major walls anyway, and since it only takes a few minutes to erect a wall, this was the best way for me. You'll have to consider your own situation and decide if it makes more sense to sheath the wall after it's erected.

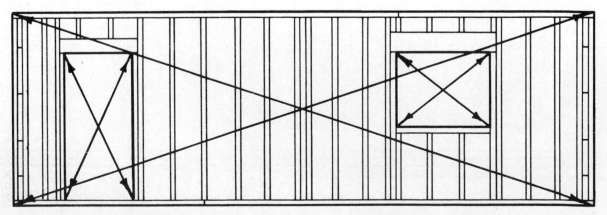

Fig. 13-6 *Making sure the wall and the openings are square.*

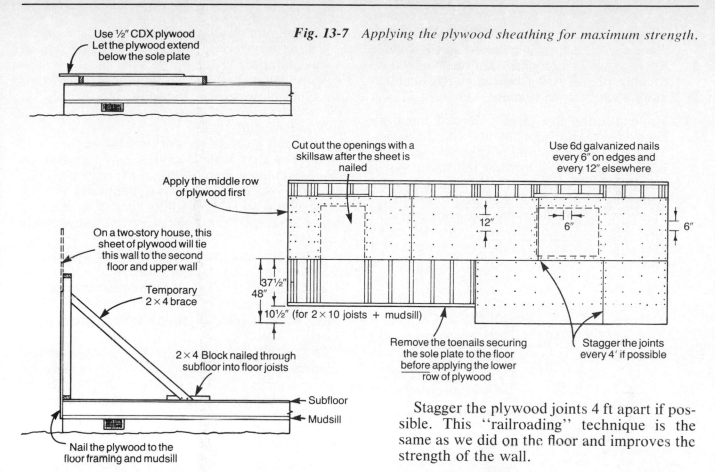

Fig. 13-7 *Applying the plywood sheathing for maximum strength.*

Use ½" CDX plywood
Let the plywood extend
below the sole plate

Cut out the openings with a
skillsaw after the sheet is
nailed

Use 6d galvanized nails
every 6" on edges and
every 12" elsewhere

Apply the middle row
of plywood first

On a two-story house, this
sheet of plywood will tie
this wall to the second
floor and upper wall

Temporary
2 × 4 brace

12"

6"

6"

37½"
48"
10½" (for 2 × 10 joists + mudsill)

2 × 4 Block nailed through
subfloor into floor joists

Remove the toenails securing
the sole plate to the floor
before applying the lower
row of plywood

Stagger the joints
every 4' if possible

Subfloor
Mudsill

Nail the plywood to the
floor framing and mudsill

Stagger the plywood joints 4 ft apart if possible. This "railroading" technique is the same as we did on the floor and improves the strength of the wall.

8. The plywood sheathing (½-in CDX plywood) should be laid perpendicularly to the studs so it ties many studs together into one rigid structure. After seeing how relatively easy it is to rack or distort a stud wall, you'll notice an amazing strength to the wall after it's sheathed.

Allow the plywood to extend beyond the sole plate so when the wall is erected the plywood can be nailed to both the floor framing and to the mudsill. (See Fig. 13-7.) With the mudsill bolted to 20 tons of concrete, it'll take a heck of a windstorm or earthquake to rattle this house off its foundation.

If insulating board is used in place of plywood sheathing, apply plywood or let-in diagonal bracing at every corner to ensure adequate lateral stability.

9. Nail the plywood to the wall framing with 8d bright box nails or 6d galvanized (hdg) nails every 6 in along the edges of the plywood and every 12 in elsewhere. (See Fig. 13-7.)

Just nail the plywood right over the window and door openings and then saw them out with a skill saw before the wall is erected. Mark the opening carefully so you don't saw the window in the wrong place.

10. Erect the wall. Remember to remove the toenails into the floor that held the wall square or it will be very difficult to lift the wall. If you're electing to erect the wall, then sheath it, nail a long diagonal to the wall to keep it square while it's tilted into position.

Sweep the floor where the wall will sit. Have some 2 × 4 braces and blocks all ready to attach to the wall, and a couple of sawhorses standing at full alert. Rally your troops. It takes two strong people to tilt up a 20 ft wall and a 30 ft wall will take three or four folks. Use all your influence to keep the wind from blowing while you do this.

Skid the wall over so the plywood overhangs the edge of the floor and the ends of the wall align with their intended final resting place. One tough part is just getting your hands under the wall to lift it. A crowbar and some blocks of wood aid in this.

The first lift should be to about waist height, then shove the sawhorses under to hold it. When you lift, keep your butt down and your back straight. For the final push, crouch down and use your legs for the power. Don't push too hard now; if the wall topples off the floor it'll be a bearcat getting it back into position again.

Immediately brace the wall securely. Try to place the braces so they aren't in the way of the next wall to be built. The wall should be roughly plumb when braced; we'll plumb it more accurately in a minute. (See Fig. 13-7.)

11. Make sure the sole plate is aligned along the floor edge. Make adjustments with the sledge hammer. Check the end positions to see if the wall is the same length as the floor. When everything's copacetic, <u>nail the plywood sheathing to the floor framing and mudsill. Then face nail through the sole plate and sub-floor into the floor framing with 16d nails every 16 in.</u> (Make sure each nail penetrates a floor header or joist.) (See Fig. 13-8.)

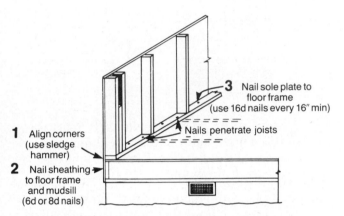

Fig. 13-8   *Nailing the wall to the floor.*

If there is a gap between the sole plate and the wall in places, it means that the floor isn't level, the sole plate isn't straight, or the wall is sitting on a nail or something. Diagnose the problem and attempt to correct it. Shims may help.

12. <u>Plumb the wall at each brace.</u> The braces have to be loosened one by one as each place is plumbed. With two people, this job will go fast. Rebrace the wall securely so a high wind can't blow it over. (See Fig. 13-9.)

Poof! A wall!

You can go right to work on the second wall now. The first wall always seems dangerously floppy standing there all by itself; once a second wall is in place it'll seem less precarious. It's not until all the walls, second floor joists, and ceiling joists are in place that the structure seems really sturdy.

When framing the second and following walls, you have to start getting creative about how to frame the wall right around the braces holding the first wall. Sometimes certain pieces of sheathing have to be applied after the wall is erected because of this.

When all four exterior walls are erected, plumbed, and braced, *check the diagonals between their top corners* to ensure that the second floor and ceiling will be square. Then the second top plates can be added, joining the walls together. Also, join the walls at the corners by nailing with 16d nails every 16 in.

Notice in Fig. 13-10 that the plywood sheathing on the second wall should overlap the end of the first wall. Here again certain pieces of sheathing may have to be applied after the second wall is erected. As an alternative, the wall can be

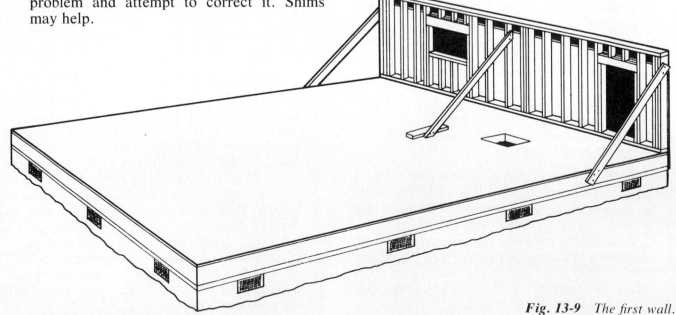

Fig. 13-9   *The first wall.*

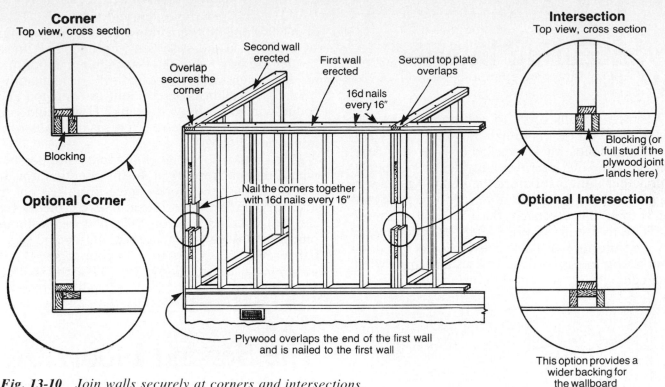

Corner
Top view, cross section

Blocking

Optional Corner

Overlap
secures the
corner

Second wall
erected

First wall
erected

Second top plate
overlaps

16d nails
every 16"

Nail the corners together
with 16d nails every 16"

Plywood overlaps the end of the first wall
and is nailed to the first wall

Intersection
Top view, cross section

Blocking (or
full stud if the
plywood joint
lands here)

Optional Intersection

This option provides a
wider backing for
the wallboard

Fig. 13-10 *Join walls securely at corners and intersections.*

sheathed with a 4-in end overlap, then erected and slid into position.

If there are any large heavy items (cast iron bathtubs, pipe organs, and the like) that go into the house, they should be hauled in before the last exterior wall goes in place. After that last wall goes up, everything that comes in must fit through the door or windows.

Just keep on building walls until the place starts to look like a house. (See Fig. 13-11.) Interior walls that are bearing walls are built and erected just like the exterior walls, but without plywood sheathing. If lateral stability is a concern, let-in diagonal bracing can be used.

In the sample house, there's a central bearing wall and a short wall by the stairs that will support the second floor. Both walls are well constrained by the rest of the structure so lateral stability isn't a problem. Keep all bearing walls well braced until all structural elements of the house are complete.

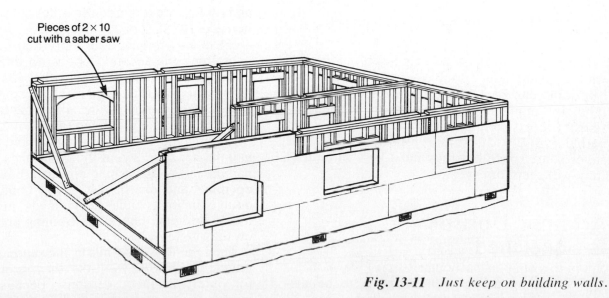

Pieces of 2 × 10
cut with a saber saw

Fig. 13-11 *Just keep on building walls.*

Nonbearing Interior Walls

The interior walls serve to partition the space in the house into rooms. Since they don't carry any structural loads, their location is somewhat flexible. In fact, the interior walls can be postponed altogether until just before the plumbing and wiring is installed if you're in a rush to get the roof on. I took that approach, but I found it was more difficult to install the walls after the ceilings were in. The interior walls also provide a place to stand and a way to support the trusses and joists while the ceilings and second floor are built.

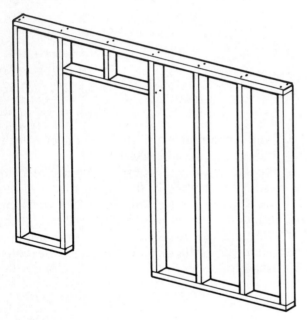

Fig. 13-12 *A nonbearing wall (partition).*

Figure 13-12 shows the framing of an interior wall. I usually add a double top plate so the interior walls can be securely connected with each other by lapping the top plates at the intersections. Closets and cabinets can be partitioned with 2 × 3, 2 × 2, or ¾-in plywood walls. This saves some space but they must be well built or they will seem flimsy.

A Minor Uprising Is Squelched

It's in this chapter particularly that I feel a dozen topics tugging at my pen, wanting to be discussed. "What about arched windows, Frank?"

"And what about me, curved walls?" "How about framing all the window and door openings with 2 × 12 headers, so you won't need all those header studs?" "Frank, you didn't mention me at all, single-wall construction!" (indignantly) "Yeah, you didn't forget about us, octagon and round windows, did you Frank? How could you slight us so, when we're what makes your own house so unique?"

Hey, settle down all you topics! Remember what I told our readers in the introduction: this is a path through the woods and I promised not to lose them in the underbrush. There's plenty for them to contend with on their first trip without plunging through a wilderness of unruly and semi-dangerous topics. Now back to your cages! Don't go confusing these folks! (*Clang!*) Having snuffed the rebellion, we'll continue through the forest.

The Second Floor

The sample house has a unique design feature here. If you've come up hard against some severe time or financial problems, there's the option of eliminating the second floor altogether. It's possible to put a truss roof over the entire first story as it stands, nail in some aluminum-framed windows and have the place tight to the weather in a couple of months. That would make the house a lot smaller (768 sq ft) and certainly change its style, but it could be done if there was a pressing need to finish up in a hurry.

However, for a small additional time investment, (300 hours of the total 2,500 hours), the second story adds 480 sq ft of living space. All the work on the foundation, floor, plumbing, heating, and roofing are required regardless of whether the house is 768 sq ft or 1,248 sq ft. A second story is a cheap way to gain a lot more house.

Framing the second floor is no different than framing the first floor except that it's farther to fall if you slip. (See Fig. 13-13.) Use a plank between two sawhorses and a stepladder to work from. Follow the procedure given in Chapter 11, "Building the Floor." In a nutshell, place and nail the headers, layout the floor joists, place and nail the joists (all arching upward), build the floor openings and overhangs as you come to them, double the joists under heavy loads, install blocking or bridging, check the framing and then nail on the subfloor.

The second-story walls of the sample house are unique in a few ways. First, they're only 5 ft tall for purely aesthetic reasons. Because of their height, the windows bisect the top plates, which

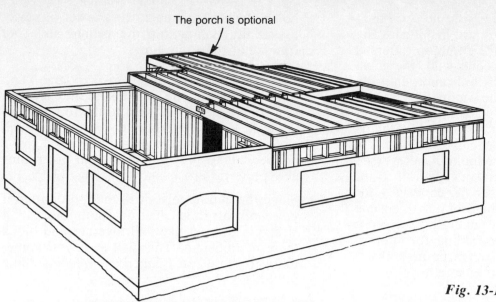

The porch is optional

Fig. 13-13 *Framing the second floor.*

calls for slightly creative wall framing, and the top plates must be securely fastened to the studs next to the dormer windows. In addition, the double top plates extend 12 in to support the roof overhang. (This is optional. The lookout rafters will support the overhang, but I found it's easier to build the overhang this way.) Since the top plates don't hold the corners together, be sure to nail the corner studs and sheathing securely to make the corners strong.

There are a number of ways to handle roof overhangs, but the ones I like are shown in Figs. 14-2 and 14-4.

Notice the two alternatives shown in Fig. 13-14 for framing the gable end walls. For the far wall, the gable part is framed lying down and then the entire wall is erected. The framing must be done

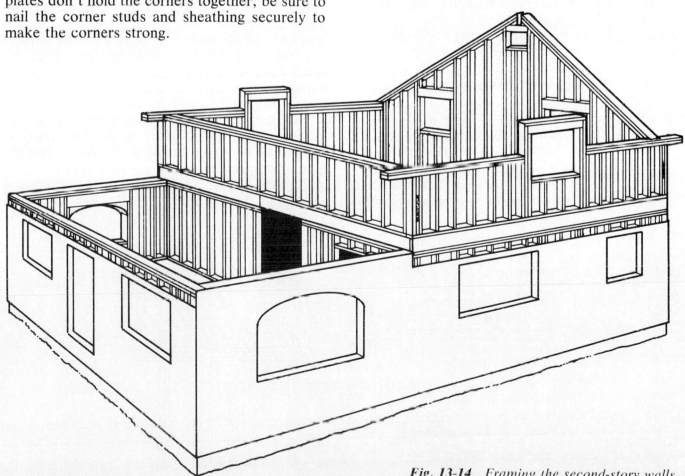

Fig. 13-14 *Framing the second-story walls.*

carefully so the top plate supports the overhanging lookout rafters when they are installed. (The size of the bird's mouth in the common rafters is critical here.) On the closer end wall, the gable part of the wall has been left unframed until the overhang is installed; then the wall will be framed in place. (The double-top-plate extension supports the fly rafter for a short time while this is done.) The top plate of the gable wall will be nailed to the underside of the lookout rafters (see Fig. 14-4) and then each stud will be cut to fit. This method ensures a better fit, especially for beginners. Don't get up on the roof until the gable wall is in place, though. In general, always use self-standing ladders or scaffolding for support when building high walls or roof structures. Don't get on the structure until you're sure it's structurally sound.

Ceilings

The type of ceiling support depends upon the roof design. With trusses, the ceiling sheetrock is nailed to the lower cross-member of the truss. With most custom-framed roofs, the sheetrock is nailed to the ceiling joists or collar ties, which also act to hold the top of the walls and the rafters together. By installing the ceiling joists, the entire structure becomes more solid and sturdy.

STEP-BY-STEP PROCEDURE

1. Lay out the truss or ceiling joist/rafter positions on the second top plate of the wall. Usually this will be every 16 in on center, though 24-in spacing is common too. Remember to allow for chimneys and skylights. The crane on the delivery truck places a bundle of trusses where they are well supported by the house walls. Then each truss is set in place, essentially completing the ceiling and roof framing at the same time.

2. With ceiling joists, cut the joists to length and sight down each one. Make sure they all arch upward, then toenail them to the top plate of the wall. Use three 8d bright box nails at each end. The ends of the joists may need to be bobbed off to prevent them from protruding above the rafters as shown in Fig. 13-15.

3. Since the ceiling joists will tend to domino if you walk on them, nail on some long 2 × 4 stringers to hold them all together and block them at midspan. This will give you something to stand on (gingerly) while installing the rafters.

4. Now is a good time to provide backing for the ceiling wallboard where interior walls run parallel to the ceiling joists. These blocks also provide a way to secure the top of the interior walls. Look around the ceiling and make sure there is something to nail the wallboard to every 16 in or so.

With the ceiling joists in place the house finally starts feeling like a house. It probably feels like a *small* house, but don't worry about that. As you build, the house will seem to contract and expand several times. When comparing room sizes, always compare a 10 ft × 12 ft space in your new house with a 10 ft × 12 ft room in a finished house, otherwise you'll be fooled frequently.

Notice that the sample house doesn't have any ceiling joists. On the first floor, the ceiling is formed by the lower chord of the trusses and the bottom edge of the second floor joists. In the second story, the ceiling is formed by the rafters and collar ties, which will be installed in Chapter 14.

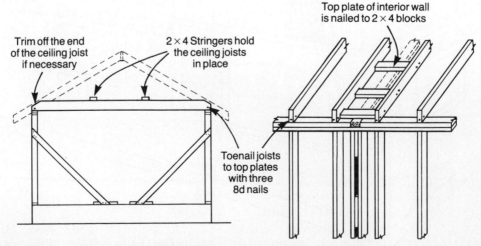

Trim off the end of the ceiling joist if necessary

2 × 4 Stringers hold the ceiling joists in place

Top plate of interior wall is nailed to 2 × 4 blocks

Toenail joists to top plates with three 8d nails

Fig. 13-15 *Installing ceiling joists.*

Food for Thought

Last Chance for Major Design Changes

I remember thinking that standard operating procedure for building walls was to get out the plans, quickly lay out the walls, and start sawing and nailing like crazy. According to Uncle Barney, I should be able to frame the entire house in three days. Well, that's a lot of baloncy. Take your time so you can think and feel your way, here's why.

When you get to the walls, it's the first time that you actually can stand on the floor and start to see what the house will be like. Is there a particular view from the dining table of an interesting old cedar tree or a sunset that should be taken advantage of? Is there a dark area that calls for a skylight? Is that hallway a little too cramped? Could the entry be enhanced by a sidelight near the door? Yes, yes, I know. All of this was supposed to be figured out during the design stage. But there's nothing like spending a lot of time at the site and standing at the kitchen window looking out to really give you a feel for what the house *could* be like. Does the house make the most of the surroundings and of your imagination? Here's one last chance to make modifications as each wall is built. You won't be able to frame a house in 3 days this way but it's an opportunity to make your house special that shouldn't be overlooked.

One fellow I know redesigned his entire house after the floor was built because he discovered how much noise emanated from a nearby road. It would have been better had he noticed that during the design stage, but at least he recognized the problem before he got the walls built. Once the walls go up, there's very little chance that they will ever be changed.

Rough Opening Dimensions—A Wolf in a Sheepskin

A problem that crept up on me during wall framing was how to decide on the proper size for window and door openings when I hadn't yet purchased the windows and doors. This was a difficult problem indeed. All of the books act like you automatically know the proper rough opening sizes from the blueprints. That's OK for cus-

tom-framed windows or when you're prepared to call up the window company and order $3,000 of preframed window units. Then you just make the openings 1/2 in taller and wider than the frames. But we wanted a certain style of wooden window. We couldn't buy them new and we were getting frustrated by the crummy batch of mismatched windows we were finding in the surplus stores.

Finally I built the window openings all slightly larger than necessary, figuring that I could always make them smaller but making them larger would be nearly impossible. Eventually we found some windows that met our stringent specifications. Then we tackled the time-consuming job of making the openings smaller to fit the windows. If I were to do it again, I'd make a major effort to locate most of the windows and doors before the walls were framed. I would still buy them second-hand (unless my financial status allowed for new ones), and I would still custom frame the non-opening double-pane windows. This procedure will be discussed in Chapter 15.

By reading Chapter 15 before framing the walls, you'll know the basic options for windows and doors. Then you can locate and purchase them, store them in a safe place, and be in the envious position of being able to frame the walls correctly on the first shot.

Salvaging Technique

This is a good time to talk about locating salvage materials. It's always a big unknown, and a definite risk. How can you know that you'll be able to find something that you like at a good price? Windows are a good example. People who have already built their houses say, "No sweat, just keep looking and something will turn up." And it's easy for me to say this too because I eventually found my windows at a good price. But it's always a risk and before you find what you're looking for you'll get frustrated and dejected. The best I can do is to give you some tips on how to increase your odds for success when salvaging.

1. Start looking early. As soon as you know you're going to build a house it's a good idea to start frequenting the surplus stores. You may buy some things that you never use in the house, but often you can use them in a shed or sell them at a garage sale.

2. Find the good sources and check them frequently. Let the proprietor know what you're looking for. I have one source that has never failed me for good doors at a reasonable price. Windows were a bit more difficult to

find, but I finally hit a bonanza and bought enough windows for two houses.

3. As you get near the time that you need something, hustle a little harder. Answer the ads, drive a little further to find surplus stores, look at more garage sales, and ask your friends to keep their eyes open for you.

4. When you start getting desperate, put out vibrations. I know this sounds a bit far out, but it works. It scored the windows and a set of French doors for us. We didn't find them, they found us.

Be Safety Conscious

The quickest way to fail at building a house is to hurt yourself. You really can't afford to have an accident. A broken arm or leg will not only delay the project 6 months, it may cause a financial disaster that will lead to abandoning the whole scheme. I don't know if there's anything I can say that will help you be safety conscious, but I'm going to try.

Perhaps it will help to read about some accident situations so you can recognize one as it presents itself. Usually the person is in a hurry for some reason that seems very important to him or her. The most serious problems seem to be falls from high places and tangles with power tools.

Wrench is a long-time do-it-yourselfer and usually is very safety conscious. Part of this stems from an incident in his youth when he expedited the painting of a two-story house by placing his ladder on top of the conveniently located dog house. I guess I can see how I might try that too. The ladder was a mere foot short of reaching that last spot on the gable and he was running behind schedule that day. Just as he was on the top rung, stretching to apply those last few strokes, the dog house slowly tipped up on edge, then rolled over. He lived through it alright, though he left ten fingernail grooves down the end of the house and splashed a lot of paint around. And after that day, the family dog was seldom seen sleeping in his dog house.

Steve did some unintended research into the change in the static coefficient of friction of cedar shakes as a function of their exposure time to weather. Apparently the cedar shakes were quite safe to walk on at first, but became very slippery after only a few months of weathering. His eaves are about 12 ft from the ground. He landed with such a thump that it startled the horses a hundred yards away, and gave himself two huge black eyes where his knees met his face. He was lucky; it easily could have been the end of his building for a while, if not worse.

A woman we know was not so lucky. While painting under the eaves of a two-story house, the ladder slipped from under her and down she came. Six-foot falls can cause broken bones, but 20-ft falls guarantee them. She landed on her elbow, shattering her upper arm and pushing parts out through the shoulder. It's not a pleasant story, but it might remind you to take you time and *be careful!*

Enough stories; you'll hear plenty when you start building. Here are some dangerous situations to avoid:

- Fingers in line with power saw blades.

- Knees or electric cords under plywood while sawing. (Build a work bench and have plenty of sawhorses around to help avoid this one.)

- Ripping boards on radial or table saws without an antikickback device. Also ripping warped wood.

- Twisting or jamming a skill saw. (Sometimes it jolts out of the wood and runs loose.)

- Carrying plywood on a roof when it's windy.

- Lifting heavy objects from an awkward position, or with inadequate mechanical advantage. Always keep your back straight and your butt down when lifting heavy things.

- Standing in line with a rope or cable that is under severe tension (for example, a truck pulling a log). If the cable or nylon rope snaps, the lashback is deadly.

- Dropping something heavy on a foot. (Steel-toed boots can be a toe-saver here.)

- Breathing fumes from paint, varnish, Swedish finish, paint remover, plastic plumbing cement, and a variety of other chemicals.

- Being on the downhill side of a large unstable object such as a log or a timber.

- Working on jury-rigged scaffolding, steep roofs (sans safety line), or unstable ladders. Ladders placed on wet plywood floors can be a serious hazard. Nail a short length of 2 × 4 behind the ladder legs so it can't slip.

- Getting objects in your eyes when sawing, drilling, chipping, nailing, grinding, etc. (It helps to have two pairs of safety glasses—one at the radial saw and one in your apron.)

- Using an electric nail gun on wallboard and missing the stud. The nail becomes a bullet and an unpleasant surprise for anyone on the other side of the wall.

- Working on an energized circuit at any time is foolhardy. A 110-volt circuit can kill you just as dead as a 440-volt circuit if conditions are right.

- Setting fire to the wood behind the copper pipe joint that is being soldered.

- Cutting your hands on jagged sheet metal.

- Spraining ankles on cluttered floors.

- Stepping on nails left protruding from boards.

- Stacking lumber sloppily can present a fatal hazard for small kids.

- Stacking asphalt roofing all in one place on the roof can cave in the roof.

With the proper eye, you can see accidents shaping up before they happen. It's worthwhile to develop that kind of eye.

Sorry, one more story. I had been carefully fitting the floor joists together for the octagonal dining room. None of them were nailed yet, due to fitting difficulties. Then I ran out of 2 × 10's and went on to something else. The next day, Dad had some spare time and was helping me on another part of the floor. He was carrying a couple of long boards and nonchalantly walking from joist to joist as only a seasoned carpenter can do. I heard a slam and turned to see him tip-toeing across the loose joists in the dining area. I froze in panic and a terrible feeling welled up inside me as I watched. Each joist fell over as he stepped on it, of course, so he just kept right on going, abandoning one loose joist for another all the way across the floor. Slam, slam, slam, slam, slam, slam! Boy, did Dad give me a disgusted look when he got to the other side. I took my heart out of my mouth and complimented him on his skill and agility (at 60-plus years old). I still get a shudder whenever I think about it.

After an accident (or near accident) occurs, it's easy to see the "dumb mistake." To be safety conscious a person must be constantly thinking of what *might* happen. In the case of the loose joists, I was used to thinking in terms of working alone. I knew they were there and their hazard potential never crossed my mind.

Since accidents are invariably accidental, it's smart to take out some medical insurance and to get a homeowner's policy that covers injuries to visitors and fire damage to the house. The best insurance, however, is to be safety conscious. Work smart, not fast. *That's* faster in the long run.

CHAPTER 14

Roof Framing and Roofing

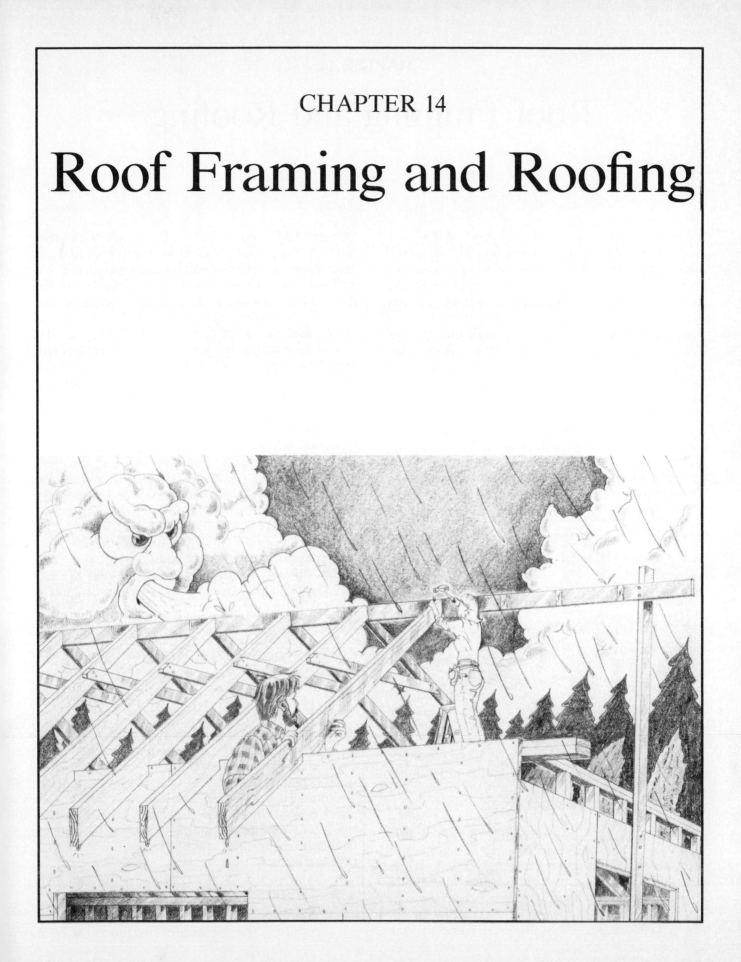

Roof Framing and Roofing

Working in the rain is depressing. Maybe it's God's way of keeping humans humble, if such a thing is possible. At any rate, forcing yourself to go outside into the mud and wetness to pound nails held with freezing fingers, feeling the rain run down your neck and the slow dampness on your back as the rain soaks through three layers of clothes, kneeling in puddles on wet floors, tip-toeing across wet and slippery floor joists, trying to ignore this drizzly, grizzly precipitation and wondering how in the world your lumber is ever going to dry out . . . it's all part of building. Having the experience yourself is a great way to realize how tough our ancestors were.

William Clark of the Lewis and Clark Expedition wasn't fond of the Pacific Northwest rain either, though he devotes many of the entries in his diary to the subject. On December 17, 1805 for example:

> "The most of our Stores are wet. Our Leather Lodge has become so rotten that the Smallest thing tares it into holes and it is now scarcely Sufficient to keep off the rain off a Spot Sufficiently large for our bead."

Another entry:

> ". . . all our Stores and bedding are again wet by the hard rain of last night . . . we are wet and disagreeable. . . . Four men complaining of violent coalds."

Geez, it makes me feel like a wimp for complaining. I could always pack it up and go home for a hot shower, put on some dry clothes, and stick a frozen dinner in the oven.

The low point in *my* diary seems to be the morning I found a lake covering the entire first floor of the house. It was exactly 1½-in deep—just deep enough to float the scraps of wood, saturate the boxes of nails, and get my apron sopping wet. I cut out the sole plate in the doorway with my handsaw and watched the water pour out like it was a reservoir spillway. I came to look

forward to having a roof on the house with great anticipation, relishing each little dry spot as I got the rafters up and covered them with plastic. I'd pull up a sawhorse and eat my sandwiches in the dry, as I peered out at the rivulets running off the "roof" and tried to imagine what the house would feel like when it was warm and dry inside.

The motivation for getting the roof built is tremendous, but it's not a stage to focus on speed. More than ever, careful attention to detail and safe methods are required. It took me nearly 5 wet, winter months to frame, sheath, tar-paper, flash, and roof the 23 different roof surfaces I'd designed into my house. That's a lot more effort than most sane house designs would require. The professional crew-built tract house is roofed in a week (2 days to install the trusses and sheathing, 3 days to apply the roofing). The roof of the sample house will take a beginner about 6 weeks. The truss part is good for 2 weeks and the custom-framed part (over the second story) will take about 4 weeks. Considering the 100-year life expectancy of this house, your 6 weeks of careful labor will be amply rewarded.

Framing Truss Roofs

Roofs framed with manufactured trusses are quite simple to built. Normally a truck with a crane delivers the trusses to the roof. Then two people place the trusses. One person plumbs it and holds it in position while another person secures it. Prior preparation is necessary so everything will proceed smoothly. The manufacturer usually provides a brochure that includes installation instructions.

STEP-BY-STEP PROCEDURE

1. Preparation. Mark the truss positions on the top plate of the walls on 16-in centers (or 24-in

centers if so designed). Remember the plywood now; the first spacing is 15¼ in or as necessary to accommodate the overhang. By installing metal framing anchors, the trusses will be more easily placed without errors. Install 2 × 6 scabs if necessary to prevent the end truss from toppling off the roof. Have some lumber ready to tie and brace the trusses and a ladder or some scaffolding on hand.

If the trusses are delivered by a truck with a crane, brace the wall where the bundle of trusses will be set. Interior walls will be needed to support the trusses until they are placed. If the trusses must be stored on the ground, don't allow them to become twisted.

2. Placement. Install one gable end first, then place and plumb each truss. Start at the end of the roof farthest from the bundle of trusses. Check each truss for the proper amount of eave overhang as it's placed.

Use care when handling the trusses. Sometimes when you pick them up flat, their own weight will cause them to pull apart. Trusses are strong in the upright position and precariously floppy in the flat position. Figure 14-1 shows one way of erecting trusses; the photos at the end of this chapter shows another method.

3. Ties and bracing. The trick here is to plumb and secure each truss as it is placed. The constant nightmare is that they'll all fall over like a line of dominos. Make sure the first truss is plumb and well secured. (See Fig. 14-1.) Then as each successive truss is placed, plumb it and tie it to the previous truss. A temporary diagonal tie will add to its lateral stability.

Be careful! This whole arrangement is a house of cards until the trusses are permanently secured and the plywood sheathing is applied.

4. Check everything. Check the roof surfaces for squareness, check each truss for plumb, check the overhang dimension on both sides, check for straight eaves (the overhanging ends of the trusses can be stringlined), and check the spacing so the plywood will fit.

5. Secure the trusses by nailing them permanently to the framing anchors and sheathing them with plywood. (Instructions for sheathing come later in this chapter.) Where special framing is required, as for the gable overhang or openings for chimneys or skylights, check the manufacturer's brochure to learn their recommended methods for use with their trusses.

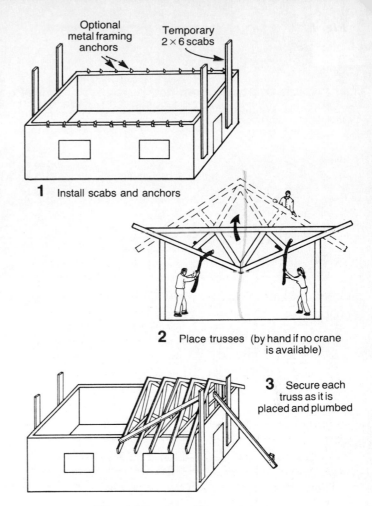

1 Install scabs and anchors

2 Place trusses (by hand if no crane is available)

3 Secure each truss as it is placed and plumbed

Fig. 14-1   Installing trusses.

In the sample house, the two special framing situations are with the gable overhang and the optional chimney opening. My preference would be to secure the first truss to the end wall of the second story and temporarily tie all the trusses together as they are placed. Then I'd immediately sheath as much of the roof as possible (removing the temporary ties as I went) but leave the sheathing off those places where special framing is required. This way I could work on the special framing at my leisure, without the nagging concern of being careful not to bump the trusses out of position or worrying about a windstorm coming along. Figure 14-2 shows the framing without the sheathing for the sake of clarity.

Custom-Framed Roofs

I call roofs that are built rafter by rafter "custom-framed roofs." They used to be the norm, but

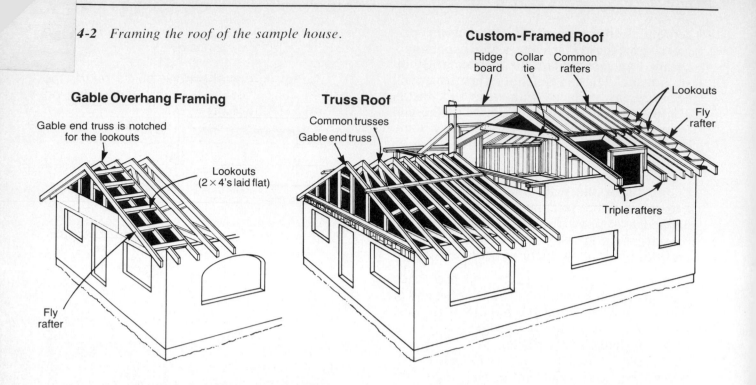

Gable Overhang Framing

Gable end truss is notched
for the lookouts

Lookouts
(2 × 4's laid flat)

Fly
rafter

Truss Roof

Common trusses

Gable end truss

Custom-Framed Roof

Ridge
board

Collar
tie

Common
rafters

Lookouts

Fly
rafter

Triple rafters

Fig. 14-3 *Custom framing a roof.*

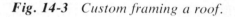

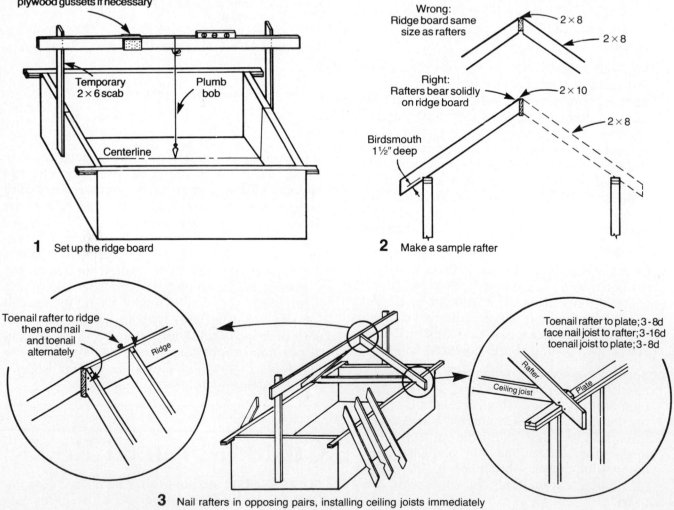

Splice ridge board with
plywood gussets if necessary

Temporary
2 × 6 scab

Plumb
bob

Centerline

1 Set up the ridge board

Wrong:
Ridge board same
size as rafters

2 × 8

2 × 8

Right:
Rafters bear solidly
on ridge board

2 × 10

2 × 8

Birdsmouth
1½" deep

2 Make a sample rafter

Toenail rafter to ridge
then end nail
and toenail
alternately

Ridge

Toenail rafter to plate; 3 - 8d
face nail joist to rafter; 3 - 16d
toenail joist to plate; 3 - 8d

Rafter

Ceiling joist

Plate

3 Nail rafters in opposing pairs, installing ceiling joists immediately

today far more roofs are built with trusses. Custom-framed roofs allow more flexibility to the designer and builder, but they usually take longer to build. The rafters are installed in pairs. The ceiling joists can be installed either before the rafters, (see Fig. 13-15), or as each pair of rafters is placed. (See Fig. 14-3.)

STEP-BY-STEP PROCEDURE

1. The ridge board goes up first. This may be a two-person job depending on the roof.

   First find the precise location of the peak of the roof and mark a line along the floor and at the top of the end walls. Plumb and nail a long, straight 2 × 6 scab to the end wall to support the ridge board. (See Fig. 14-3.)

   Select a straight ridge board. If necessary, splice two boards together with a plywood gusset or metal splice plates. The ridge board should normally be one size larger than the rafters to provide a full bearing for the rafter tops. For example, if the rafters are 2 × 8's the ridge board should be a 2 × 10. A 1 × 10 might be used, but the 2 × 10 will provide

more support while the rafters are being installed, and help to keep them in a straight line.

Before raising the ridge board, mark the rafter locations on it using a combination square. Do I need to remind you that the first spacing is 15¼ in? (Nah.) Also mark the rafter locations on the top plate of both side walls. Openings for chimneys and skylights are framed just like the floor openings with the rafters doubled on both sides of the opening. (See Fig. 14-4.) For the sample house, the 5-ft wide opening for the dormers is framed with *triple* rafters on each side of the opening, since the 48-in maximum in Fig. 14-4 is exceeded. The triple rafters are nailed together and function as a beam, carrying the load of the dormer and the roof. (See Fig. 14-2).

Finally, raise the ridge board and nail it in place. Check the height of the ridge board above the top of the walls. This will determine the slope of the roof. Locate the ridge exactly over the line on the floor using a plumb bob. The ridge seems kind of precarious right now and you may have to devise additional ways

Fig. 14-4 Special roof framing.

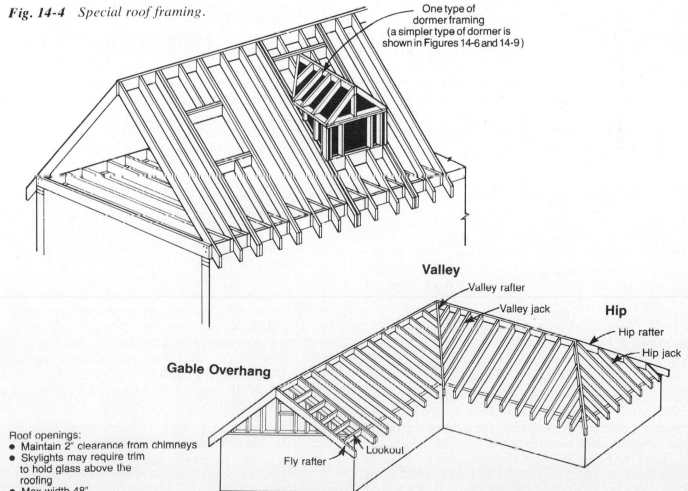

One type of
dormer framing
(a simpler type of dormer is
shown in Figures 14-6 and 14-9)

Valley

Valley rafter

Valley jack

Hip

Hip rafter

Hip jack

Gable Overhang

Roof openings:
- Maintain 2" clearance from chimneys
- Skylights may require trim to hold glass above the roofing
- Max width 48"

Fly rafter

Lookout

to support it. When the rafters are installed, the structure will become stiffer.

If you're into ceremonies, here's an opportunity for one. Raising the ridge is an important milestone and traditions ranging from pine boughs to incantations to celebrations with secretly-formulated beverages pervade various cultures. You may wish to get a couple pairs of rafters up to secure the ridge before getting heavily into the celebration.

2. Make a sample rafter. Find your straightest board and place it between the ridge board and the top of the wall. A helping hand will be useful here. Carefully mark and cut this board so it fits tightly to both the ridge and the top of the wall. Use your T-bevel here. It's a bit tricky and requires some trial-and-error, boatbuilding-type fitting. (The other option is to go at it using a carpenter's square, which is time consuming.) *Check this rafter to be sure it fits on both sides of the roof and in every rafter position.* This is the sample rafter and all other rafters will be made just like it. (If you mess up on the first try—cutting it too short before it fits right, for example—use it to help make a new one. This one can be used in short pieces to make lookout rafters.) (See Fig. 14-3.)

Once you have a sample rafter in hand, the world lies at your feet. It should fit to the ridge board and to the top plate at every rafter position. If it doesn't, check to see if the ridge board or wall is not straight and make adjustments as needed. Then use the sample rafter as a templet to trace all the other rafters with a sharp pencil. In order to avoid making a cumulative error, don't nail the sample rafter in place until all other rafters are installed.

3. Nail the rafters in place in opposing pairs. This helps to keep the ridge straight. To keep the weight of the rafters from pushing the walls out or over, *nail the ceiling joist or collar tie to each pair of rafters as soon as they're installed.* (See Fig. 14-3.)

Full Alert!

To your stations! The stray topics have broken out again. Good grief! This time there are roof overhangs, valleys, hips, dutch hips, octagons, hexagons, habrehedrons, and six species of dormers all clamoring for attention. (Din of the battle for several long minutes.)

They're fierce fighters and they want to get in this book. I'd hate to kill any of them; I just want to minimize their disruptive influence. In a negotiated settlement on the battlefield (and possibly at a weak moment), I've agreed to illustrate some of these aggressive critters here and possibly involve some of the others at another time. They're open game for any careful, patient, beginning roof builder. In fact, to build the sample house you'll have to hunt down and capture the (dangerous if provoked) gable overhang. Go get 'em folks.

While the references at the end of Chapter 5 (Wagner's *Modern Carpentry,* in particular) describe some of these topics in detail, your main resource will be your native ability.

Sheathing the Roof

You may recall that I'm a proponent of going as far as possible with each stage of the construction on a small scale to see what sort of problems are going to crop up. Add to this knowledge the fact that I needed a dry place to store things and you won't be surprised to learn that I didn't finish framing the roof before I sheathed and tarpapered a portion of it. This forced me to figure out some of the details on flashing and gutters early in the process and saved me some problems later on.

Applying plywood sheathing is a lot like sheathing the walls and nailing down the subfloor. Make sure the roof surface is square, then lay the plywood perpendicular to the rafters with the end joints railroaded so all of the joints don't land on the same rafter. The end of each sheet must land on the center of a rafter. Avoid using small pieces.

A handy way to get the plywood on the roof is to set up two sawhorses as shown in Fig. 14-5. Put six or eight sheets of plywood on the sawhorses, then go up on the roof and haul them up.

Another helpful trick is to put some planks between sawhorses so you can stand at a comfortable position to nail on the lower row of sheathing, the facia, the lower rows of roofing, and the gutters. On two-story houses, you may want to rent some free-standing scaffolding or build your own. Be cognizant of how much weight you put on the scaffolding. You're cruisin' for a bruisin' if you start storing a lot of heavy roofing materials on the scaffolding.

STEP-BY-STEP PROCEDURE

1. Start at any corner of the square roof surface and nail on the plywood (½-in CDX) with 6d

galvanized box nails spaced every 6 in along the edges and every 12 in elsewhere. By starting at a lower corner, the first piece of plywood provides a place to stand to bring up plywood for the rest of the roof. Make sure the plywood is positioned squarely on the rafters and that the ends fall on the center of a rafter. Railroad the end joints. The lower edge of the plywood should match the end of the rafters in most situations. (See Fig. 14-6.)

2. Continue applying sheathing along each row until you come to a hip, valley, or end of the roof. At the gable ends, ridges and hips, allow the plywood to project a few feet while it is

Fig. 14-5 *Getting plywood to the roof.*

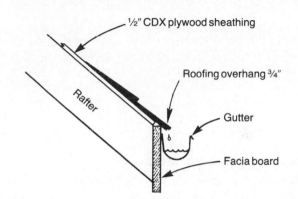

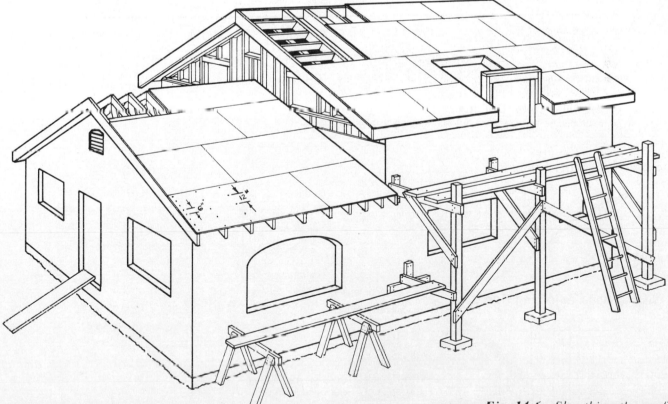

Fig. 14-6 *Sheathing the roof.*

Roof Framing and Roofing / 163

nailed on. Then snap a chalkline and saw off the excess. When you come to a valley, measure carefully from the end of the last piece of plywood to the valley rafter, saw the plywood to fit and nail it down.

Sheath right over any roof openings, then cut them out with a skill saw. The sheathing must clear metal chimneys by 2 in and masonry chimneys by ½ in. The flashing or chimney collar will cover the gaps.

3. Check the entire roof for a complete nailing pattern. Then nail on the facia boards. The damp conditions around facias promote rot, so choose a rot-resistant wood like cedar or redwood. Use galvanized nails.

Now, take a moment to view the house from several hundred feet away. It's really starting to look like a house isn't it?

Flashing and Gutters

I always thought flashing had something to do with a guy in a trench coat. It turns out that flashing refers to pieces of galvanized sheet metal or asphalt roofing that are artfully bent and secured to keep the rain out around chimneys, plumbing vents, dormers, skylights, doors, windows, and such. Some applications for roof flashing are illustrated in Fig. 14-7.

If you don't like the bright metal look of flashing, you can paint it. Metal flashing has an oily residue left from the forming process. By washing it down with vinegar, the oil is removed and paint will adhere to the metal.

In most cases, flashing is installed at the same time as the roofing and woven in with the shingles. If you try to wait until later to install it,

Fig. 14-7 *Typical flashing applications.*

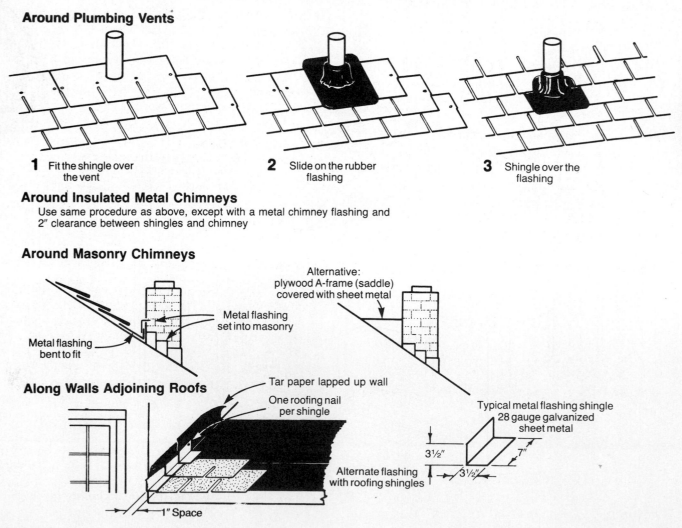

portions of the roofing will have to be removed and the chances for leaks will be increased. This means that before applying the roofing, all chimney and plumbing vent locations should be accurately determined.

Locating Round Metal-Fireplace Chimneys

To decide on the chimney location, you need to know the type of fireplace or stove and the type of flue required, or at least make a good guess. Chapter 16 discusses fireplace installation in more detail, but for now your main concerns are the inside and outside diameter of the chimney and the location of the boot where the chimney will attach to the stove. Notice that insulated chimney pipe must be used wherever a chimney goes through a roof or wall. Further notice that certain types of insulated chimney (triple wall, specifically) may be too large to fit between the rafters without a specially-framed opening.

A specific example: My Franklin fireplace requires a flue with an 8-in inside diameter. I could buy either triple wall or Metabestos insulated chimney locally. The major difference is that 8-in triple wall has a 12-in outside diameter while 8-in Metabestos has a 10-in outside diameter. To pass the chimney between 16-in O.C. rafters (14½-in space) and maintain the required 2-in space from wood, the Metabestos chimney would fit and the triple wall wouldn't. I discovered this when I got my triple wall chimney home and started working it all out. Consider yourself forewarned.

Another factor is the location of the stove itself. The stove must have between 18-in and 36-in clearance from the interior wall, depending on the stove design and whether heat shields are installed. Don't forget to allow ½ in for the wallboard which will be installed later. (Fig. 16-27 shows details.)

Locating Masonry Chimneys

Masonry chimneys require ½-in clearance from combustibles. Normally a specially-framed floor and roof opening are required due to their large size. Using the chimney foundation as a guide, hang a plumb bob from the rafters to locate the chimney. Your plans for the chimney design have to be pretty firm before it can be located.

Flashing for Chimneys

When the chimney is accurately located and the size is known, cut a hole for it in the roof sheathing with a saber saw. For round metal chimneys, a standard chimney flashing can be obtained from a mail-order store or a sheet-metal shop for about $15. Use this flashing to trace the hole on the sheathing, then saw the cutout. Install the flashing when the roofing is applied the same way that vent flashing is installed. (See Fig. 14-7.) Flashing for masonry chimneys is also shown.

Locating and Flashing Plumbing Vents

You probably have a basic plan for your drain-waste-vent (DWV) plumbing system, but now you need to locate the vents accurately. (Plumbing design was discussed in Chapter 7.) The main vent will probably be located close to the toilet and go up through the house in a place that won't compromise the structure. The end of the vent(s) shouldn't be near any windows or other places where the offensive smell can be a problem. In the sample house, I would locate the 3-in main vent in the water heater closet, where I could get at it to install an upstairs bathroom some day. Upstairs, it will be concealed either in a non-bearing wall or in a post made of 1 × 6's. The clothes washer is so far from the main vent that it has its own 1½-in vent that goes up in the wall by the stairs and is also concealed by a post upstairs. Consequently, no major structural elements of the house are compromised.

Just cut a hole in the roof sheathing for each vent, leaving about ¼-in clearance around the pipe. (Remember that a 3-in pipe has an outside diameter of 3½-in.) When you get to that place while roofing, just weave the rubber vent flashing in with the shingles as shown in Fig. 14-7. The vent pipe will be pushed through later.

Gutters

Gutters are important to the preservation of the house. Without gutters, rain water from the roof will splash on the ground and will splatter mud and water on the base of the house and promote rot. The gutters won't be installed until after the roofing, but it's good to think about them now.

The common type of gutter is shown in Fig. 14-8. They are available in both metal and plastic. The metal ones tend to rust after about 20 years.

To install, just nail the gutters to the facia making sure they slope downhill (about 1 in in 12 to 16 ft) in the direction of a downspout. The ferrules hold the gutter shape while you pound the spike through the hollow ferrule into the facia. Use caulking at the joints to prevent leaks.

Common Gutter

Typical Custom Gutter

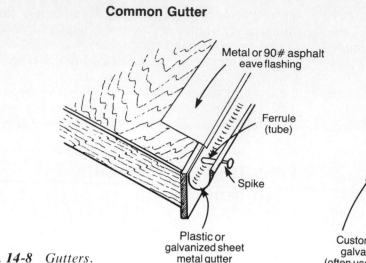

Metal or 90# asphalt
eave flashing

Ferrule
(tube)

Spike

Plastic or
galvanized sheet
metal gutter

Custom formed 24 gauge
galvanized sheet metal
(often used on low slope roofs)

Fig. 14-8   Gutters.

The custom-made type of gutters shown in Fig. 14-8 are used in situations where it looks better for the gutter to be hidden from view. They're more expensive because they have to be custom made by a sheet metal shop. Mine cost $300 and took quite a lot of work to install, but I do like their appearance.

Take particular care to provide a means of routing the downspout water away from the house. If you installed drainpipes for this in Chapter 11, the downspouts can just be connected to them. I use plastic drainpipes buried about 6 in to take the rainwater to a small gravel-filled pit on the downhill side of the house. This way I have no problems with water getting into the crawl space. Incidentally, if you have moisture on the windows of your new house, it usually is due to either excess moisture in the crawl space or lack of a plastic vapor barrier over the dirt.

Uncle Barney Scores

Poor ol' Uncle Barney. I'd been studiously ignoring all of his suggestions since my problems with the foundation. It's not necessarily that his advice was wrong. It's just that using another person's way of doing things is a lot like borrowing his or her trousers; sometimes the acquired item just doesn't fit you.

Barney thought that single-wall construction (where the tar paper is applied directly to the wall studs and the siding nailed over that) made good economic sense. I decided on double-wall construction (plywood sheathing, then tar paper, then siding) for strength reasons. Barney thought

the roof ought to be open-sheathed (1 × 6 stringers spaced 9 in apart with cedar shingles applied directly), arguing that the ventilation ensured longevity. I decided on plywood sheathing covered with tar paper and asphalt shingles, primarily for the added strength and better fire resistance. And so it went. Barney suggested method A, I chose method B. I wasn't rude or anything, I just heard him out and then did it my way.

Every time Barney left, I didn't expect him to come back. After all, when I did something different from his suggestion he took it as a personal affront, a ding in his integrity and a smirch on his gold-plated experience. But no, Barney was my

most faithful spectator. He'd stop by at least twice a week.

This particular day Barney was complaining of an injury he'd sustained in a confrontation with an obstinate pull tab. It didn't slow his stream of advice.

"You know, Frank, it's too windy to be working up on that roof today. Why don't you join me for a cool one down at the Duchess Tavern?"

"It's not that bad up here, Barney. I can handle it." I was getting used to heights and figured there would never be a day calm enough for Barney.

The words had scarcely left my mouth when a gust of wind caught the sheet of plywood I was carrying, spinning me around abruptly. I foolishly tried to hang on to it, lost my balance, and sprawled onto the sheet as it slid down the roof. I'd never seen Barney move so fast as I tobogganed toward him. I flew off the drop at the clerestory windows, then landed on a roll of tar paper which provided a roller bearing to accelerate my sleigh off the eaves at a high rate of speed. In terms of both height and distance the flight qualified as "sky" by any ski jumping standard.

The next thing I knew, Barney was admonishing me to wake up and was pressing a bottle of brandy to my lips. My head hurt, one arm was numb and my skill saw lay smashed to smithereens on the ground. I recovered slowly, nursing on the brandy. It was good stuff. Finally Barney, observing the level in the bottle dropping, reckoned I had recuperated adequately and retired the bottle to his truck.

"Barney, you're right. It's too windy today!" We got in his truck and headed for the Duchess Tavern.

Roofing

Once the sheathing is successfully nailed down and the flashing and gutter system is mentally worked out, the next task is to install the roofing and flashing so not even the most devious raindrop can gain access to the house. There are a broad array of types and styles of roofing. The discussion on roofing design in Chapter 5 may help you decide which type to use. By the way, shop around for roofing. I found that prices varied widely.

When installing roofing, it's very important to follow the manufacturer's instructions. The instructions here encompass most of the general principles and are for regular 3-tab asphalt shingles and the common type of gutter.

STEP-BY-STEP PROCEDURE

1. Cover the entire roof with 15-lb felt tar paper. Do this when the roof is dry. Otherwise the moisture will be held in forever, promoting rot.

   Start at the lower edge and roll it out, stapling it down every few feet as you go. (See Fig. 14-9.) You can also use roofing nails, which hold better than staples but take longer. Align the edge of the tar paper with the edge of the roof. Cut it with a utility knife or an old pair of scissors.

   Roll out the second strip of tar paper overlapping the edge of the first strip by 2 in to 4 in. Use the lines on the tarpaper as a guide. A roll of tar paper is usually 3 ft wide and about 100 ft long. When you come to the end of one roll and start the next, overlap the ends by 4 in.

   Be sure to put the tar paper on straight. When you get to the ridge, the tar paper should be perfectly parallel with the ridgeline. Then when the shingles are applied they can be lined up with the lines on the tar paper. No mistake is quite so apparent as a roof with the shingles on crooked.

   Where the tar paper meets an adjoining wall or other vertical surface, lap the tar paper up the wall about 8 in. Press the tar paper tightly into the corner so it won't rip when the flashing shingles are applied.

   Finally, avoid walking on the tar paper any more than absolutely necessary. I found that wearing tennis shoes helped to avoid damage and gave my toes a better grip on the roof.

Fig. 14-9   Typical roofing procedure.

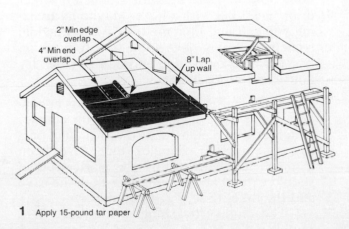

2. Lay the eave flashing. I use an 18-in strip of 90-lb roll roofing laid along the eave and over-hanging the edge by ½ in. (See Fig. 14-9.) In cold climates a 36-in strip is needed so the flashing is well inside the wall line. This way a backup of ice or snowmelt on the colder eaves can't get under the eave flashing.

Note: Sometimes a metal drip edge is installed before (or in place of) the eave flashing. It helps the drips to drop free of the roof and facia, and is especially important on low slope roofs.

Nail the eave flashing down with ⅞-in galvanized roofing nails. If the soffits won't be covered, use ½-in nails here so they won't protrude through the sheathing and be visible underneath.

What's a soffit? It's a small, soft, furry animal that lives up under the eaves of houses. When you look up under there sometimes you can see them through the screening. (See Fig. 15-9.) Keep looking, and don't feel silly. There are some nice ones, though a lot of builders seem to ignore them these days.

3. Lay the valley flashing. The valleys of a roof see a lot of rainwater, just like their name-sakes in the mountains. Consequently the valley is often the first place to leak. By laying a double layer of 90# roll roofing or galvanized metal in the valleys, the inevitable leaks can be postponed many years. Metal can be painted to avoid that bright shiny appearance.

Press the roll roofing tightly into the valleys. Use ⅞-in nails on the first layer, 1½-in nails on the second. Space the nails every 6 to 8 in. *Do not nail within 6 in of the center of the valley!* Drips just have a way of finding nail holes in the valley. (See Fig. 14-9, step 3.)

When roofing, put yourself in the place of the most devilish, devious, determined drip that you can imagine. Get yourself in a 50-mph wind or gang up with a bunch of other drips and a kid with a hose. Maybe try freezing yourself and then melting. Form a puddle somehow so you have some time to work on it. Your mission is to get inside that nice, dry house and make it wet.

4. Install the starter course. Start at a lower corner of the roof. Nail on the starter course, which can be a 9-in strip of roll roofing or a row of inverted shingles that have been trimmed to a 9-in width. The starter course should extend about ¾ in over both the eave and the rake. (See Fig. 14-9, step 4.)

Nail each inverted shingle with three 1½-in roofing nails. Use shorter nails for uncovered

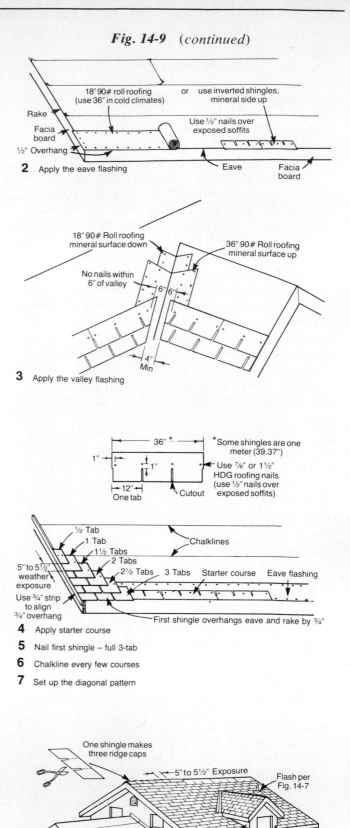

Fig. 14-9 *(continued)*

2 Apply the eave flashing

18″ 90# roll roofing (use 36″ in cold climates) or use inverted shingles, mineral side up
Rake
Facia board
½″ Overhang
Use ½″ nails over exposed soffits
Eave
Facia board

3 Apply the valley flashing

18″ 90# Roll roofing mineral surface down
36″ 90# Roll roofing mineral surface up
No nails within 6″ of valley
6″ 6″
4″ Min

36″ * *Some shingles are one meter (39.37″)
1″
1″
Use ⅞″ or 1½″ HDG roofing nails (use ½″ nails over exposed soffits)
12″ One tab
Cutout

4 Apply starter course
5 Nail first shingle – full 3-tab
6 Chalkline every few courses
7 Set up the diagonal pattern

½ Tab
1 Tab
1½ Tabs
2 Tabs
2½ Tabs 3 Tabs Starter course Eave flashing
Chalklines
5″ to 5½″ weather exposure
Use ¾″ strip to align
¾″ overhang
First shingle overhangs eave and rake by ¾″

8 Apply shingles and flashing
9 Apply ridge caps

One shingle makes three ridge caps
5″ to 5½″ Exposure
Flash per Fig. 14-7
Flash per Fig. 14-7

soffits. The starter course ensures that the entire roof is covered with a double thickness of roofing.

Note: Roofing nails come in several sizes. For the best holding power, the nails should penetrate through the sheathing by 1/4 in or more. If the sheathing thickness is 3/4 in or more, penetration isn't so important. All roofing nails should have large heads, be galvanized (preferably hdg), and have barbs. Pound the nails flush, but don't sink them into the roofing or cause any tears.

5. First course, first shingle. Position the shingle so it aligns with the starter course. That is, with a 3/4-in overhang. This overhang keeps the drips away from the wood and allows them to drip into the gutter. A larger overhang tends to droop after a few hot days. Nail the shingle (and all shingles) with four 1½-in hdg roofing nails as shown in Fig. 14-9. (Use short nails over exposed soffits.) When you've completed the roof, no nail heads should be visible.

6. Layout the position of the courses on the roof. Check the manufacturer's instructions for the amount of shingle that should be exposed to the weather. This is usually about 5 in and often the depth of the cut-out is the weather exposure. This makes a ready guide for aligning the shingles as they are nailed in place.

   You want the courses to be equally spaced right up to the ridge cap. A skinny row of shingles by the ridge looks kind of funny. Measure the distance from the eave to the ridge and mathematically determine the proper exposure to get an even number of courses. For example, 160 in (from eave to ridge) divided by 5 in (exposure) gives 32 courses. A 161-in measurement would require about a 5 1/32-in exposure.

   Layout the courses accordingly and snap chalk lines for every third course. These lines should be parallel to the lines on the tar paper.

Note: Experienced roofers often don't bother with the chalk lines, they just eyeball the course from the tar paper lines.

7. Set up the diagonal roofing pattern in a lower corner of the roof as shown in Fig. 14-9, step 7. Cut the shingles as straight as possible using a straightedge and a utility knife (or large tin snips). It helps to nail a 3/4-in strip of wood along the rake to align the overhanging edge of the shingles. This can also be done along the eaves.

Notice that each shingle is offset by half a tab. This locates the cut-outs midway between the cut-outs on the shingle below. So no devilish drip can get through to the roof.

Note: Some professionals offset the shingles by 5 in, rather than half a tab. This way, as you look up the roof, the cut-outs will not line up and any mistakes or inconsistencies are not apparent.

With the diagonal pattern set up, the first shingle is a full three tabs, the third is two tabs, and the fifth is one tab (and is the part of the third shingle that was cut off). Nail these shingles just like the first one: four nails, 1 in from the ends and 1 in above each cut-out.

8. Take a bunch of full shingles and start nailing them on. Usually the easiest procedure is to continue the diagonal pattern. That is, one shingle on the first course, one on the second course, and so on, then repeat. This technique minimizes your movement on the roof and the amount of time spent kneeling on installed shingles. If you are working from a scaffolding, however, it may be easier to nail on the first three to six full courses, then get up on the roof and continue.

   When you come to a valley or to the other end of the roof, trim the shingles as necessary. Remember not to nail within 6 in of the center of the valley and to allow a clear chute (4 in minimum) for rainwater and debris.

   After applying six courses across the roof, come back and begin another diagonal pattern. Then apply the next six courses across the roof.

   When you come to an adjoining wall, chimney, vent, skylight, or such, remember to weave in the flashing as was shown in Fig. 14-7. Leave 1 in to 1½ in of clear flashing. A narrow channel of flashing is less noticeable from the visual standpoint, but it tends to get clogged with debris, affording another opportunity for leaks.

   As you near the ridge, double check to be sure the shingles are parallel to the ridge. Readjust the weather exposure if necessary so the last course will have about the same exposure as the other courses.

9. Apply the ridge caps along the hips and ridges as shown in Fig. 14-9, step 9. Start at the bottom of any hips and work up to the ridge.

10. Sometimes there are places that can't be sealed with roofing and flashing alone. The top edges of flashing around masonry chim-

neys or places where octagonal turrets meet planar roofs are good examples. Caulk any of these unusual places with asphalt cement or silicone, acrylic, or polysulphide caulking.

Use a good grade of exterior caulking. I found out real quick that these three-for-$1 tubes of caulking were a waste of time. The stuff dries out and cracks about 6 months after it's applied and then has to be picked out of the cracks before another caulking can be effective.

Logistics

The slope of the roof will determine some of the logistics. On a medium or low slope roof, the roofing can be placed on the roof at the time it is delivered. Be careful to distribute the roofing—don't pile it all in one place or you may cave in the roof. Roofing is heavy stuff and just getting it up on the roof can be a trial. Many suppliers will deliver it to your rooftop with a special scissors-jack truck or a conveyor.

On steep roofs, the difficulty of keeping roofing, tools, and yourself on the roof becomes a major problem. I built my 6 in 12 slope roof by just being cautious and careful, and with a safety rope nearby as my security blanket. On the other side of the house the slope was 12 in 12 (45 degrees) and was a different safety story altogether. I tried a number of safety methods, the most successful of which was a sort of harness around me attached to a rope, which was tied to an eye-bolt at the top of the roof. There are various types of roof scaffolding that can be very helpful also.

Dry at Last

It's a great feeling to get the roof on the house. You are hereby allowed 3 days off for having surmounted the worst of this house building business. It may take a few days to realize it, but once you've gotten this far you are going to make it. There's still a lot to do, but working out of the rain will give you a whole new outlook on the project. (Review Fig. 14-10.)

Even if you're flat broke, exhausted, and ready to quit you can always steal some plastic to cover the windows, get some old blankets from the Salvation Army and kiss your landlord goodbye. Sure, it's kind of primitive, but it would look pretty good to ol' William Clark, mushing along in his wet leather buckskins. I bet he'd have traded his diary and his last pair of dry socks for

a nice dry shelter that's near to that dandy outhouse. No deal William, I don't wear those kind of socks.

For More Information
Sunset Books Editors. *Do-It-Yourself Roofing and Siding.* Menlo Park, Calif.: Sunset-Lane Publishing Company, 1981.

Fig. 14-10 *Framing, sheathing, and roofing the house.*

CHAPTER 15

Windows, Doors, and Siding

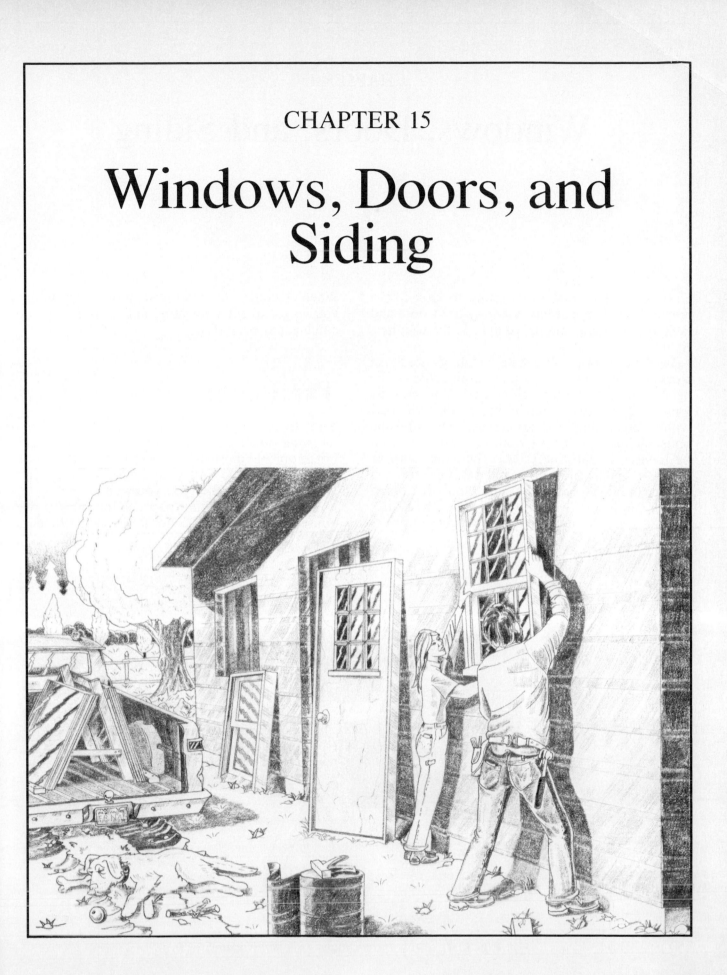

Windows, Doors, and Siding

Hooo boy! This place is starting to look like a house. Just slam in a few windows and doors and tack up the siding and we're tight to the weather, right?

Well, yes and no. Windows and doors are finish work, as opposed to the rough framing that we've been doing, and they take time. You can no longer rely on that catchall excuse that this part will be covered up and no one will see it after the house is finished. Those windows and doors will be looked at constantly. So when the hammer misses the nail and puts a big smile in the trim next to the front door, I bet you'll pull that board off and put it somewhere at the back of the house. I did. I'm no perfectionist but I've got my pride.

When it comes to windows, doors, and siding you've got to set your mind and body to be two notches more careful.

Tar Paper

You already know about tar paper. It has an interesting smell; it's kind of sticky on hot days and brittle on cold days. It's easy to work with if your stapler is reliable and your knife is sharp. The 15-lb tar paper is the type you want.

When applying the tar paper to the sides of the house, things go much faster if you have a helper.

Fig. 15-1 *Apply the tar paper so it sheds water.*

Start at the bottom of the wall and let the tar paper extend about ½ in below the sheathing so any water will drip off the tar paper and not dribble down onto the mudsill and into the crawl space. Overlap each strip by 4 in so it sheds water. If it's windy, nail strips of wood over it so it won't blow off. (See Fig. 15-1.)

Wrap the corners of the house to make them wind tight. At window and door openings, staple the tar paper right over them for now. Later, cut a big X with a knife and fold the flaps into the opening.

Important: The flaps will be folded under the window sills so they shed water. Don't slice the flaps off until after the window or door is installed.

Another place for potential leaks is at the tops of windows and doors. Here the tar paper will be lapped over a drip cap, so again the flap is needed. To sum it up in one sentence: Apply the tar paper so no water can get behind it.

Installing Windows

There are a multitude of window systems. We'll discuss the most common ones, which are illustrated in Fig. 15-2.

Aluminum Frame Windows

Aluminum frame windows are quick and easy to install. If the rough opening has been framed to the proper size, just set the window in the opening (from the outside), level it, and nail the lips to the wall. (My older brother used to threaten me quite frequently with that last step.)

Make sure the window is right-side-up (usually the top is indicated) and cut the tar paper so it overhangs the top lip of the window. If the window is the opening type, check its operation after it's nailed in place. Sometimes they get twisted and won't open and close right.

That's all there is to it. Your first window may take an hour to install, but it'll take only about ten minutes after you get some practice. The siding and trim are fitted around the window and over the lips so no water can intrude. The wood trim can be placed to cover the aluminum frame and give it a more distinctive appearance. On the inside, the window can be finished with wood trim or with wallboard and paint.

Preframed Window Units

By preframed units, I mean windows with a sill, frame, and pane such as the one illustrated in Fig. 15-2 (*center*). They come in so many styles that it is imperative to read each manufacturer's instruc-

Aluminum Frame Window   **Pre-Framed Window Unit**   **Custom-Framed Window**

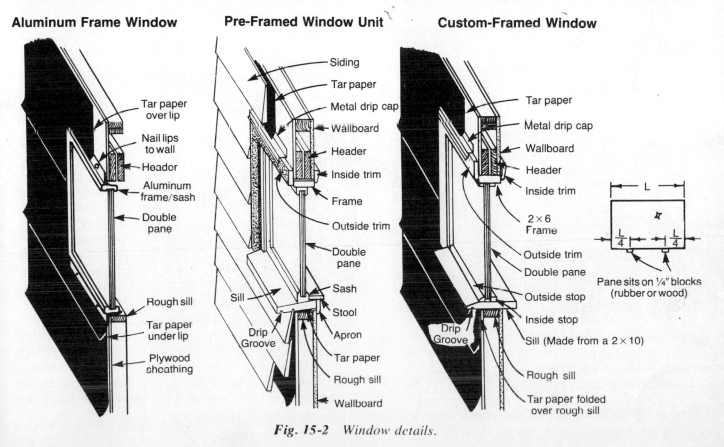

Fig. 15-2 Window details.

tions. The description that follows is for the general case.

Place the window in the rough opening from the outside. The sill often has two grooves in the underside; the outer one is a drip groove (to keep water from running back under the sill) and the second groove accepts the siding. Often the window unit has outside trim attached to it so you can just push the window in until the trim is snug against the tar paper. At the top of the window, unstaple and cut the tar paper so it overlaps the drip cap or top of the window. It usually takes a few minutes of careful study to see how all the inside and outside trim, drip cap, and siding fit together. Don't rush it.

One of your many assistants can hold the window (or nail a strip of wood across it) while you go inside to level and plumb it. Use shims or blocks as shown in Fig. 15-3. When that's done,

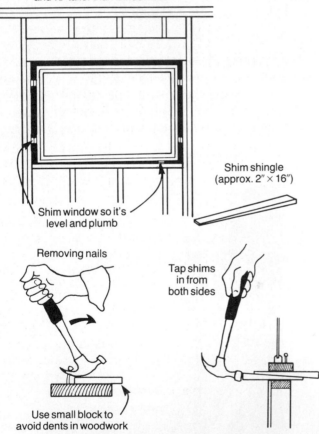

Rough opening ½″ wider and ½″ taller than window unit

Shim shingle (approx. 2″ × 16″)

Shim window so it's level and plumb

Removing nails

Tap shims in from both sides

Use small block to avoid dents in woodwork

Fig. 15-3 *Shimming window and door frames.*

go back outside to check everything. Then tack a few nails through the trim to hold the window tight to the wall. Back inside again, recheck the level and plumb and nail through the window frame and shims into the rough opening frame. Don't drive the nails all the way home yet. You may find some adjustments to make when you

start putting on the siding. If you have to pull the nails, put a block under the hammer head so it doesn't leave a dent in the woodwork.

When absolutely sure the window is properly installed, secure it by nailing it in several places. Use 8d finish nails. The gap between the window frame and the rough opening will be stuffed with insulation and eventually covered with the inside trim after the wallboard is installed. The drip cap, outside trim (if necessary), and a few siding boards should all be installed before the window is secured, particularly on the first window you install. You don't want to discover any mistakes after you've installed 20 windows improperly.

Custom-Framed Windows

These are single- and double-pane windows installed in frames built to fit the opening.

All this talk about preframed windows is nice, but have you ever priced those things? Yeah, they're expensive! Aluminum windows are cheaper, but the frames sweat on cold days and the appearance may not be exactly to your taste. I ended up using aluminum windows (disguised with wood trim) in the kitchen and bathroom where I wanted screened, opening, double-pane windows. On most other windows I built my own frames, see Fig. 15-2 (*right*), and installed either double-pane glass or wooden sash windows.

The sill can be bought in a lumber yard or made from a 2 × 10 or 2 × 12. I ripped the slope on 2 × 10's by running them through the table saw on edge. Use a new blade as a dull blade could be murder on a rip like this. Then I ripped a drip groove on the bottom of the sill (again with the table saw) and sanded it with a belt sander. The frame was made of selected 2 × 6's that I'd been setting aside for this purpose, see Fig. 15-2 (*right*).

I used ½-in base shoe for the lower outside window stops because I wanted the water to drain off of it, not just sit there in a puddle. For the inside stops I used ½ in × 1¹⁄₁₆-in bull nose stop. Almost any type of molding is satisfactory for the inside stop as long as it holds the window firmly in place. You can get a brochure from the lumber yard showing all the styles of moldings. A few common ones are shown in Fig. 15-4.

If the window rough openings are square and the rough sill is level, the sill and the 2 × 6 frame can be nailed with 8d or 10d casing nails directly to the opening. If the rough opening isn't square, build a square frame (at the work bench) and shim it into the opening. Stain, varnish, or paint the frame now; it's easier than doing it after the glass is installed.

Moldings

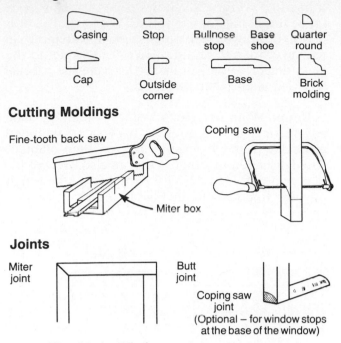

Cutting Moldings

Joints

Fig. 15-4 *Window and door moldings.*

Order the glass for the window pane ½ in shorter and narrower than the inside of the frame. The pane will sit on ¼-in blocks as shown in Fig. 15-2. When ordering glass, be aware that the terminology differs from shop to shop. Normally **single pane** means a single sheet of glass, which may be either **single strength** (³⁄₃₂-in) or **double strength** (⅛-in). **Double pane,** which may also be called **insulated glass** or by a brand name (for example, Twindow or Thermopane), is two sheets of glass with a sealed vacuum space between them. If the seal is imperfect or broken, condensation may form in the space. Considering the minor cost difference and the high rate of heat loss through single-pane glass, you'll probably want double-pane windows. For windows in doors and for windows larger than 12 sq ft, use double-strength, double-pane windows. Your local wind conditions may affect the thickness of the glass that is necessary, so check the code before you buy. The Uniform Building Code, Chapter 54 Table 54-A, specifies glazing requirements.

I was surprised to learn that most glass dealers don't care what size window pane you order. A 36 in × 48 in pane was no cheaper than a 35¹¹⁄₁₆ in × 47⅝ in pane. It seemed that these double panes are just made up specially for the customer. This turned out to be a blessing, because many of my windows had odd shapes and dimensions. The odd shapes (such as trapezoids and arches) do cost a little extra.

After the window frames are installed, the sid-

ing can be applied. There is some advantage, particularly on two-story houses, to nail the siding on before putting glass in the frames because you can use the windows for access to the scaffolding. For the sake of continuity, though, we'll take the process for custom-framed windows step by step without interruption.

STEP-BY-STEP PROCEDURE

1. Install the frame in the rough opening so it is level, plumb, and square.

2. Nail the outside stops in place with 4d finish nails. The corners can be mitered, butted, or coping sawed, but they must fit perfectly. (See Figs. 15-2, 15-4.) The stops must be straight so the window will fit tightly against them without bending.

3. Saw the inside stops making sure they fit snugly. Just set them near the window until step 8.

4. Place two ¼-in thick pads of neoprene rubber or wood for the window to rest on as shown in Fig. 15-2 (*right*). These pads must be no wider than the thickness of the window. For example, some double panes are ¾-in thick so the pads should be ¼ in × ¾ in × 2 in.

5. Check the fit of the pane (or wooden sash window) in the opening. The glass is placed from *inside* the house. Handle it carefully and watch out for stray nails on the sill. If it fits, remove it from the opening.

6. Apply caulking where the window will push against the outside stop. Use a caulking gun and a good grade of butyl rubber or silicone caulking. Carefully caulk windows that face a prevailing wind.

7. Make sure you have your hammer, some 4d finish nails, and the inside stops all within easy reach. Then, carefully place the pane of glass on the two pads. Place the bottom of the pane snugly against the bottom stop, then tilt the pane in place, squishing the caulking against the outside stops.

8. Nail the inside stops in place, pushing them firmly against the window pane. Start with the bottom stop, then do the sides, then the top. With each stop, put the first nail in the center and work toward each end.

Warning: You may find that if the pads are too wide you won't get a snug fit at the bottom of the pane. Also, be careful not to drive the nail at an angle toward the pane—you'll break the glass. To avoid bumping the glass with the hammer, hold the hammer head lightly against the pane as you hammer, and concentrate!

9. Finally, go around and clean up the excess caulking before it dries. Nail on the outside trim, making sure the tar paper at the top is free to overlap the drip cap. There are many types of trim and your choice will make a major effect on the appearance of the house. Look carefully at the details on some older houses when you get to this step. Use 8d or 10d *galvanized* finish or casing nails to avoid future rust streaks.

Galvanized sheet metal drip caps are available in lumber yards in 8-ft to 10-ft lengths. Cut them to length with tin snips, wipe then with vinegar to remove the oil, then paint them the color of the trim. Nail the drip cap over the trim at the top of the window to intercept the rain coming down the siding or the tar paper. Wood drip caps are also available.

You can see that making your own window frames takes a lot more time than installing aluminum or preframed windows. I figure it takes a total of 6 to 10 hours per window. The main advantage is that you get the exact appearance that you want, you minimize the cash outlay, and you can do any size or shape of window you desire.

Windows Ad Infinitum

Wooden Sash Windows. Usually found in second-hand stores, wooden sash windows can be stopped into custom-built frames the same way as glass panes. They are easier to work with than panes and are less fragile. Check for rotted wood and the condition of the putty before you buy them. The putty (glazing compound) can be replaced but it takes about 1 to 5 hours per window. Mullioned windows take the most time to repair and paint, but are very attractive. Incidentally, by making a wood grid that snaps in front of a pane of glass from the inside, the window will appear to be mullioned.

Triangular, Trapezoidal, Hexagonal, Octagonal, and Other Oddly Shaped Windows. Other shapes are also installed in custom-built frames like glass panes. I was surprised when I told my glass dealer that I wanted a double-pane trapezoidal window and he said "Sure. What are the dimensions?" just as if he did it all the time. It was hardly any more expensive than a rectangular window.

The safest way to get a good fit is to build the frame first. (Fig. 15-5 shows rough frame shapes for several kinds of windows.) Then make a cardboard (or plywood) templet ½ in shorter and ½ in narrower than the frame. Give the templet to the glass dealer so he can make the pane the precise size and shape that you need.

Round Windows. Round windows can be made by making a frame of ⅛-in veneer wrapped around a round form (like a tire), then stapled, glued or nailed together. You can also use wet ¼-in plywood for this. Nail the frame into the rough opening like the one in Fig. 15-5 and make sure

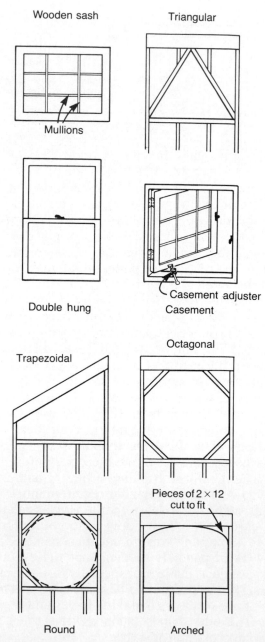

Fig. 15-5 *Window types and rough framing details.*

it's very close to round. Make a cardboard templet to have the glass cut. (An option is to shop for round glass coffee table tops in salvage stores.) For stops, use some 1-in wide strips of 1/8-in veneer wrapped around the inside of the frame. Install the outside stop, apply caulking, plunk the glass in place, then install the inner stop.

I know this sounds pretty easy, but you'd better plan on about 20 hours for each round window.

Arched Windows. Arched windows are framed into rough openings like the one shown in Fig. 15-5. A couple of pieces of 2 × 12 are cut to shape with a saber saw (or band saw) and nailed up against the header. The arched part of the frame is made with a flexible board, or laminated with thin strips of wood. Some woods like oak and ash can be steamed to fit the arch but that's more trouble than most people will want to get involved in. After the window is framed, make a cardboard templet the precise shape of the glass and take it to a glass dealer. Don't forget that the glass pane will sit on 1/4-in pads. The glass must be 1/2 in shorter and narrower than the frame.

Casement Windows. Use a wooden sash window and build a frame like you would for a glass pane. Most casement windows should open out, because nothing can be left on the sill of a window that opens in. However, when planning window opening casements that open out, make sure the window doesn't bump into an overhanging eave.

Make the inside stops and nail them in place. Attach rubber, felt, or foam weatherstripping to the stops so the window closes against it. Hang the window on hinges. Install the latch and the casement adjuster, a simple device that holds the window open so it won't blow shut.

Double-Hung Windows. This type should be bought with a frame, since the frame is difficult to build. Install them as described for preframed window units. With used windows, check the condition of the wood and the glazing putty before you buy. Most double-hung windows get painted shut at some point in their lives. With diligence, they can be put into working order. When you get around to painting them, you'll see why most of them get painted shut. Plan on 20 to 40 hours to strip, sand, stain, varnish (inside), paint (outside), and install one set of double-hung, mullioned windows. They look very attractive when completed.

Handling, Transporting, and Storing Windows

In a word, be careful. Don't allow the edge of a glass pane to be bumped. Store glass and windows on edge on blocks of wood, styrofoam, or the fiberboard that glass dealers use. The weight of a pane setting on a small sand granule or nail can crack it.

Handle glass with clean hands or clean gloves. Don't lay a pane flat to transport it or it's goodbye window at the first chuckhole. You can build an A-frame to sit in a truck bed that will hold the glass upright during transport. Separate the panes with cardboard or styrofoam, then drive carefully, avoiding bumps and railroad tracks. I got caught in a pinch once and transported a single pane by laying it flat on a mattress. That worked, but it's better to keep glass upright. Better yet, get the company to deliver the glass.

Doors

When you pay $800 for a fancy front door, it usually comes with instructions on how to install it. But the kind of doors that you get from a garage sale or a surplus store come without instructions or promises and there's a few things you'll need to know.

Notes on Buying Doors

- Make sure the door is absolutely flat. A warp in the door is not immediately noticeable and makes the door practically useless. Check the door by sighting it like you would a board.

- If you buy a prehung door complete with the hinges, doorknobs, frame, and stops you can save about 5 to 10 hours in installation time.

- If you buy a prehung door, make sure it opens in the direction you want it to open. There are right-handed and left-handed doors, and they aren't easily interchangeable.

- The outside dimensions of the door frame should be a 1/2 in or more shorter and narrower than the rough opening. If you custom-build the frame, the door should be 3/16 in shorter and narrower than the frame. Solid-core doors can be trimmed smaller, but hollow-core doors can usually be trimmed only about 1/2 in. Knock on the door to tell if it's hollow or solid. Also, lift it. Solid-core doors are heavy.

Installing a Prehung Door and Frame

Since this is the easiest situation, we'll look at it first. Getting the proper size unit is the toughest part. A common size for a main entry door is 36 in × 80 in, which is large enough to accommodate a refrigerator or a clothes washer on a mover's appliance dolly. Other exterior and interior doors are commonly 32 in × 80 in.

As with windows, it's a good idea to have the door unit in hand before framing the wall. One point that requires special care is the height dimension. Since the final finish floor should be flush with the door sill, and the door sill is typically 1½ in thick, the ¾-in subfloor normally must be removed to accommodate the sill. The sill rests on the floor joists. (See Fig. 15-6.)

I found that removing that little piece of subfloor took some determination, a skill saw, and a sharp chisel. Watch out for nails!

STEP-BY-STEP PROCEDURE

1. Make sure the rough opening is the correct size for the door unit.

2. Carefully remove the subfloor where the sill sits. If the angle of the sill requires a strip of wood for support, make and install that strip.

3. Set the door unit in place and make sure the sill rests solid and level on the floor. This may require that you plane or chisel the sheathing or the floor framing.

4. Shim the door unit until it's plumb in both directions. Shim just as you did for the windows. (See Fig. 15-3.) Normally the outside edge of the door frame will be aligned with the plywood sheathing and the exterior trim will cover the gap between them.

5. Nail an 8d or 10d casing nail through the door frame, through the shims and into the trimmer stud. Don't drive the nail home until all adjustments are made and the door is tested.

6. Test the door to be sure it opens and closes easily. If the door swings open or closed on its own, the door frame (specifically the jamb) is not plumb.

7. When you're satisfied with the operation of the door, drive the nails in the door frame home. Allow for the weatherstripping to be fastened to the door stop. (See Fig. 15-8.) When insulating the house, stuff insulation in the gaps between the door frame and the rough opening.

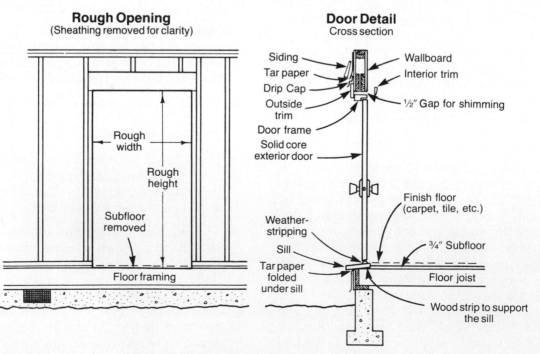

Fig. 15-6 *Installing a prehung exterior door.*

8. Normally at this point the trim around the doors can be nailed in place. Use galvanized finish or casing nails on the outside to avoid rust streaks.

Installing a door is no cakewalk, particularly the first time. It takes concentration. Some doors are more complicated than the one I've described here, so be sure to read the instructions whenever they're provided.

Hanging Doors in Custom-Built Frames

Doors bought without frames or hardware save you money, but will result in a lot more work. If you have more time than money, it's a good way to go. You have to build the door frame and the sill, install the sill and door frame in the rough opening, mount the hinges on the door and on the frame, hang the door and get it working right, then attach the door stops and weatherstripping. All of this requires careful workmanship. If possible, practice on a shed door or the back door to the house.

1. Buy a door sill (sometimes called a threshold) or make your own from a 2 × 10. Using the sill and 2 × 6's (or regular door frame stock), build the door frame 3/16 in wider than the door. The height of the frame will depend on the type of weatherstripping used at the bottom of the door.

2. Remove a piece of subfloor to accommodate the sill per Fig. 15-6. Then place the door frame, including the sill. The sill must sit solid and level. The frame must be shimmed, plumbed, squared, and tacked in place. Don't drive the nails home yet. Wherever possible, place the nails so they will later be covered by the door stops. Check the door for fit in the frame, being careful not to scratch it on the protruding nails.

3. Locate and mount the hinges on the door. The inside of the door should be flush with the finished inside of the wall. The hinge pins must be inside the house and should be clear of the frame and the door by about 1/4 in. The

Door Detail—Top View
Cross section

Fig. 15-7 *Installing a custom-framed door.*

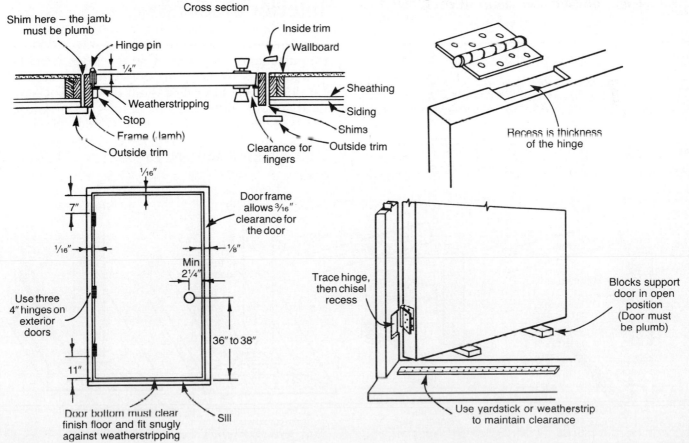

recesses for the hinges should be exactly the thickness of the hinge leaf. The recesses can be chiseled carefully or made with a router and a templet. (See Fig. 15-7.)

Pilot drill two holes for each hinge and screw the hinges to the door. We'll put in the other screws later when we're sure that the hinges are on straight and the door works properly. This was illustrated for the outhouse door in Fig. 9-20, step 11.

4. Install the doorknob. Since doorknobs come in a thousand varieties, use the instructions that come with the one you buy. Avoid putting the knob or handle so close to the edge of the door that it will be a knuckle smasher.

5. With the hinges and doorknob installed, set the door in the frame. It helps to put a yardstick flat on the sill to support the door with adequate clearance. Mark the hinge locations on the edge of the door frame. Then open the door and set it on some small blocks of wood to support it, keeping it plumb and at the right height. Trace the hinge positions on the frame with a knife point, remove the door, then carefully chisel (or rout) a recess for each hinge. Replace the door, then screw two screws to secure each hinge leaf to the door frame.

Note: Some people pull the hinge pin and just mount the hinge leaf on the door frame. Then the door is positioned and the pin replaced. This avoids wrestling with the door, but care must be taken to mount the hinge leaves precisely plumb.

6. Gently check the operation of the door. Is the inside surface flush with the door frame? Is there adequate clearance to open and close the door? Does the door swing open or closed? Make adjustments to the shims behind the door frame or to the hinge positions as necessary. When satisfied, drive the frame nails home and put the rest of the screws in the hinges.

7. Install the door stops with the door in the closed position. Allow for weatherstripping. (If you're using regular door frame material, the stop is an integral part of the frame, so ignore this step.)

8. Chisel a recess for the latch plate and the strike plate where the doorknob latches. Install the weatherstripping.

9. Install the outside trim with galvanized casing nails.

Now you thought I was kidding when I said that a prehung door complete with hardware could save up to 10 hours of installation time. Remember this when you're in the surplus store; it pays to look for prehung doors.

Interior Doors

Since there is no need for a door sill, locking doorknobs, or weatherstripping on interior doors, the installation is easier. Use the same basic instructions as were given for prehung doors and for doors hung in custom-built frames. The standard interior door size is 32 in × 80 in. Closet and special doors come in a variety of sizes, styles, and hanging systems (for example, bifolds, pocket doors, and so on).

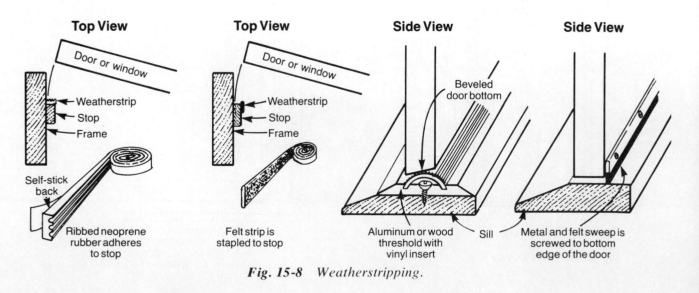

Fig. 15-8 Weatherstripping.

Weatherstripping

A bewildering array of weatherstripping options awaits you at your building materials store.

The sales person pointed me to aisle 19 where the shelves were spilling over with goodies. There were automatic bottoms, interlocking thresholds, wool felts, spring metals, foams, rubbers, and tubes. For a moment I thought I'd gotten lost in the birth control aisle, but upon closer inspection I found that these materials were all for use around doors and windows. As I frequently do when confused, I bought one of each and tried them all on various casement windows and doors around my house. My conclusion is that they all seem to fulfill their intended purpose of sealing out the rain and wind, provided they are installed correctly. The best ones are those that are easy to install and inconspicuous. Figure 15-8 shows my favorites.

Things work out best when you have the weatherstripping in mind when the door or window is installed. Otherwise there's a good chance you'll end up trimming the door bottom or adjusting the door stops later when you get around to weatherstripping.

Soffits and Vents

I know you're getting anxious to put the siding on, but there's one more job to do. Unless you've decided to leave the overhangs unfinished, the soffits must be installed now. And you'll need vents to the rafters regardless of whether or not soffits are installed. (See Fig. 15-9.)

Soffits are primarily for the sake of appearance. They cover up the underside of the roof overhang. The vents allow air to circulate to the attic, which reduces moisture problems and keeps the attic cooler on hot summer days. The Uniform Building Code Sec. 3205c requires the area of the vents to be at least $\frac{1}{150}$th of the area of the attic, with certain exceptions and conditions.

I think that soffits make a pronounced effect on the appearance of a house. Unfortunately, they can be awkward and time-consuming to install. Most builders of tract homes don't bother with them.

Besides the vents in the soffits, most houses also have gable vents which are easy to install. Just saw through the tar paper and sheathing and set the vent in place. The siding fits around the vent. An assortment of vents is available at the lumber store or they can be custom built.

Siding

At last! Take a couple of pictures now because once the siding is applied, the house will look basically finished from the outside. Structurally,

Fig. 15-9 *Soffits and attic ventilation.*

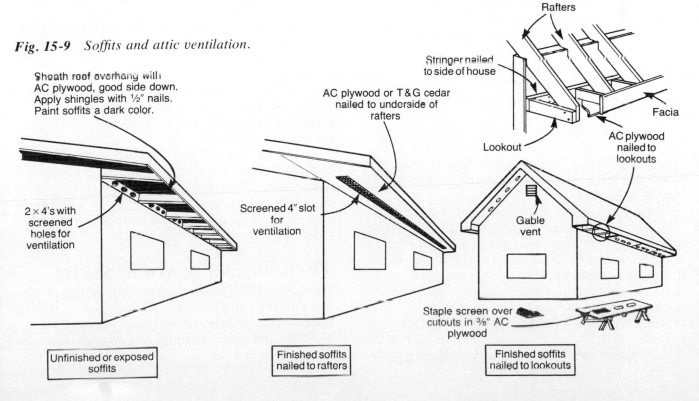

Sheath roof overhang with AC plywood, good side down. Apply shingles with ½″ nails. Paint soffits a dark color.

AC plywood or T&G cedar nailed to underside of rafters

Rafters

Stringer nailed to side of house

Facia

Lookout

AC plywood nailed to lookouts

2 × 4's with screened holes for ventilation

Screened 4″ slot for ventilation

Gable vent

Staple screen over cutouts in ⅜″ AC plywood

Unfinished or exposed soffits

Finished soffits nailed to rafters

Finished soffits nailed to lookouts

the house is already complete. You can nail up anything you like for siding: shingles, boards and battens, bevel siding, aluminum siding, vinyl siding, T-111, bricks, or buffalo chips if you like. Since the siding is the most visible part of the house, take some time to make a decision you'll be happy with 20 years from now.

You'll probably select a siding that you've seen on some house that you've liked. Now is a good time to check out that house more carefully. Looking at drawings in books is helpful, but actually seeing how the siding was applied is better. Look at the way the door and window trim joins with the siding. Look at the corners and up in the gable. Check the nailing patterns. Are the boards warping? Notice shingling patterns and check the difference in appearance between single and double-course shingles. Notice the different types of texturized plywood. Note how corner boards, masonry facings, and shutters can immeasurably enhance the appearance of a house. Often, painting the window and door trim a color that contrasts with the siding gives a previously drab house a distinctive, stylish appearance. And finally, beware of those doggone color chips the paint companies have; they'll fake you out every time. Sometimes it's worth the price of a quart of paint just to paint a part of the house to see what it's going to look like.

I found it best to stain or paint the siding before I nailed it up. Siding with a smooth surface can be painted or stained. On rough surfaces, stains work better. Paints tend to peel with age while stains tend to fade. Both seem to last about 10 years.

The method of applying the stain or paint is a matter of personal preference. I set the boards on sawhorses and applied the stain with a roller and a brush. (It was during the perpetual rainy season and this method allowed me to work inside.) Take care to stir the paint or stain well and often to prevent variations in the color. Also, as you get to the bottom third of one can of stain, mix it with the next can. The pigments tend to settle fast and cause the boards stained from the bottom of the can to be darker.

The fastest method of staining or painting is with a compressed-air or airless spray gun. While it takes me 2 or 3 hours to apply 1 gallon of paint using a roller, I've sprayed on 8 gallons of paint in the same amount of time. Of course it has to be done outdoors, and if the siding is already on the house the windows and doors must be masked. Normally trim is hand painted after the spraying is done. You'll need to rent the spray equipment as it's quite expensive to buy. As with any rental

situation, the odds of getting a functional piece of equipment are about fifty-fifty.

Notes on Applying Siding

- Siding should be stained or painted to preserve the wood. To keep that attractive wood-tone appearance, use a clear or semi-transparent stain or preservative

- Use hot-dip galvanized (hdg) nails or aluminum nails. Other nails, including electro-galvanized (eg) nails, will leave rust streaks on the siding after a few years.

- Siding must be applied so it sheds water. This sounds simple but there are a lot of places where people make mistakes. One place is the joint in the siding where Z-metal flashing should be used. (See Fig. 15-10.)

- Install drip caps above all windows and doors where there is any chance of windblown rain beating against the wall. (See Fig. 15-10.)

- Siding should extend about 1 in below the top of the foundation, so the rainwater can't drip onto the ledge and get into the crawl space.

- Use scaffolding to apply the siding when necessary.

When applying the siding I frequently found that I needed one hand to hold the board, one hand to hang onto the ladder, and two hands to start the nail. My usual solution was to hold the board with one hand and start the nail with the other. Grip the hammer with the nail between your knuckles and the nail head against the hammer handle. Hanging onto the ladder or scaffolding is a luxury I seldom enjoyed.

Another solution to this "four-handed job" problem is to rent or buy a power nailer (or power stapler). Some are electric and others pneumatic (that is, operated by compressed air). The cheapest run about $60.

To orient yourself to types of siding, flip back to the photos at the start of Chapter 2. Figure 2-2 shows a house with wide-bevel siding, Fig. 2-4 shows board-and-batten siding with shingles in places, and Fig. 2-5 shows a narrow-bevel siding. (Also see Fig. 15-11.)

Texturized Plywood Siding

Plywood siding comes in several different textures and in thicknesses of 3/8 in to 3/4 in. It's applied like plywood sheathing, except that the window and door openings are often cut out before it's nailed to the wall. Measure carefully.

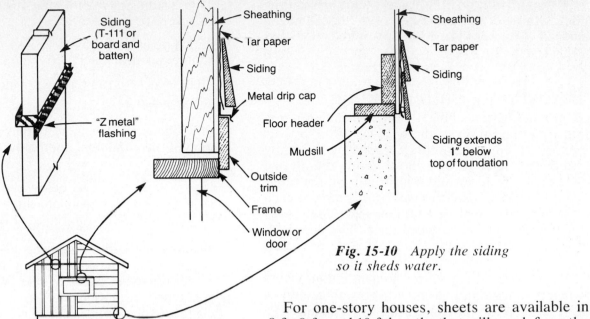

Siding (T-111 or board and batten)

"Z metal" flashing

Sheathing

Tar paper

Siding

Metal drip cap

Floor header

Mudsill

Outside trim

Frame

Window or door

Sheathing

Tar paper

Siding

Siding extends 1" below top of foundation

Fig. 15-10 *Apply the siding so it sheds water.*

Nail it in place with 6d or 8d hdg or aluminum siding nails spaced every 6 in along the edges and every 12 in along the stud locations.

For one-story houses, sheets are available in 8-ft, 9-ft, and 10-ft lengths that will reach from the foundation to the eaves. In the gables and on two-story houses, there will be a joint where Z-metal flashing must be used. (See Fig. 15-10.) Let the bottom of the plywood extend 1 in over the top of the foundation.

Fig. 15-11 *Siding types and installation details.*

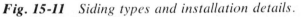

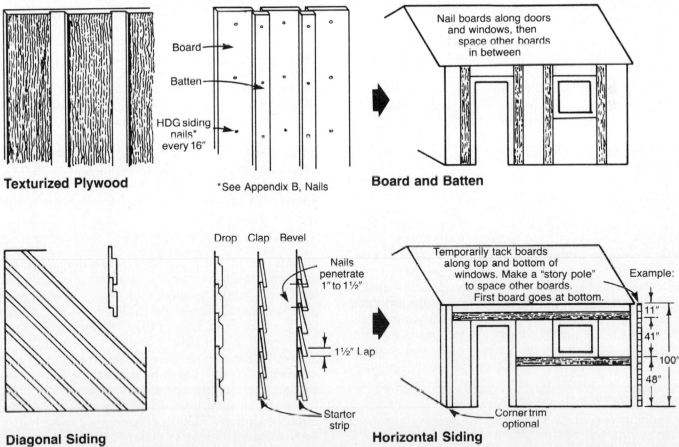

Board

Batten

HDG siding nails* every 16"

Texturized Plywood

*See Appendix B, Nails

Nail boards along doors and windows, then space other boards in between

Board and Batten

Drop Clap Bevel

Nails penetrate 1" to 1½"

1½" Lap

Starter strip

Diagonal Siding

Temporarily tack boards along top and bottom of windows. Make a "story pole" to space other boards. First board goes at bottom.

Corner trim optional

Example:

11"

41"

100"

48"

Horizontal Siding

Plywood siding is the easiest and fastest type that I know of. The pictures of the house at the end of Chapter 14 show texturized plywood (T-111) siding.

Board and Batten Siding

Board and batten siding is easy to install, but you must pay attention to the spacing of the boards next to windows and doors. It looks kind of funny if a batten lands next to the window trim on one side of the window and 6 in from it on the other side. I found it easiest to nail boards next to each window and door first and then space the rest of the boards to fit in between them. The board exposure may vary some (from 5½ in to 6½ in on my house), but the variation isn't noticeable.

By the way, there's a phenomenon that occurs when building that you should know about. When you first build something that has a minor flaw (like a variation in the spacing of the siding, for example), that flaw jumps out at you every time you walk by. Strangely, most other people can't see it, even if you point it out. The trick is to keep your mouth shut about your minor mistakes.

The boards and battens are nailed on with 7d hdg siding nails spaced every 18 in. The boards are put on first. If the boards are a lumber species or wetness that indicates they may crack, nail the heart side in and nail along the center of the boards rather than along the edges. This allows the boards to be free to shrink as they dry out. I used well-dried cedar 1 × 8's and nailed them at both edges where the nail heads would be covered by the 1 × 3 battens. There hasn't been any cracking, but knowledgeable people have told me that I might have had problems if I'd used lumber that wasn't completely dry.

Horizontal Siding

This category includes bevel siding, drop siding, clapboards, Victorian siding, and several other horizontally applied types of boards, aluminum, or vinyl.

The board spacing is the toughest part of applying horizontal siding. You should avoid notching the boards at windows and doors by temporarily tacking long siding boards above and below the windows. Then use a "story pole" as shown in Fig. 15-11 to determine the proper board spacing. See the following for the proper way to calculate the spacing:

1. Soffits to 1 in below the foundation top measures 100 in

2. 1 × 8 siding is 7½ in wide and requires a 1½ in overlap

3. Each board will have 7½ less 1½ in to leave 6 in exposure

4. 100 in ÷ 6 = 16⅔ rows, which implies 17 rows, each 5⅞ in

Now check the distance between the top and bottom of the windows, and below and above the windows:

48 in ÷ 6 in = 8 rows below the window
41 in ÷ 5⅞ in = 7 rows beside the window
11 in ÷ 5½ in = 2 rows above the window

This spacing will give us the 17-row total and avoid the need for any notching of the boards. The discrepancy between the widths of the boards won't be noticeable.

Make a story pole accordingly and mark the bottom position of each board at the corners of the house and along the trim by each door and window.

Nail on the siding beginning at the bottom. A starter strip is needed for bevel siding. (See Fig. 15-11.) Use 7d hdg siding nails as shown in Fig. 15-11 and be careful not to leave hammer dents in the boards. The corners of the house can be finished with corner board trim or with metal corner caps.

Diagonal Siding

Some types of horizontal siding can be installed diagonally, though it's not as weatherproof that way. Other siding is specifically made to be installed diagonally and usually has a larger overlap. Install the boards so the ends overhang the house corners and openings slightly, then trim off these ends with a skill saw. Otherwise you'll spend a lot of time cutting and fitting each board.

Shingles

Here's one of the most attractive sidings, but it's time-consuming to apply. To my eye, shingles look best when applied with a 7½ in exposure and spaced about ⅛ in to ³⁄₁₆ in apart. When shingles are fitted tightly together the appearance is similar to bevel siding. Double course shingles, shown in Fig. 15-12, give more of a shadow line or visual texture. Look at some shingled houses to decide what appeals to your eye.

Undercoursing is usually a lower grade, cheaper shingle. The outercoursing can be No. 1- or No. 2-grade shingles. There are also fancy butt

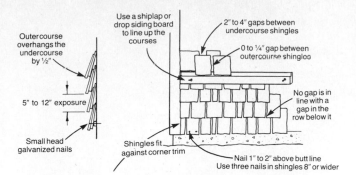

Fig. 15-12 *Applying double course shingles as siding.*

Labels in figure:
Outer course overhangs the undercourse by ½"

5" to 12" exposure

Small head galvanized nails

Use a shiplap or drop siding board to line up the courses

Shingles fit against corner trim

2" to 4" gaps between undercourse shingles

0 to ¼" gap between outercourse shingles

No gap is in line with a gap in the row below it

Nail 1" to 2" above butt line
Use three nails in shingles 8" or wider

---

shingles that are used to lace geometric designs into walls, and can be very attractive. Uncle Barney thinks that buffalo chip siding looks even better though, and he's insisting that I include instructions for that too.

Buffalo Chip Siding

This is by far the most difficult siding to install. The grain of each chip must be examined and the chip placed so the grain will shed water. Careful attention to maintain the one-third chip overlap is essential, and the spacing around windows and doors is tricky. Special aluminum alloy ringed nails (7d) must be used since the zinc on galvanized nails reacts with the chip to form a toxic gas. Although experts disagree on which texture and application technique is superior, there's general agreement that the chips are scarce and getting scarcer. I've heard it takes a herd of buffalo nearly 2 years to produce enough chips to do one house.

Really now, would I put you on?

Chitchat

No kidding, sports fans, this place is going to be a home. The roof is on, the windows and doors are in, and with the siding nailed up anyone passing by will see a finished, bona fide house.

Looking back, this was an involved, complicated project. While it's impossible to keep everything straight in your mind all at once, no single step is insurmountable when attacked with concentration. Once you gain the confidence that building a house can be broken down into small, manageable steps, there's nothing that can stop you.

I like building. Finally I've found a job in which my inherent forgetfulness, impatience, and profanity are not tremendous liabilities. They're not assets either, but at least when I cuss out a board it doesn't turn around and fire me. And when I get enamored by the sales girl and forget to buy a paint brush to go with the paint, there's no one to remark about my loose brain circuit when I get back (except the cats, whom I've taught to keep prudently tight lips in such situations).

But for all the foibles that you can get away with, you have to make up for them in two other departments. First, you have to be able to figure things out, but not necessarily with intuition or just by looking. It's entirely fair to ask questions of friends, inspectors, and store clerks, to use the yellow pages to get brochures, and to read a book or two on the subject. Other construction sites are fair game too, as are the details of friends' houses. When I got really desperate I would talk to Uncle Barney. But at each step you've got to come up with at least one way to accomplish the task.

Second, and this is crucial, you've got to have grit. You can't be a quitter. When you've tried six stores to find the right window or door and struck out, you have to try a few more. You may find you'll have to build it yourself. And when it has rained 12 days in a row and you just can't stand it any longer, you have to bear down, get tough, and keep going. Remember, it can't rain forever. Just keep making some progress every week, no matter how small. One step at a time toward a house that will last a hundred years.

Yep, that's the formula. The ability to figure things out by hook or by crook plus a heavy dose of grit. You'll find that these two ingredients will get you through any difficulty.

For More Information

Care and Handling of Fir and Hemlock Doors. Fir and Hemlock Door Association, Yeon Building, Portland, Oregon 97204.

Roof and Wall Shingling Made Easy. Red Cedar Shingle and Handsplit Shake Bureau, Suite 275, 515 116th Ave. N.E., Bellevue, Washington 98004.

Wagner, Willis. *Modern Carpentry.* South Holland, Ill.: Goodheart-Willcox Co., 1979.

Wood Siding—Installing, Finishing, Maintaining. U.S. Department of Agriculture Home and Garden Bulletin 203, 1973. U.S. Government Printing Office, Washington, D.C. 20402.

CHAPTER 16

Installing the Plumbing, Wiring, and Heating

Installing the Plumbing, Wiring, and Heating

Wrench compares the plumbing, wiring, and heating of a house to the dessert course of a meal. The house has been created and stands there as it will for the next hundred years. Equipping the structure with its inner workings is the fun and exciting part. Soon a deft flip of a switch will turn darkness into light; grasp a faucet handle and pure, clean water will flow; nudge a thermostat and effortlessly warm the house; twist a knob and a 60-piece orchestra performs.

These systems transform a mere shelter into an honest-to-God home and provide another opportunity to give free rein to your imagination. Install a shower out in the trees. Provide a way to switch the heat on in the bedroom before getting out of bed in the morning. Wire special lighting for a large carving by the front steps. Once you can plumb and wire there are a lot of possibilities open to you.

As you dive into this chapter you'll notice that the water is substantially deeper than you've been in so far. This chapter touches on the basics and you'll have to do some study beyond what is covered here. By visiting plumbing and electrical stores, reading more specialized books, and concentrating on each specific task one at a time, you'll soon gain knowledge that will serve you a lifetime.

The sequence in which to tackle this three-course dessert is our first decision. For the sample house, and most other houses, plumbing is my first choice because the slope of the drain-waste-vent (DWV) pipes is critical and their location is not very flexible. The heat ducts and wiring can be routed around the DWV pipes normally, but not vice versa. Identify situations where the heat ducts and drain pipes are competing for the same space and resolve how each should be routed. Depending on the season, the supply plumbing may have to be drained after it's installed to avoid a freeze-up (and burst pipes). The sample house has electric heat, so there won't be any heat until both the heating system and wiring are installed.

The sequence in which each subject is discussed here is a logical order for their installation, but nearly any sequence is feasible.

Plumbing
Putting Your Childhood Training to Work

The basic challenge of plumbing is to get clean water to all the necessary locations in the house and to get the used water back out to the sewer or septic tank, with no drips in between. If you were brought up with Tinker Toys in the toy box, plumbing will be a natural extension of your childhood skills. The theory of fitting all the pieces together is nearly identical, but the logistics of routing the pipes to avoid structural damage and to keep the waste water flowing downhill makes things a bit tougher. If you designed the foundation to provide a 24-in or larger crawl space and made holes in the concrete in the appropriate places, you'll benefit from that now. If not, well, as Uncle Barney would say, "You lose, Buckwheat. I warned you about this back in Chapter 5." Slithering around in a 12 in crawl space is a tough job.

Installing a Septic System

My outhouse was sort of an open-air affair, making it pleasant in the summer. I mused away many hours sitting there looking at the blossoms and watching the birds. In the winter though, the wet leaves and frost had to be brushed away and the temperature of the seat made it hard for me to relax. That prompted me to install the septic system early. (If you're hooking up to a sewer, this section can be skipped.)

In my locality, the Health Department wanted

a design like the one shown in Fig. 7-3 and issued a permit for the installation. Briefly, here's the procedure. Specifics can be obtained from Fig. 7-3.

STEP-BY-STEP PROCEDURE

1. Submit a septic system design to the Health Department and get a permit, if necessary.

2. Dig the septic tank hole and drainline trenches. I suggest hiring a backhoe for this.

3. Put the septic tank in the hole and level it. Fiberglass tanks can be placed by hand; concrete tanks are delivered by a truck with a crane arrangement.

4. Install the distribution box downhill from the tank outlet and level it carefully.

5. Install the tightlines (4-in plastic PVC pipe) between the house, septic tank, and drainlines. Cut the pipe with a handsaw, dry fit it together, then glue it when satisfied. Seal the connection to the concrete distribution box with cement or topping mix.

6. Install level grade boards in the trenches.

7. Put the gravel and drainlines in the trenches. Sounds easy, doesn't it? But now that there are mounds of dirt next to the trenches, which have turned to mud in the rain, where do you run the wheelbarrow? I shoveled a ramp up each mound, flattened off the top so planks could be placed, then ran the wheelbarrow along the planks. Occasionally I'd slip in the mud and wheelbarrow, gravel, shovel, and I would crash into the trench, obliterating my level grade boards. When I did make it to the desired location, dumping was another trick. I spent more time in the trench with the wheelbarrow than I'd planned, but eventually I got the trenches filled with gravel to the top of the grade boards.

   The drain lines are 4-in perforated plastic pipe that often come in 10-ft sections that snap together. Lay them on top of the grade boards and secure them with baling wire. Mine had a yellow line along the top to ensure the drain holes were properly positioned. (See Fig. 7-3.)

8. Fill the septic tank with water and test the system. Have it inspected if necessary.

9. Pile the last 6 in to 8 in of gravel around and over the drainlines. Cover it with about four pages of newspaper. This keeps the dirt from sifting into the gravel. The dirt can lead to roots of trees and rowdy vegetation taking over and clogging the drainfield. The newspaper eventually biodegrades and allows the system to aerate, a necessary function I'm told.

10. Finally, shovel the dirt back in the trenches and plant grass seed to combat the mud.

There, it's done! Now you know why the Indians didn't bother with flush toilets.

Installing the Drain-Waste-Vent (DWV) Plumbing

Review the plumbing section of Chapter 7 to refresh yourself on the design principles of DWV plumbing. House drains must be sloped at $1/4$ in per foot toward the sewer or septic tank and vents must be provided close to each fixture trap. This discussion is for plastic pipe and fittings (ABS type), which are nearly universally accepted now and included in the Uniform Plumbing Code (UPC Sec. 401). If you don't want to design the system, sometimes plumbing stores have drawings of code-accepted systems that can be copied, or you can hire a plumber to design it.

STEP-BY-STEP PROCEDURE

1. Make a drawing of the drain-waste-vent (DWV) system and get a plumbing permit. (See Fig. 16-1.) If you can, get the plumbing inspector to review the design with you when you get the permit and follow his suggestions. Check the local codes; Uniform Plumbing Code is the most universally used code.

2. List and acquire the necessary materials. (See Fig. 16-1.)

3. Determine the location of all plumbing fixtures and appliances (toilets, sinks, tub, shower stall, clothes washer, and so on).

   Mark the drain pipe location and vent location right on the floor or sole plate. I like to visualize things, so I set the sinks on sawhorses, pushed the tub in place, and positioned the toilet to see how everything fit together. If the toilet or tub drain comes out directly over a floor joist, either relocate the fixture or reframe the floor. Also check to be sure the drain pipes can be sloped to the main drain. Remember, $1/4$ in per foot is the minimum slope.

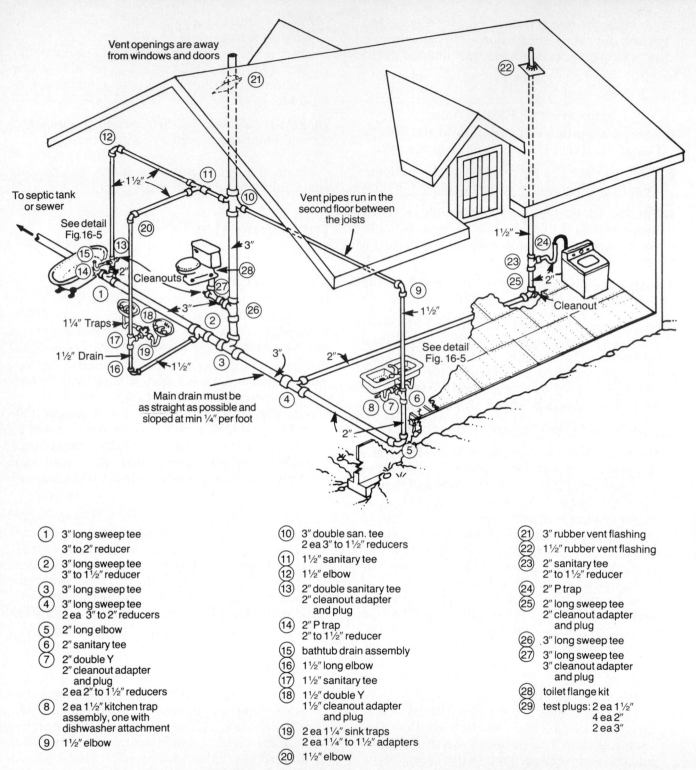

Vent openings are away from windows and doors

To septic tank or sewer

See detail Fig. 16-5

Vent pipes run in the second floor between the joists

Cleanouts

1¼" Traps

1½" Drain

Main drain must be as straight as possible and sloped at min ¼" per foot

See detail Fig. 16-5

Cleanout

Fig. 16-1 *Drain/waste/vent plumbing system for the sample house.*

① 3" long sweep tee
    3" to 2" reducer

② 3" long sweep tee
    3" to 1½" reducer

③ 3" long sweep tee

④ 3" long sweep tee
    2 ea 3" to 2" reducers

⑤ 2" long elbow

⑥ 2" sanitary tee

⑦ 2" double Y
    2" cleanout adapter
    and plug
    2 ea 2" to 1½" reducers

⑧ 2 ea 1½" kitchen trap
    assembly, one with
    dishwasher attachment

⑨ 1½" elbow

⑩ 3" double san. tee
    2 ea 3" to 1½" reducers

⑪ 1½" sanitary tee

⑫ 1½" elbow

⑬ 2" double sanitary tee
    2" cleanout adapter
    and plug

⑭ 2" P trap
    2" to 1½" reducer

⑮ bathtub drain assembly

⑯ 1½" long elbow

⑰ 1½" sanitary tee

⑱ 1½" double Y
    1½" cleanout adapter
    and plug

⑲ 2 ea 1¼" sink traps
    2 ea 1¼" to 1½" adapters

⑳ 1½" elbow

㉑ 3" rubber vent flashing

㉒ 1½" rubber vent flashing

㉓ 2" sanitary tee
    2" to 1½" reducer

㉔ 2" P trap

㉕ 2" long sweep tee
    2" cleanout adapter
    and plug

㉖ 3" long sweep tee

㉗ 3" long sweep tee
    3" cleanout adapter
    and plug

㉘ toilet flange kit

㉙ test plugs: 2 ea 1½"
              4 ea 2"
              2 ea 3"

4. Determine the location of the main stack and main drain. Use a plumb bob to align with the roof flashing. The flashing is installed per Fig. 14-7.

5. Saw the holes for the pipes. Start with the holes for the main stack and drain. Large holes can be cut with a saber saw, or even better with a $130 reciprocating saw. Others can be drilled with a hole saw and a large drill. Avoid devastating the structure. (See Fig. 6-15 for notching guidelines.)

The pipes should not fit tightly in the holes or they will squeak forever more, in response to temperature changes. The gaps will be stuffed with insulation later. Where pipes go through sole plates or top plates of a wall, protect the pipe against wallboard nail punctures by installing steel nail plates. (See Fig. 16-4.)

I found it useful to get the main drain and stack in place by supporting the drain with baling wire wrapped around nails in the floor joists. Get it straight and at a minimum slope of ¼ in per foot. Dry fit the drain and the stack; don't glue them until you're certain of their position.

When sawing the hole for the toilet flange, be careful not to make it too large. The screws must go through the flange and into the subfloor. The hole for the tub drain is the opposite situation; it must be large enough to allow room to use a pair of wrenches to tighten the drain fittings.

6. Measure, cut, and dry fit the pipes together. Don't glue any sections together until you're satisfied. The fit is very tight when it's dry, but the glue melts the ABS and allows them to slide all the way together. For this reason, always measure pipe lengths from fitting seat to fitting seat so you don't cut the pipe too short based on the dry fit.

Cut the ABS pipe with a handsaw, making sure the ends are square so they will seat tightly into the fitting. Pare off all burrs with a knife or file. (See Fig. 16-2.)

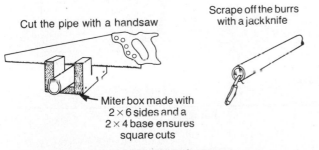

Fig. 16-2 Cutting plastic pipe.

Assemble the pipes and fittings just like Tinker Toys. Sanitary tees must be assembled so the tee sweep goes in the direction of the flow. Where a tee serves both water and air flow, always position for the water flow.

I ensure the proper orientation of each joint by marking the pipe and fitting with a white grease pencil during the dry fit, as shown in Fig. 16-3. Proper orientation is critical. When the joint is glued you'll only have a few seconds to get the fitting oriented right (to get the proper drain slope), and once the glue dries

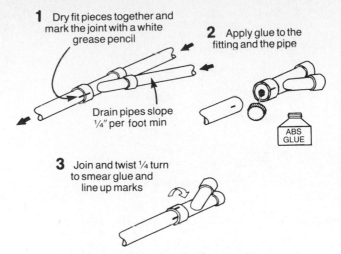

Fig. 16-3 Gluing plastic pipe.

there's no way to make adjustments short of sawing the whole section out and doing it over again.

7. Glue the pipes and fittings together. Here the benefits of living in the 1980s are readily apparent. Just dip the dabber in the glue can, swab a liberal amount of glue all around the inside of the fitting and around the end of the pipe, and push the pipe end into the fitting. Give the pipe a one-quarter turn until the grease pencil marks line up. This will ensure a water tight joint. Poof! You're a plumber!

8. Test the system. When the entire DWV system is assembled and glued, glue test plugs in the pipe ends where they come out of the floor or walls to meet the fixtures. The test required by most inspectors is to fill the DWV system with water by running the garden hose in the roof vent. The sewer or septic tank end of the system is plugged also, so the system is subject to the static pressure of the water. If there are any leaks, you'll find out this way. Make repairs as necessary and then have the inspector look at it. All pipes must be supported adequately and protected from nail punctures where necessary. (See Fig. 16-4.)

After the inspector's signature is on the dotted line, remove the plug at the sewer end of the main drain. If it's glued in, just smack it with a hammer and stand back. Leave the other test plugs in place until the fixtures are hooked up. This keeps sewer gases from escaping into the house if you don't install all of the fixtures right away.

The fixtures will be connected after the wallboard and finish floors are installed, but it's helpful at this point to know how it's done. Figure 16-5 shows some typical fixture connections,

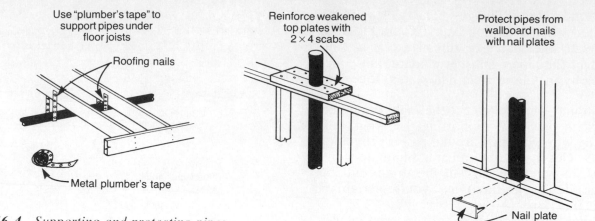

Use "plumber's tape" to support pipes under floor joists

Roofing nails

Metal plumber's tape

Reinforce weakened top plates with 2 × 4 scabs

Protect pipes from wallboard nails with nail plates

Nail plate min ¹⁄₁₆" thick steel

Fig. 16-4 *Supporting and protecting pipes.*

though recognize that there are several different acceptable ways to connect any fixture and your plumbing inspector may prefer other arrangements.

Installing the Supply Plumbing

The supply plumbing is more of a free-spirited thing than the DWV system. Just run a pipe to deliver water from point A to point B any way

Fig. 16-5 *Typical plumbing fixture connections.*

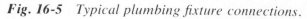

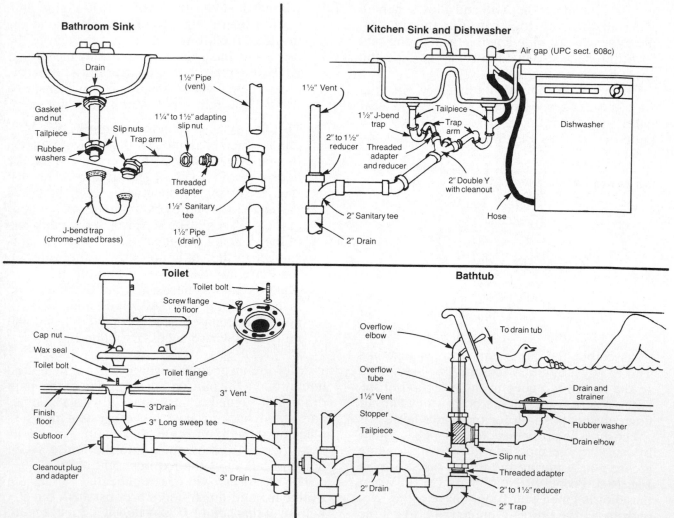

Bathroom Sink

Drain
Gasket and nut
Tailpiece
Rubber washers
Slip nuts
Trap arm
J-bend trap (chrome-plated brass)
1½" Pipe (vent)
1¼" to 1½" adapting slip nut
Threaded adapter
1½" Sanitary tee
1½" Pipe (drain)

Kitchen Sink and Dishwasher

Air gap (UPC sect. 608c)
1½" Vent
1½" J-bend trap
Tailpiece
Trap arm
2" to 1½" reducer
Threaded adapter and reducer
2" Double Y with cleanout
Dishwasher
Hose
2" Sanitary tee
2" Drain

Toilet

Toilet bolt
Screw flange to floor
Cap nut
Wax seal
Toilet bolt
Toilet flange
Finish floor
Subfloor
Cleanout plug and adapter
3" Vent
3" Drain
3" Long sweep tee
3" Drain

Bathtub

Overflow elbow
To drain tub
Overflow tube
1½" Vent
Stopper
Tailpiece
Drain and strainer
Rubber washer
Drain elbow
Slip nut
Threaded adapter
2" Drain
2" to 1½" reducer
2" Trap

that suits your fancy. I kind of hate to take the fun out of it by giving instructions, but there are a few things that need to be mentioned.

STEP-BY-STEP PROCEDURE

1. Work from a drawing like Fig. 16-6. Chapter 7 discusses the considerations necessary to make such a drawing.

2. List and acquire the necessary materials and tools. Figure 16-7 shows the common copper fittings and pipes. Figure 16-8 shows most of the tools needed. You'll also need a heavy duty drill like the one shown in Fig. 16-6.

The valves and faucets for the system are a bit expensive. Study what is available in the plumbing supply store and buy the ones that best suit your needs. The sample house has a shut-off valve where the water comes into the house. Normally, there is also a valve at the street water meter or at the well tank. In conjunction with the shut-off valves at every fixture and appliance, these valves will make the system easy to repair or modify. The garden standpipe has a valve with a drain cap that allows it to be drained to avoid winter freezing. The house system can be drained through a yard faucet if it is put at the lowest point in the system. Figure 16-9 shows some common valves and faucets.

There are several ways of making connections between plastic pipe (like the polyethylene pipe from the water meter), copper pipe (like the house system), and galvanized steel pipe (like the garden standpipe). The best way

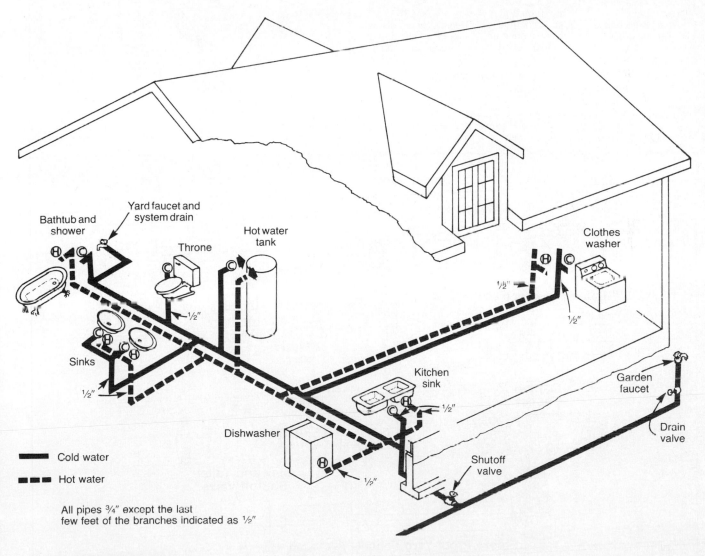

Fig. 16-6 *Supply plumbing system for the sample house.*

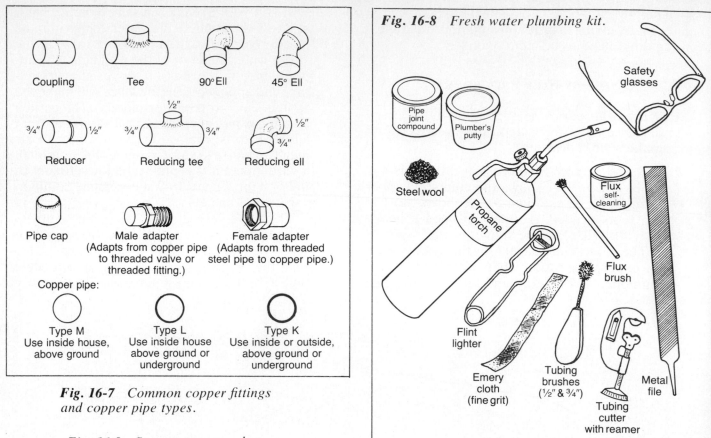

Fig. 16-7 *Common copper fittings and copper pipe types.*

Coupling Tee 90° Ell 45° Ell

Reducer Reducing tee Reducing ell

Pipe cap
Male adapter
(Adapts from copper pipe to threaded valve or threaded fitting.)
Female adapter
(Adapts from threaded steel pipe to copper pipe.)

Copper pipe:
Type M
Use inside house, above ground
Type L
Use inside house above ground or underground
Type K
Use inside or outside, above ground or underground

Fig. 16-8 *Fresh water plumbing kit.*

Pipe joint compound
Plumber's putty
Safety glasses
Steel wool
Flux self-cleaning
Propane torch
Flint lighter
Flux brush
Emery cloth (fine grit)
Tubing brushes (½" & ¾")
Tubing cutter with reamer
Metal file

Fig. 16-9 *Some common valves and faucets.*

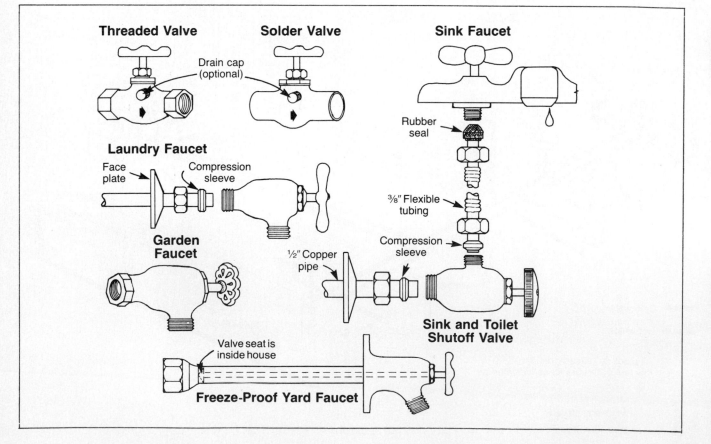

Threaded Valve **Solder Valve** **Sink Faucet**

Drain cap (optional)

Rubber seal

Laundry Faucet

Face plate
Compression sleeve

⅜" Flexible tubing

Compression sleeve

Garden Faucet

½" Copper pipe

Sink and Toilet Shutoff Valve

Valve seat is inside house

Freeze-Proof Yard Faucet

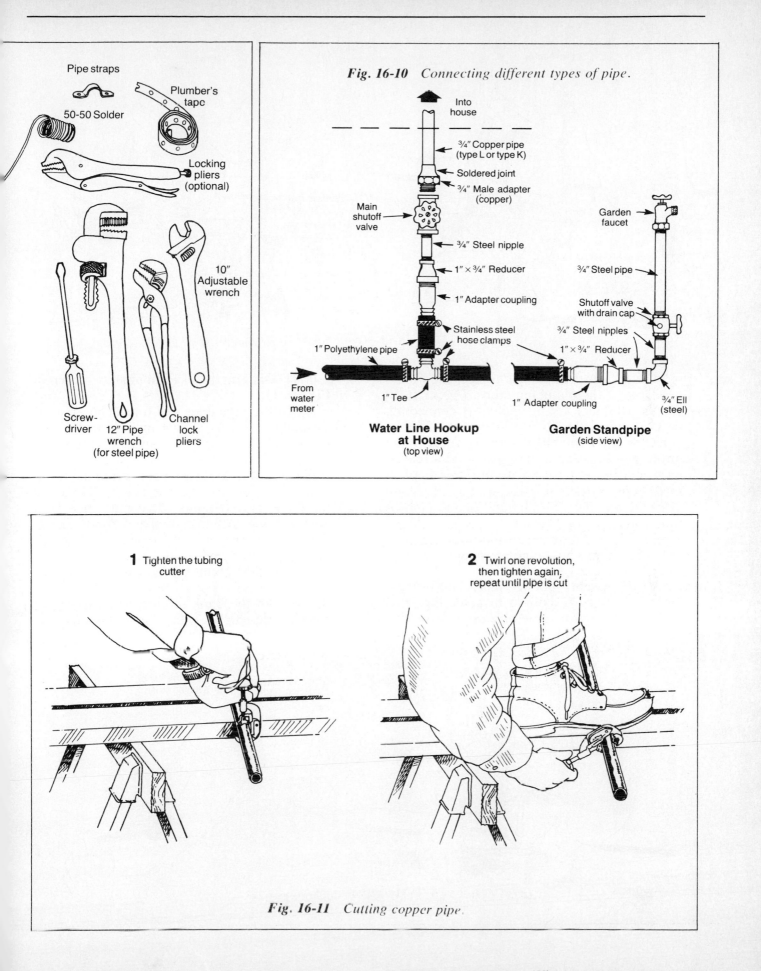

Pipe straps

50-50 Solder

Plumber's tape

Locking pliers (optional)

Screwdriver

12" Pipe wrench (for steel pipe)

Channel lock pliers

10" Adjustable wrench

Fig. 16-10 *Connecting different types of pipe.*

Into house

¾" Copper pipe (type L or type K)

Soldered joint

¾" Male adapter (copper)

Main shutoff valve

¾" Steel nipple

1" × ¾" Reducer

1" Adapter coupling

Stainless steel hose clamps

1" Polyethylene pipe

From water meter

1" Tee

Water Line Hookup at House
(top view)

Garden faucet

¾" Steel pipe

Shutoff valve with drain cap

¾" Steel nipples

1" × ¾" Reducer

1" Adapter coupling

¾" Ell (steel)

Garden Standpipe
(side view)

1 Tighten the tubing cutter

2 Twirl one revolution, then tighten again, repeat until pipe is cut

Fig. 16-11 *Cutting copper pipe.*

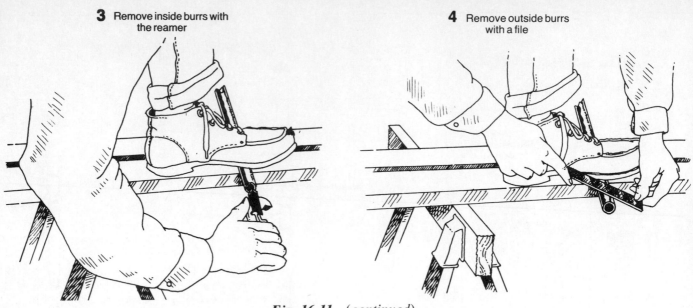

3 Remove inside burrs with the reamer

4 Remove outside burrs with a file

Fig. 16-11 (*continued*)

is to work out the connections from what's available at the plumbing store. Figure 16-10 shows a typical arrangement for connecting different pipe types.

By the way, you might want to work out the difference between a male part and a female part here in the privacy of your own book. I made the mistake of asking the plumbing salesgirl one day. She looked at me kind of funny and finally told me I'd better ask my mother. It still took me a minute to figure it out. Me oh my, how embarrassing!

3. Learn to "sweat" copper pipes and fittings. Here's a short $2 course in soldering, also called sweating for some obscure reason. Cut

four 11¼-in lengths of ½-in pipe as shown in Fig. 16-11.

Clean the pipe ends and four ½-in elbows, assemble them as shown in Fig. 16-12, flux the joints and solder ("sweat") them together.

Heat up one of the joints, moving the torch around the joint to heat it evenly. Hold the torch upright. It tends to go out when held upside down. After about five seconds, touch the solder to the pipe and see if it melts, keeping the torch flame on the elbow but on the opposite side from the solder. The solder melts all at once and is sucked in and around the joint in the blink of an eye. Remove the torch immediately. There should be a narrow

Fig. 16-12 *Soldering copper pipe.*

1 Clean the pipe ends with emery cloth

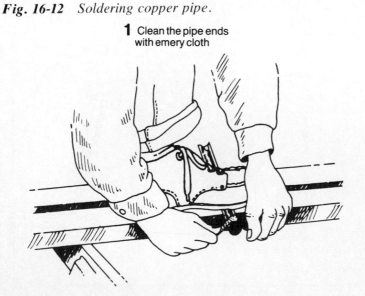

2 Clean the fitting with a tubing brush

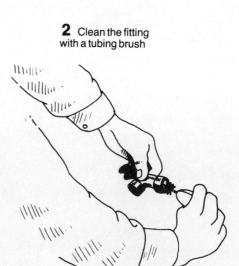

collar of solder all around the pipe. Sometimes uneven heating, incomplete fluxing or
inadequate cleaning will result in a partial collar around the joint. Such a joint will leak.
Fortunately, the joint can be reheated and
pulled apart with pliers, recleaned and re-
soldered. That's an advantage to working
with copper pipe that we didn't have with the
plastic pipe for the DWV system.

You can solder a copper wire to this practice project and use it as a frame for a picture
or a mirror tile.

4. Plumb the house in accordance with your
drawing. This is a big step, but mostly it's
enjoyable work. I call it "Tinker Toys with a
torch." Start at the water line from the water
source and install a branch into the house.
Install all cold water pipes leading to every
necessary location, including the hot water
tank. From the hot water tank, install pipes to
every place that hot water is needed. When
you come to a fixture or appliance location,
"stub out" the pipe as shown in Fig. 16-13
and cap it. After the wallboard is installed, the
stubout and cap are cut off and a shut-off
valve is installed. (See Fig. 16-9.) The stubout
should be 4 in long so there is room to swing
the tubing cutter and enough pipe left to connect the compression sleeve.

Some approximate dimensions for various
fixtures are shown in Fig. 16-14. While not all
fixtures will fit these dimensions, I found that
working without a rough estimate was more
troublesome than working with the good
guesses shown here.

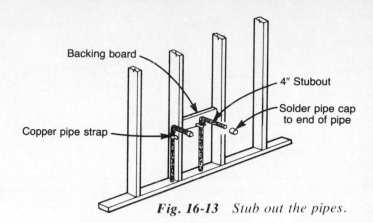

Fig. 16-13 *Stub out the pipes.*

When soldering the stubouts, you'll often
be working right next to a stud or backing
board. Here it's handy to have a 6-in × 6-in
piece of asbestos or metal to put between the
wood and the torch when soldering in close
quarters. To burn the house down at this
point would be a bummer.

One other tip: Many plumbers don't bother
with the time-consuming job of cleaning the
pipes and fittings before applying the flux.
The flux is supposed to do the cleaning. As a
rank beginner, I preferred to clean each joint
rather than take a chance with leaks. The reward was only one leaker in 150 joints. I'll tell
you about the leaker in a minute.

Finally, hook up the hot water tank as
shown in Fig. 16-15.

*Important: Don't put a shut-off valve on the hot
water pipe coming out of the hot water tank. It
could cause a dangerous pressure build-up.*

3 Make sure everything fits together

5 Solder each joint

4 Flux each joint

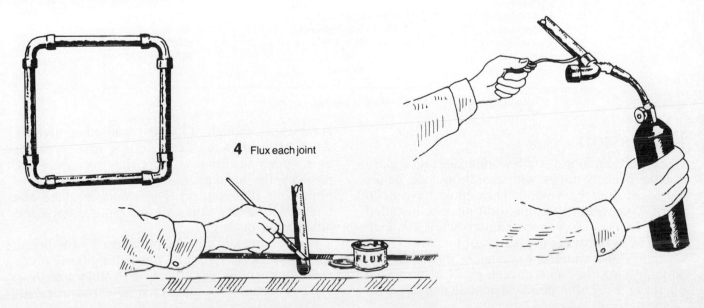

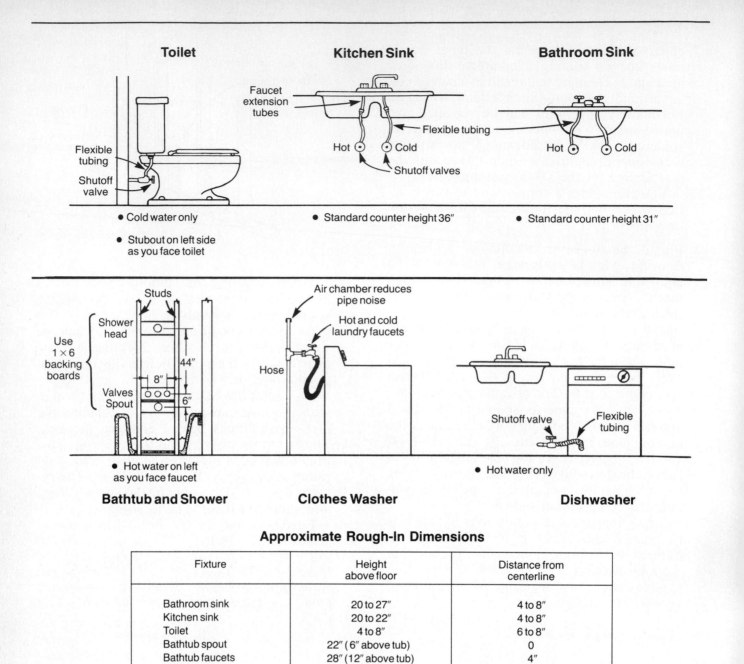

Toilet

Flexible tubing

Shutoff valve

- Cold water only
- Stubout on left side as you face toilet

Kitchen Sink

Faucet extension tubes

Flexible tubing

Hot   Cold

Shutoff valves

- Standard counter height 36″

Bathroom Sink

Hot   Cold

- Standard counter height 31″

Studs

Shower head

Use 1 × 6 backing boards

44″

8″

Valves Spout

6″

- Hot water on left as you face faucet

Bathtub and Shower

Air chamber reduces pipe noise

Hot and cold laundry faucets

Hose

Clothes Washer

Shutoff valve

Flexible tubing

- Hot water only

Dishwasher

Approximate Rough-In Dimensions

Fixture	Height above floor	Distance from centerline
Bathroom sink	20 to 27″	4 to 8″
Kitchen sink	20 to 22″	4 to 8″
Toilet	4 to 8″	6 to 8″
Bathtub spout	22″ (6″ above tub)	0
Bathtub faucets	28″ (12″ above tub)	4″
Shower head	72″ (44″ above faucets)	0
Clothes washer	38 to 42″	0 to 10″
Dishwasher	6″	15″
Hot water heater	48 to 60″	0 to 10″

Fig. 16-14   Water supply connections.

Digression

Deb's parents found some wonderful leaded glass double French doors with matching side panes hiding in their basement. They must have been there 30 years or more, just waiting to be installed in our house. They were painted white, so Deb spent weeks braving the fumes of paint remover, scraping, and sanding her fingers into oblivion, and applying six coats of high-gloss spar varnish on these beautiful doors. I installed them with great care, though it almost caused a divorce when I ripped ⅛ in of one door to make them fit in the opening. Finally, there they were, gracing possibly the most stunning laundry room/future library in the county. The doors became the highlight of every tour with guaranteed oohs and ahs from everyone.

Along comes Frank the plumber. By the time I got to the laundry room I was getting fairly proficient and could put 10 or 20 joints together, cleaned and fluxed, and then soldered in rapid-

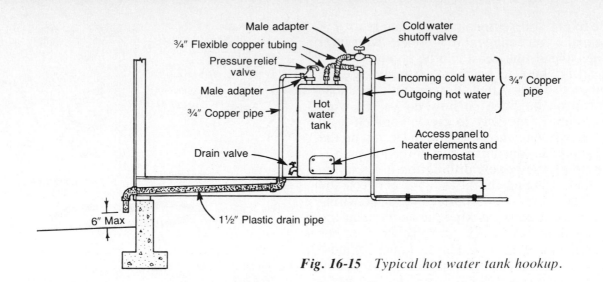

Fig. 16-15 *Typical hot water tank hookup.*

fire succession. My carefulness was wearing thin and I was into *speed*. I polished off the laundry room in record time, ran through my mental checklist, and went outside to the main valve. With the apprehension of a kid on the high dive for his first time, I slowly opened the valve.

Everything was quiet, then there was a slow whooshing sound as the water rushed through to pressurize the pipes. Then quiet again. I strained to hear any telltale noises of water spilling onto the floor—nothing. Good. I got up and walked toward the house.

BLAM!! A gun went off blasting a hole through the leaded glass doors! Glass tinkled and the shot hit the ground behind me. Instinctively I dove behind the house; my heart was pounding.

Scrambling along the side wall I grabbed a big monkey wrench and prepared for the worst. I couldn't see anything from the dining room window, so I snuck around to where I could see the laundry room. All was still, except there was water spraying across the room and onto the doors.

Hmmm . . . a light begins to dawn. I go back to the valve to turn off the water. The bullet hole in the glass is about waist high. I look in the grass about 20 ft from the house and there it is . . . a copper plumbing cap. First I feel a wave of comic relief, then a wave of dread—Deb will be home shortly.

Epilogue. It appears that I had tapped on that particular cap with a wrench because the pipe had not been adequately burred. Unfortunately I missed soldering that cap during the attempt to set a speed record. Since the water pressure was high (about 90 to 100 psi, before the pressure reducer was installed), the air in the end of the pipe was slowly compressed until, finally, the cap blew off.

I spent the afternoon mopping up the water, assessing the damage and making up a good story for Deb. I devised a dandy story and modified my T-shirt with some catsup to fit it. She was so relieved I was still alive that she didn't kill me outright. The truth eventually came out though, as it has a tendency to do. Now *I'm* included in every house tour, as the butt of the story, of course. Everyone leaves with visions of me peering through the windows, monkey wrench in hand, fearfully stalking my assassin as water fills the room. (When you're building, some days go better than others.)

Installing Plumbing Fixtures and Appliances

After the rough plumbing is installed and signed off by the inspector, continue straight away to wire the house. The fixtures and appliances are normally installed after the wallboard and finish floors are completed. There is a wealth of information available on fixture and appliance installation, both from the manufacturers of fixtures, appliances, shower valves, shut-off valve kits, and sink faucets, and from literature such as that cited at the end of the plumbing section in Chapter 7.

Wiring

You'll be starting to hit your stride now, a skilled student at acquiring new trades. Just as you're getting proficient at foundation work, carpentry, roofing, installing windows, or plumbing it's time to move on to the next step and become a beginner all over again. This process doesn't give you a chance to put yourself on autopilot and relax, but it does make you confident in yourself. No matter what lies ahead, the beast can be conquered.

Wiring is likely to be more intimidating than any subject so far . After all, you can't see those little electron buggers, so how do ya tell what they're doing? One way to cover yourself is to find an electrician who is willing to let you do the boring and time-consuming part of the job such as nailing up all the boxes, drilling the holes in the studs, and stringing the housewire between the boxes. The electrician's part consists of wiring the service entrance, making connections at the panel, checking your work, and placing his or her name on the panel sticker. This last part is almost guaranteed to make the inspection go smoothly and seems to be a popular way among do-it-yourselfers.

Knowing that the electrician-assisted option is available, briefly acquaint yourself with the subject here, testing the water, so to speak, and see if you can handle the current. As with plumbing, this is a subject that requires some study. Visit the electrical store, check the wiring displays there, look at construction sites, and pick up a book or two devoted specifically to wiring. This section and the electrical design section in Chapter 7 cover the basics. Other information sources are referenced in Chapter 7.

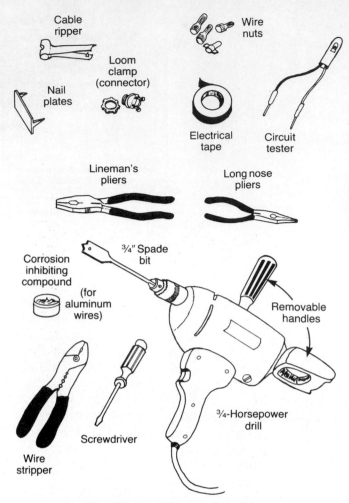

Fig. 16-16   *Electrical tool kit.*

STEP-BY-STEP PROCEDURE

This is a brief description to give the whole picture. We'll get into the details in the next three sections.

1. Make a layout of the electrical system, if you didn't do it in Chapter 7. (See Fig. 7-5.) Also reread the section on electrical design.

2. Get a permit and check your local codes and requirements, which may differ from the rules discussed here. The National Electric Code (NEC) is the code most used. Also call the power company and see what requirements they have regarding height of the meter and where they will connect the wires at the service entrance.

3. List and acquire the necessary materials and tools. The service entrance parts are often available as a kit (panel, main breaker, meter base, weatherhead, and conduit). There are only a few new tools required (shown in Fig. 16-16) and they're mercifully cheap.

4. Install the service entrance. The power com-

pany will provide and install the meter when they hook up the power.

Aside from the electrician-assisted option, the easiest way here is to copy the display at the electrical store. There are only three wires to deal with. Figure 7-8 shows the basic setup. Figure 16-17 shows the panel wiring in more detail.

The means of connecting the wires and arrangement of the bus bars varies by brand of panel. Usually the insulation is stripped off the end of the wire with a sharp knife. Be careful not to nick the copper. Then insert it in a hole where a machine screw holds it in place. With aluminum wires, dip the end in goop used to inhibit corrosion (like Penetrox) before connecting it.

The housewires won't be connected to the circuit breakers until all of the circuits have been rough-wired. (The housewires run from box to box and to the panel.) Connect one circuit breaker now for practice. (There's no power to the panel yet, right?) If you can con-

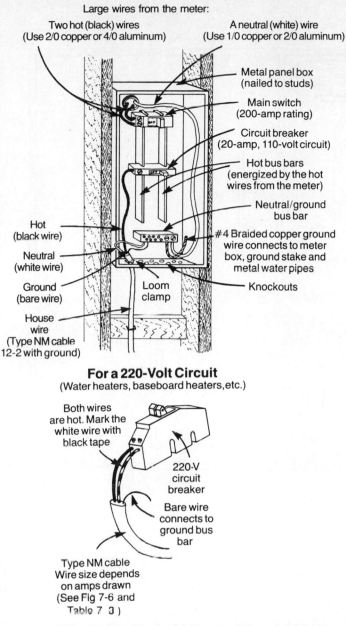

Large wires from the meter:

Two hot (black) wires
(Use 2/0 copper or 4/0 aluminum)

A neutral (white) wire
(Use 1/0 copper or 2/0 aluminum)

Metal panel box
(nailed to studs)

Main switch
(200-amp rating)

Circuit breaker
(20-amp, 110-volt circuit)

Hot bus bars
(energized by the hot
wires from the meter)

Neutral/ground
bus bar

#4 Braided copper ground
wire connects to meter
box, ground stake, and
metal water pipes

Knockouts

Hot
(black wire)

Neutral
(white wire)

Ground
(bare wire)

Loom
clamp

House
wire
(Type NM cable
12-2 with ground)

For a 220-Volt Circuit
(Water heaters, baseboard heaters, etc.)

Both wires
are hot. Mark the
white wire with
black tape

220-V
circuit
breaker

Bare wire
connects to
ground bus
bar

Type NM cable
Wire size depends
on amps drawn
(See Fig 7-6 and
Table 7-3.)

Fig. 16-17   Typical 200-amp house panel connections.

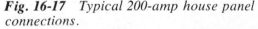

nect one circuit breaker, you can connect twenty. When you look at a completed panel, it looks like a bowl of spaghetti in there and it's pretty intimidating. Each single connection is fairly simple, however. For now, we'll just lead the housewires to the panel and in step 7 we'll connect them to the breakers.

5. Rough-wire one circuit. Basically this amounts to nailing the boxes in place, drilling holes in the studs, and pulling the wires through the holes. I'll expand on this in the next section.

6. Rough-wire all of the circuits. This is a repeat of step 5 with minor variations. If the electric

furnace circuit requires a conduit, plan where it will connect to the panel.

7. Connect the wires to the circuit breakers at the panel. (See Fig. 16-22.)

8. Have the wiring inspected.

9. After all the wallboard is installed, connect the outlets, lights, and switches.

Rough-Wire One Circuit (Step 5)

From Chapter 7 we have an electrical layout and a list of the circuits to be installed. Select one general-purpose circuit to wire first. For the sample house, I'll choose circuit 12 (Fig. 16-18), which serves the bedroom, hallway, and stairway. (Though the choice is entirely personal preference, I'd put the bathroom lights and the living room lights and outlets on one circuit, and the dining, kitchen, and laundry lights on another circuit.) Circuit 12 consists of seven outlets, four switches, and four lights. Each light is controlled by a switch, except the stairway light which is controlled by two three-way switches, one upstairs and one downstairs.

Now, let's go through the installation of the circuit step by step. (We're breaking down step 5 of the general procedure into smaller steps.)

A. Using the layout, locate all of the boxes for switches, outlets, and lights and nail them in place. Use 18.0 cu in plastic boxes since they're faster, cheaper, and safer than metal boxes (safer because they don't conduct electricity). Please study Fig. 16-19.

B. Plan how to route the wires to connect all of the boxes. Then drill ¾-in holes through the studs and joists as necessary between the panel and the first box. Minimize structural damage by drilling in the center of the face of joists as shown in Fig. 6-15.

   In circuit 12, the hallway outlet would be a logical first box. There are at least ten acceptable ways to route the wire between the outlet and the panel. Just avoid damaging the structure and make sure the wire is protected from damage. *Don't run the wire diagonally* from point to point, however. Remain either parallel or perpendicular to the joists and studs, making neat right angle corners with the wire.

C. Pull a wire from the first box to the panel. The wire is connected to the panel with a loom clamp (also called a connector). See Fig. 16-17.

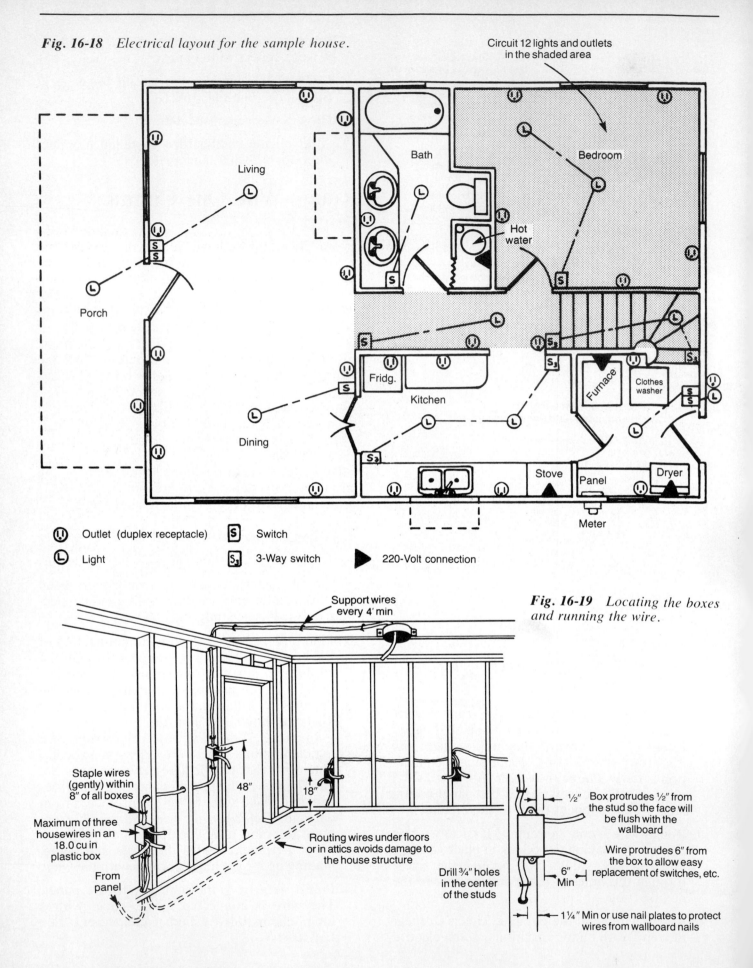

Fig. 16-18 *Electrical layout for the sample house.*

Circuit 12 lights and outlets
in the shaded area

Living

Porch

Bath

Bedroom

Hot
water

Fridg.

Kitchen

Dining

Furnace

Clothes
washer

Stove

Panel

Dryer

Meter

(U)   Outlet (duplex receptacle)   S   Switch

(L)   Light   S₃   3-Way switch   ▶   220-Volt connection

Fig. 16-19 *Locating the boxes and running the wire.*

Support wires
every 4' min

Staple wires
(gently) within
8" of all boxes

Maximum of three
housewires in an
18.0 cu in
plastic box

From
panel

48"

18"

Routing wires under floors
or in attics avoids damage to
the house structure

Drill ¾" holes
in the center
of the studs

½"   Box protrudes ½" from
the stud so the face will
be flush with the
wallboard

Wire protrudes 6" from
the box to allow easy
replacement of switches, etc.

6"
Min

1¼" Min or use nail plates to protect
wires from wallboard nails

Remove the knockout from the panel with a hammer and a screwdriver, breaking it off with pliers. The loom clamp protects the wire and clamps it. Leave at least 2 ft of wire hanging out of the panel and label it (circuit 12) with masking tape. Leave at least 6 in of wire hanging out of the first box, then cut the wire with linesman pliers. (Yes, 6 in is correct, see NEC Sec. 300-14.) All connections will be made later. Use wire staples to hold the wire loosely and keep it from sagging. *Don't dent the wire with the staple!* (NEC Sec. 370-7c, 336-5.)

D. Drill ¾-in holes and run the wire to all boxes on the circuit. To prevent wallboard nails from contacting the wires, the edge of the hole must be 1¼ in from the edge of the stud or protected with a nail plate (NEC Sec. 300-4a, 336-6, 336-7, 336-8, 333-12).

A potential problem crops up here. My large drill wouldn't fit between the studs. I solved this by removing the rear handle and drilling at an angle. Electricians use right-angle drills called hole hogs that sell for about $160.

All connections must take place in a box (NEC Sec. 300-15). No more than three housewires (that is, three three-wire cables) can come into each 18.0 cu in plastic box (NEC Sec. 370-6b elaborates). Be careful not to remove more knockouts from a box than necessary (NEC Sec. 370-8). Figure 16-20 shows a schematic of circuit 12.

Don't get heady, but you've just rough-wired the first circuit. There's still that lingering question: how do I connect the wires to the outlets, circuits, and lights? For practice, I suggest you go ahead and connect them for this circuit. Don't bother screwing the devices to the boxes because they'll have to be removed before the wall board is installed. This, again, is my practice of going through the process on a small scale to avoid perpetuating mistakes throughout the house. For the other circuits, we'll just rough-wire them, get all of the rough-wiring signed off by the inspector as "ready to cover" with wallboard, then make the connections after the wallboarding is completed.

Basic Wiring Skills

Figure 16-21 shows how to remove the plastic covering from the housewire cable and how to strip the insulation from each individual wire.

This is done to every wire at each box. The inspector will be checking for 6 in of free wire, that the individual wires are cleanly stripped (no nicks in the copper), and that the wires are stapled (gently) within 8 in of the box. He or she will also check that each wire is protected by the proper size of circuit breaker. (See Table 7-3.) (NEC Sec. 240-3, 310-15, Table 310-16.) Make sure that the stripped ends of black wires are not touching any bare or white wires. That'll cause a short if the circuit is inadvertently energized.

Figure 16-22 shows some common ways of making connections. All wire nut connections should be tested by yanking each wire to be sure

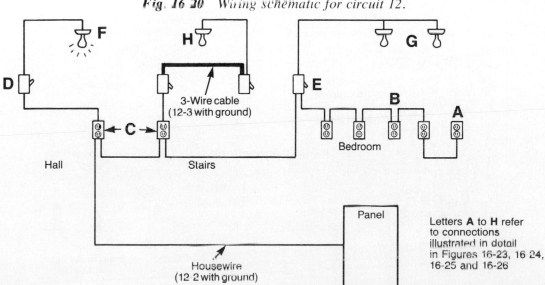

Fig. 16-20 *Wiring schematic for circuit 12.*

3-Wire cable
(12-3 with ground)

Bedroom

Hall

Stairs

Panel

Housewire
(12-2 with ground)

Letters **A** to **H** refer to connections illustrated in detail in Figures 16-23, 16-24, 16-25 and 16-26

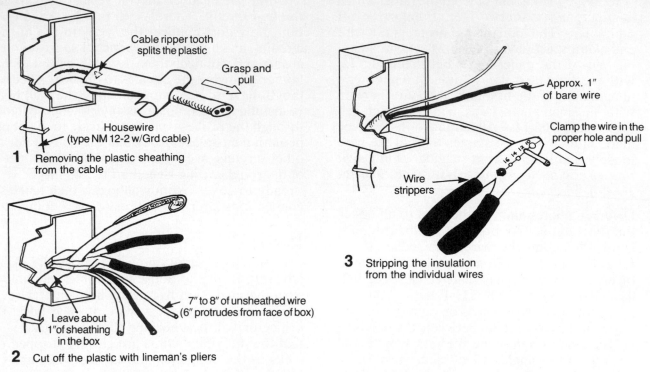

1 Removing the plastic sheathing from the cable

Cable ripper tooth splits the plastic

Grasp and pull

Housewire (type NM 12-2 w/Grd cable)

Approx. 1″ of bare wire

Clamp the wire in the proper hole and pull

Wire strippers

3 Stripping the insulation from the individual wires

2 Cut off the plastic with lineman's pliers

Leave about 1″ of sheathing in the box

7″ to 8″ of unsheathed wire (6″ protrudes from face of box)

Fig. 16-21 *Removing wire sheathing and insulation.*

Connections to Devices

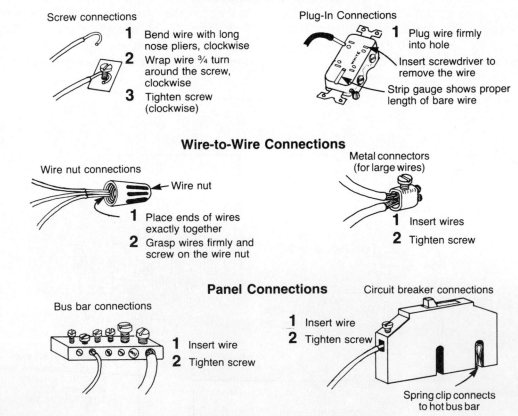

Screw connections

1 Bend wire with long nose pliers, clockwise
2 Wrap wire ¾ turn around the screw, clockwise
3 Tighten screw (clockwise)

Plug-In Connections

1 Plug wire firmly into hole

Insert screwdriver to remove the wire

Strip gauge shows proper length of bare wire

Wire-to-Wire Connections

Wire nut connections

Wire nut

1 Place ends of wires exactly together
2 Grasp wires firmly and screw on the wire nut

Metal connectors (for large wires)

1 Insert wires
2 Tighten screw

Panel Connections

Bus bar connections

1 Insert wire
2 Tighten screw

Circuit breaker connections

1 Insert wire
2 Tighten screw

Spring clip connects to hot bus bar

Fig. 16-22 *Making electrical connections.*

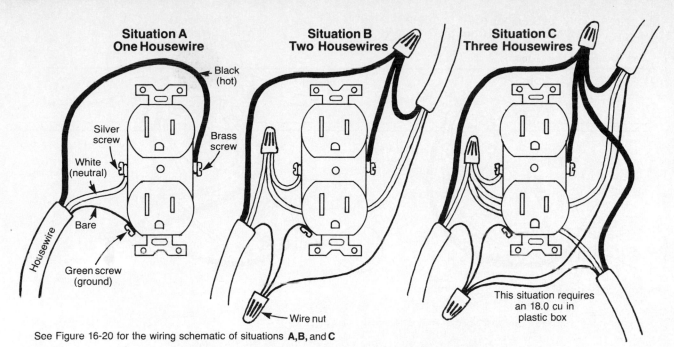

**Situation A
One Housewire**

Black (hot)

Silver screw

Brass screw

White (neutral)

Bare

Housewire

Green screw (ground)

**Situation B
Two Housewires**

Wire nut

**Situation C
Three Housewires**

This situation requires an 18.0 cu in plastic box

See Figure 16-20 for the wiring schematic of situations **A, B,** and **C**

Fig. 16-23 *Outlet connections (in plastic boxes).*

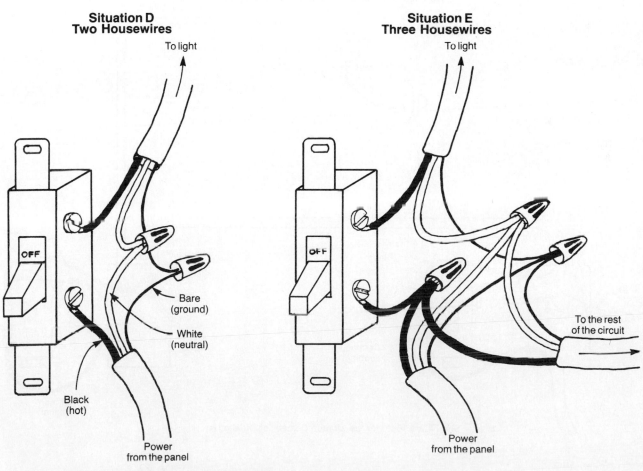

**Situation D
Two Housewires**

To light

OFF

Bare (ground)

White (neutral)

Black (hot)

Power from the panel

**Situation E
Three Housewires**

To light

OFF

To the rest of the circuit

Power from the panel

See Figure 16-20 for the wiring schematic of situations **D** and **E**

Fig. 16-24 *Switch connections (in plastic boxes).*

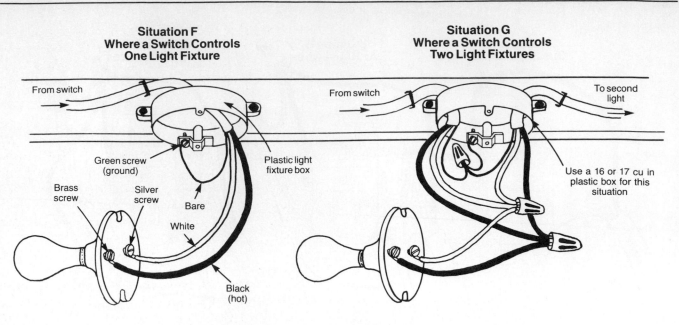

Situation F
Where a Switch Controls
One Light Fixture

From switch

Green screw
(ground)

Plastic light
fixture box

Brass
screw

Silver
screw

Bare

White

Black
(hot)

Situation G
Where a Switch Controls
Two Light Fixtures

From switch

To second
light

Use a 16 or 17 cu in
plastic box for this
situation

See Figure 16-20 for the wiring schematic of situations **F** and **G**

Fig. 16-25 *Light fixture connections (in plastic boxes).*

Situation H
One Light Controlled by Two 3-Way Switches

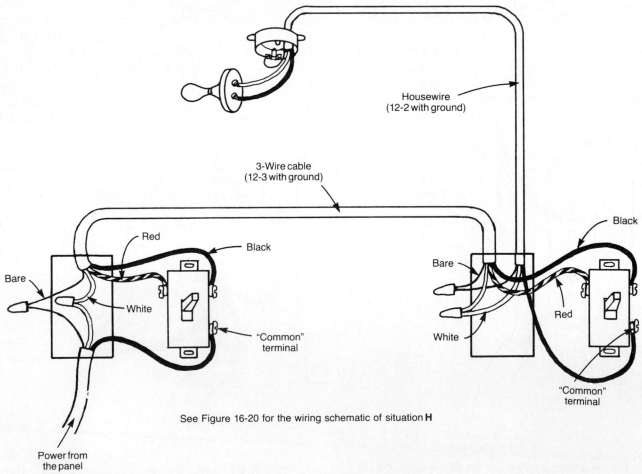

Housewire
(12-2 with ground)

3-Wire cable
(12-3 with ground)

Red

Black

Bare

Black

Bare

White

White

Red

"Common"
terminal

"Common"
terminal

Power from
the panel

See Figure 16-20 for the wiring schematic of situation **H**

Fig. 16-26 *Light fixture controlled by two switches (using 3-way switches in plastic boxes).*

they're secure. Typical wiring situations are shown in Figs. 16-23, 16-24, 16-25, and 16-26.

Three-way switches can be confusing until you devise a good way of dealing with them. There are several methods to hook them up, but here's the way I prefer: Run the housewire (12-2 with ground) carrying the power (from the panel) to one three-way switch, then run a three-wire cable (12-3 with ground) to the other switch, then run a housewire to the light(s). At the first three-way switch, connect the black wire coming from the panel to the terminal that is marked "common." The other black and red wires are connected to the two remaining terminals, it doesn't matter which to which. At the second three-way switch, the black wire going to the light is connected to the "common" terminal. Again, the other black and red wires go indiscriminately to the two remaining terminals. At the switches, the white wires are simply connected together and the bare wires are connected together. The hookup at the lights is the same as with regular switches. (See Fig. 16-26.)

Check with your inspector to see what connections he wants to inspect. In my case he wanted all the circuit breakers and service entrance wires connected, but he almost made me remove my practice connections of the outlets and lights. With his OK, the power company hooked up the power to the meter and then I could switch the power on for each circuit with the circuit breaker when I was ready.

Special Circuits

Two **small appliance circuits** serving the kitchen and dining room are required. No lights are allowed on these circuits. The kitchen outlets must be supplied by two different circuits, each roughly sharing the load (NEC Sec. 220-3b, 210-25b).

The **clothes washer circuit** may have only one outlet under most conditions (NEC Sec. 220-3c). In some localities, the dishwasher circuit may serve only the dishwasher.

Bathroom, garage, and outdoor outlets must be protected by either a Ground Fault Interrupting (GFI) circuit breaker or a GFI device at each outlet. My inspector wouldn't allow lights on the GFI circuit.

If the **water heater** is electric, the circuit usually requires "10-2 with ground" wire which is run to a metal octagonal box (4 in × 4 in × 1½ in) close to the tank. Special heat resistant wires (Type THHN 90°C, No. 10) run in an armored cable from the box to the tank. *Don't connect these to the water heater until it's full of water,* or you'll burn out the elements if the breaker is inadvertently switched on. Mark any hot white wires with black electrical tape (as was shown for the 220-volt circuit breaker in Fig. 16-17).

The **clothes dryer circuit** uses a "10-3 with ground" wire which connects to a dryer outlet. (See Table 7-3.) Check the plug on your dryer and buy an outlet to match. At the panel, connect the red and black wires to the 220-volt breaker and the white and bare wires to the neutral bus bar.

Baseboard and wall heater circuits are wired per the manufacturer's instructions. The thermostat can be placed either on the heater or on a wall. Be sure the thermostat isn't directly over the heater or near a cold draft or window.

With **range and oven circuits** the wire is large, requiring 1-in holes in the studs and J-nails, instead of wire staples, to secure it. Some range/oven units have plug-in cords, others are hardwired into the box. Check the instructions or the display at the electrical store.

The **electric furnace circuit** requires such large wires that conduit may be necessary. In the case of the sample house, two No. 2 (type T) copper wires and a No. 6 copper ground wire will be necessary (for a furnace requiring up to 95 amps as indicated in Fig. 7-6). Wires this large don't come in Type NM cables, so metal conduit (type EMT is the easiest to install) will be run from the panel down under the floor and up to the furnace. Metal conduit is required where the conduit is hidden in the walls, since plastic conduit could be drilled into accidently.

For **underground circuits,** use Type UF wire. It can be buried directly, 24 in deep. If it's protected by metal conduit, bury it 6 in deep and 18 in deep if it's in plastic conduit.

Now if you read this electrical business for the first time, you might be feeling kind of like you tried to take a drink from a fire hydrant. The sought-after material is there all right, but it comes a trifle thick and fast. By the third reading of this section, combined with some research at the electrical store, you'll start to be able to put all this knowledge to work for you.

The Final Circuit: The Inspector and You

You can tell when the inspector drives in the driveway whether you two are on the same wavelength or not. I'd given him some abbreviated instructions on how to find my place, and he pulled into my neighbor's long driveway instead of mine. On the way back out he stalled his State Motor Pool Gremlin in a huge puddle where the water was about 12 in deep. He had to get out to wade up to the engine, and when he opened the door, water poured in the car. (My neighbor told me all this later, amid gleeful hysterics.) Finally, he got the distributor or whatamakit dried out and the fearless Gremlin was running again. By the time he arrived at my door . . . barefoot, wet pantlegs rolled up, soggy clip-board in hand, I could tell right away that things weren't going to go real well.

He let me off with about three pages of corrections and stormed out of the driveway muttering something about being 2 hours late. I sat down on a sawhorse with that terrible sinking feeling. He had just gone through the house rattling off code clauses and scribbling correction notes. Of the items I could read, only a few made any sense. And some of those I couldn't believe. How could I have mounted the panel upside down? Slowly I spiraled to the bottom of a new emotional low of frustration and dashed hope. What the hell, I'm a failure. Who cares? I don't need electricity anyway. I'll use kerosene lamps and a propane stove. I'll heat water for baths in buckets like we did in Montana. I'll wash the clothes in a tub with a scrub board.

Eventually I summoned up my Last Ditch Philosophy that I reserve for such occasions: There's got to be a good side to this disaster, it's just a matter of knowing what it is. Twenty minutes later I found the answer. He hadn't told me to forget it. I just had to fix some things. He never said he wouldn't come back, though I'm glad I didn't ask him right then.

I set about reading the National Electric Code verbatum and everything else I could lay my hands on. I picked the brains of anyone who knew anything about wiring and learned everything I could about electricity. I figured out a lot of the problems and fixed them. A couple of weeks later I called the Electrical Office and asked for another inspection. (It turns out that this is a normal procedure and many places need a second inspection.) I started giving the inspector directions on how to get here but he said he remembered the place. Ah yes, I suppose so.

The second inspection went better. He was wearing some nice dry shoes and I was poised with a pad of paper and a pen. This time there were only about twenty corrections and I understood most of them.

The next week I called for a third inspection, which cost $20 (the fee for excess inspections) and left only three problems. Things were looking up.

On the fourth inspection, which cost another $20, Oscar finally hung a little tag with his signature on the meter base. That meant the power company could hook up the power. Hot damn! No kerosene lamps or scrub boards for me; this house is wired. After he left, I plugged in the radio, did a little victory dance, and graciously kissed my temporary power panel goodbye.

Heating

There's a nice reward for you at the end of this job. One day you'll walk into this fancy barn you've been working on and, suddenly, that day it'll feel more like a house than a barn. It'll be warm inside. Take off your wool hat and jacket, hang them on a nail and sit down to read the newspaper. Just like home . . . only it feels better than any place you've ever lived in. This house is your creation and almost a part of you by now; a visual, touchable, functional extension of your imagination and hard work.

Installing Metal Fireplaces and Stoves

Installing a metal fireplace or stove is pretty simple and is the quickest and cheapest way to get some reliable, inexpensive heat into the house. Most of the necessary information for installation is shown in Fig. 16-27.

The main thing to remember is that the fireplace and chimney can get extremely hot, particularly in the case of a chimney fire. Leave lots of clearance between any part of the chimney or fireplace and combustible materials. Use heat shields of at least 22-gauge metal or 1/4-in asbestos millboard in those places where you can't achieve the necessary clearance. The heat shields are mounted an inch from the wall and have space at the top and bottom to allow air circulation.

If the heat shield is made of asbestos, take care not to breathe any asbestos dust when it is cut. The way I understand it, inhaled asbestos dust will take up permanent residence in your lungs, eventually causing an assortment of medical problems. So *don't* cut it with a power saw! Set it on some 2 × 4's supported by sawhorses. Draw a line on it and use the edge of a wood file to file a 1/16-in to 1/8-in groove along the line. Carefully wipe up the dust. Then hang the sheet over the edge and snap it off. To mount it to the wall, drill holes in it with a sharp bit and fasten it to 1-in

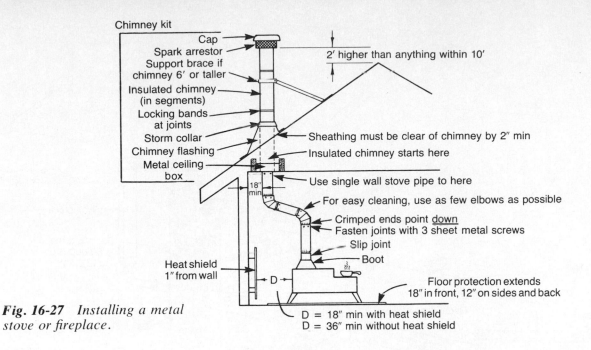

Fig. 16-27 *Installing a metal stove or fireplace.*

Diagram labels:
- Chimney kit
- Cap
- Spark arrestor
- Support brace if chimney 6' or taller
- Insulated chimney (in segments)
- Locking bands at joints
- Storm collar
- Chimney flashing
- Metal ceiling box
- Heat shield 1" from wall
- 2' higher than anything within 10'
- Sheathing must be clear of chimney by 2" min
- Insulated chimney starts here
- Use single wall stove pipe to here
- 18" min
- For easy cleaning, use as few elbows as possible
- Crimped ends point <u>down</u>
- Fasten joints with 3 sheet metal screws
- Slip joint
- Boot
- Floor protection extends 18" in front, 12" on sides and back
- D
- D = 18" min with heat shield
- D = 36" min without heat shield

noncombustible spacers with screws or nails. Handle it carefully; it's brittle and it's expensive.

STEP-BY-STEP PROCEDURE

Have you noticed how these procedures are getting more concise? With the experience you have now, I'm guessing you'll appreciate shorter explanations.

1. Read the manufacturer's instructions.

2. Determine the fireplace location, maintaining the clearances shown in Fig. 16-27.

3. If necessary, frame for the chimney. Maintain 2-in clearance between the insulated chimney and wood or other combustibles.

4. Install the roof flashing, if it wasn't installed when the house was roofed. (See Fig. 14-7.)

5. Place the fire protective floor covering; then position the fireplace.

6. Mount the ceiling box and assemble the stove pipe from fireplace to the ceiling. Use the slip joint to adjust for the ceiling height. (Pipe comes in 24-in sections, which may not work out evenly—hence the slip joint.) Point the crimped ends *down*. Otherwise, creosote drips down and out the joint, running down the outside of the chimney. Put three sheet metal screws in every joint.

7. Assemble the insulated chimney. Locking bands hold the sections together. Use caulking at the storm collar. The top of the chim-

ney should have a cap and spark arrestor and be 2 ft higher than anything within 10 ft.

8. Avoid hot fires (for example, paper and cardboard), especially at first. Fireplaces and stoves need a chance to "cure" I'm told.

Installing an Electric Forced-Air Heating System

The heating system design for the sample house is shown in Fig. 16-28. Heating design principles were discussed in the heating section of Chapter 7 and several references for more information are listed there.

STEP-BY-STEP PROCEDURE

1. Make a drawing of the system and get a permit for its installation. Check your local codes. Uniform Mechanical Code (UMC) governs most heating installations.

2. List and acquire the necessary materials and tools. Check both local sheet metal shops and mail order catalogs for prices.

3. If you've framed a place in the floor for the furnace and supply plenum, just install the supply plenum and set the furnace in place. Follow the manufacturer's instructions and maintain designated clearances from combustibles. (See the UL label for clearances.) Typical clearance for an electric furnace is 2 in on three sides and 30-in access to the front. Plenum clearance is typically 1 or 2 in. For

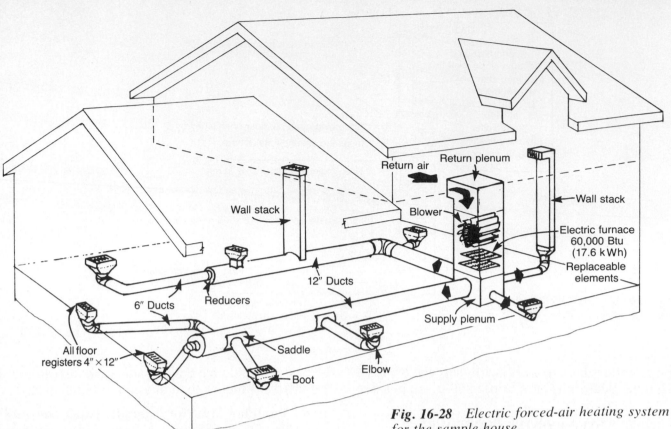

Fig. 16-28 *Electric forced-air heating system for the sample house.*

fuel-burning furnaces, clearances must be much greater.

4. Install the return plenum on the furnace. If the furnace enclosure must be cut, use a saber saw with a metal cutting blade. Yeah, it's noisy. The ductwork is thinner and can be cut with tin snips.

5. Determine the locations of the floor and wall registers. Holes for the registers will be cut in the subfloor (between the joists) about 5 in from the wall. Closer than 5 in to the wall may cause interference between the duct and the foundation.

6. Install one duct from the supply plenum to the first register.
   a. Cut a hole in the subfloor for the register. A drill and a saber saw are needed here.
   b. Nail the boot in place. I use roofing nails nailed through the boot into the edge of the plywood. (Hold the nail with pliers, not with fingers, until it punctures the sheet metal.)
   c. Cut a hole in the supply plenum for the first duct. Leave room for the other

ducts. (See Fig. 16-29.) Be careful not to slash your fingers; use gloves and pliers.
   d. Install a starting collar at the supply plenum. Use gimlet point screws and a small drill with a magnetic chuck when fastening sheet metal together. It's much faster than drilling pilot holes with an 1/8 in bit and then slowly putting in each screw with a screwdriver.
   e. Assemble the ducts and install them from the plenum to the first boot. Cut the ducts to length before snapping them together. The ducts are nearly impossible to take apart after they are snapped. The crimped ends point in the direction of the air flow. Just put one screw in each joint until the entire duct is hung and straightened. (Electric fence wire works well for hanging the ducts.) Then put three screws in each joint. Study Fig. 16-30.

7. Install all ducts using the method explained in step 6. There are lots of variations of this method that are used for reducers, saddles, wall stacks, and so on, but the primary skill involved is fitting the sheet metal together neatly and screwing it in place.

Cutting a Hole in a Plenum

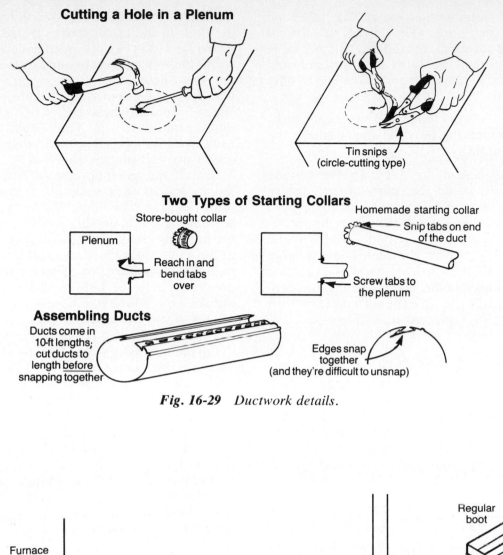

Tin snips
(circle-cutting type)

Two Types of Starting Collars

Store-bought collar

Homemade starting collar

Plenum

Reach in and
bend tabs
over

Snip tabs on end
of the duct

Screw tabs to
the plenum

Assembling Ducts

Ducts come in
10-ft lengths;
cut ducts to
length before
snapping together

Edges snap
together
(and they're difficult to unsnap)

Fig. 16-29   Ductwork details.

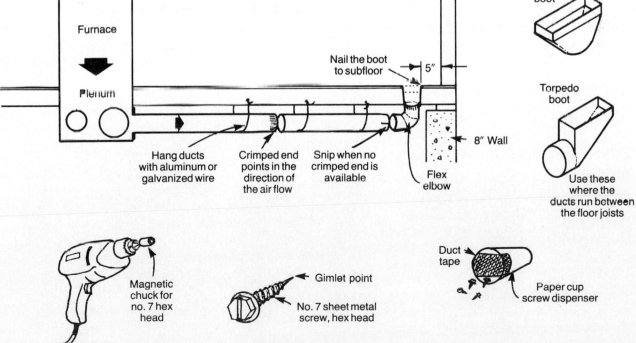

Furnace

Plenum

Nail the boot
to subfloor

5″

Regular
boot

Torpedo
boot

Hang ducts
with aluminum or
galvanized wire

Crimped end
points in the
direction of
the air flow

Snip when no
crimped end is
available

8″ Wall

Flex
elbow

Use these
where the
ducts run between
the floor joists

Magnetic
chuck for
no. 7 hex
head

Gimlet point

No. 7 sheet metal
screw, hex head

Duct
tape

Paper cup
screw dispenser

Fig. 16-30   Assembling ductwork.

8. Tape and insulate the ducts. The duct tape seals all the joints. They must be airtight (UMC, Sec. 1002c). Insulation (1 in or 2 in) is wrapped around the ducts and taped in place. It's a clumsy, itchy job and takes lots of duct tape.

9. Place the registers. This will keep your tools from disappearing into the ducts during the next few months.

10. Install the return air duct. In the sample house, this could be part of the return plenum. If the upstairs were routinely heated, you would do well to run the return air duct to draw air from the upstairs or from the stairwell. Either way, the duct should be large (about 14 in × 20 in) to avoid starving the furnace blower of air. Cover the intake opening with a grill. Be sure the furnace (or return air grill) has a replaceable air filter.

Let's Hear it for Snow

I can't remember making any giant screw-ups on the heating system. The furnace blower is a little too close to the dining room, making its noise a notable nuisance. Luckily, I didn't put the return air duct or a fresh air inlet anywhere that they might pick up objectionable odors. All in all, the installation went smoothly and the system works well.

It was a Monday when I picked up most of the ductwork. Tuesday found me banging and clanging and assembling the ducts. When I started out for lunch, I was confronted with 6 in of snow at the door of the crawl space. It had quietly piled up while I was noisily occupied with the ductwork. Quite a surprise in our climate, and a pleasant one. Snow is one of those pure miracles that was devised by someone up there who likes me. When I stuck my head out of that clang chamber into the immensely quiet and beautiful landscape, I felt like I'd emerged into paradise. If there ever was a savored silence, this was it.

One of the less appreciated features of snow storms is that they scramble everyone's habits, priorities, and schedules. All the competent folks are suddenly late, just like I always am. Suddenly the pressing goals of my life become unattainable as the snow clogs the works and holds me captive until the storm blows by. There's not much left to do except help the kids build a snowman or get out the skis. I love it.

This storm seemed particularly significant in that the snow stayed on for days, poignantly verifying the necessity of the heating system. I skied over to the house each day, banged on the ductwork for a few hours, then emerged into the pure silence of trees and hills dressed in white. It bowled me over. Changed my whole outlook. Rearranged my senses. I really like snow.

Finally the ductwork was finished and it came time to turn on the furnace. You know, I never should have done it. In an uninsulated house like that the heat all escaped, melted the snow, and we haven't had a decent snowstorm since. I should have known better.

CHAPTER 17

Insulation and Wallboard

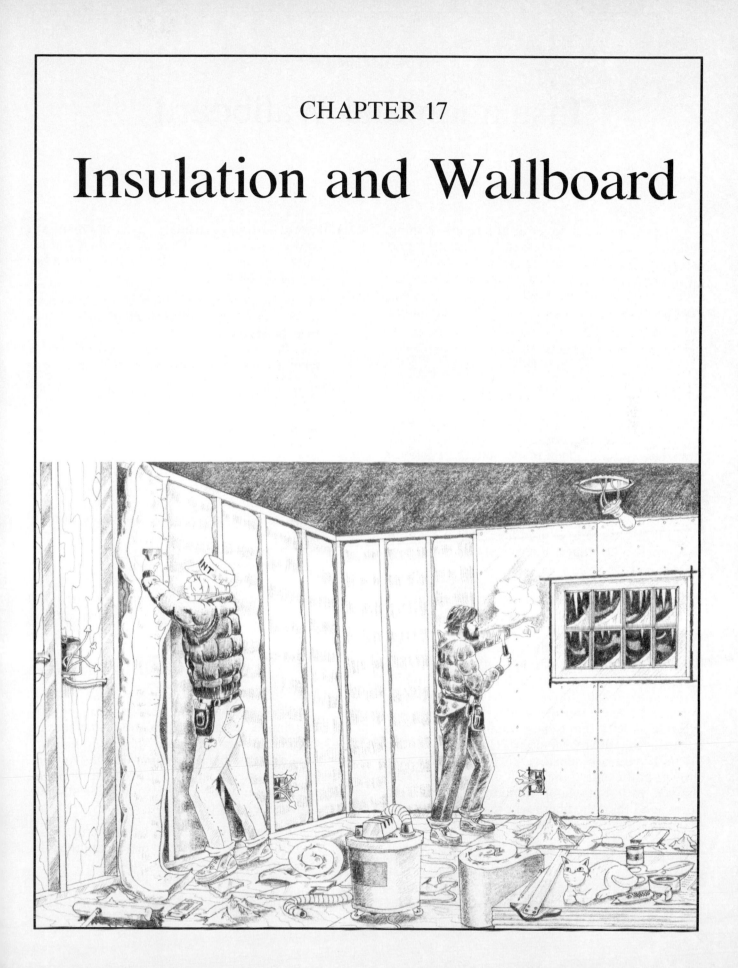

Insulation and Wallboard

In this chapter we'll cover up the rough framing, plumbing, and wiring and do the groundwork necessary to finish the interior of the house. Finishing a house, as I mentioned in Chapter 2, can be a simple task or an involved, expensive process. That's where we'll have to part company and go our separate ways.

Insulation

Insulating is an exercise in devising ways to keep the heat inside the house. Though major decisions have already been made, like the thickness of the walls, the roof design, type of windows, and portions of the house that must be heated regularly, a careful insulating job can make a major difference in your heating bills.

We'll discuss foil-covered fiberglass blanket insulation, since that's the most common and easiest method.

Fiberglass does have some drawbacks. It's real itchy stuff and if you insulate on a sunny day, pretty soon the air will be sparkly with little fibers floating by. Unless you've been forewarned to dress right and wear a dust mask, you'll have a tickle cough and be itching at your hands soon thereafter. I found that a hooded sweatshirt, gloves, dust mask, and occasionally goggles were necessary to seal out the stuff.

Some people don't have to trouble with this because the fiberglass doesn't seem to bother them. If you are sensitive to fiberglass, a cool shower at the end of the day is a good idea. Don't shower with hot water though; it opens your pores, and then the fibers will really get to you. I avoid working myself into a sweat for the same reason.

So much for the bad news. The good news is that installing insulation is a cinch and goes fast. Just whack off a piece the right length and staple it in place. There's no necessary sequence to the job, but there are a few things worth knowing.

- Before starting to insulate, take photos of all the walls. Five years from now, when you're trying to locate a pipe, wire or stud, you'll be very glad you did this.

- All the small jobs on which you've procrastinated, like wiring for the doorbell and the range hood over the stove, should be done now. Pretty soon it will be very difficult to make any wiring or plumbing additions.

- Equip yourself with the necessary tools as shown in Fig. 17-1.

- Cut fiberglass blankets to length with tin snips or a utility knife.

- The blankets have stapling flanges that are stapled to the stud or joist in such a manner that the flanges aren't a hinderance when the wallboard is nailed up. The foil vapor barrier always faces the inside of the house. It reflects radiant heat back into the house and prevents moisture from condensing inside the walls. (See Fig. 17-2.) If you use kraft-paper-lined insulation, or live in a cold climate, staple plastic sheeting over the insulation on the inside of the walls as a vapor barrier.

- Use 3/8-in staples. For holding up ceiling insulation, 1/4-in staples aren't always adequate.

- Under floors, a small problem arises. Since the foil should face the warm space, the stapling flange on regular insulation is inaccessible for stapling. There's a special floor insulation made with a kraft paper backing for this situation, or you can hold the regular blankets in place with wires or chicken fencing stapled to the underside of the floor joists.

- The standard widths of insulation blankets are 16 in and 24 in. To cut a strip to another dimension, roll up the blanket and saw it with a handsaw as shown in Fig. 17-2.

- Leave adequate space for ventilation to the attic as in Fig. 17-3.

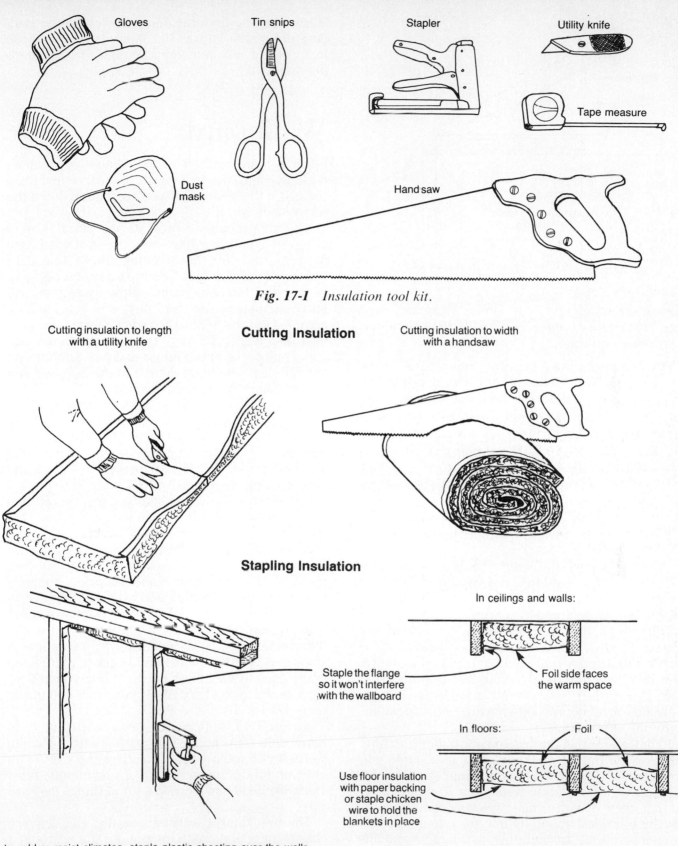

Gloves

Tin snips

Stapler

Utility knife

Tape measure

Dust mask

Hand saw

Fig. 17-1 *Insulation tool kit.*

Cutting Insulation

Cutting insulation to length
with a utility knife

Cutting insulation to width
with a handsaw

Stapling Insulation

Staple the flange
so it won't interfere
with the wallboard

In ceilings and walls:

Foil side faces
the warm space

In floors:

Foil

Use floor insulation
with paper backing
or staple chicken
wire to hold the
blankets in place

In cold or moist climates, staple plastic sheeting over the walls, ceilings, and under the floor as a continuous vapor barrier. In all cases, the dirt in the crawl space must be covered with plastic sheeting.

Fig. 17-2 *Installing fiberglass blanket insulation.*

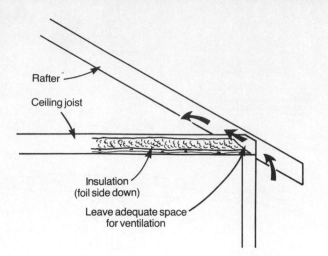

Fig. 17-3 *Allow ventilation to the attic.*

- Be thorough and neat. Cover all areas where heat could escape. Stuff insulation behind the switch boxes.

- Keep insulation 3 in away from recessed light fixtures (NEC Sec. 410-66). Regular light fixture boxes can touch the insulation since the heat from the light bulb is inside the room, not in the ceiling.

- Where wires or pipes go into the crawl space, the attic, or upstairs, stuff the holes and spaces with insulation. This will help slow the spread of flames in the event of a fire (NEC Sec. 300-16). Also stuff the gaps around window and door frames with insulation.

Normally only the walls, ceilings, and floors that adjoin the outdoors are insulated, but you may wish to insulate some interior walls as well. A room or two served by a fireplace or baseboard heater might be insulated from the rest of the house to allow those rooms to be heated separately. Bathroom walls can be insulated to muffle various tinkling and mouse-on-a-motorcycle noises. And if you've got an Uncle Barney around, insulate for rip-snorting bull-moose-in-rutting-season noises.

That's about all there is to say about insulating. It took me only 2 weeks in spite of a staple gun which developed a debilitating ailment, causing it to jam with infuriating frequency. I got adept at unjamming it with a nail, spewing thousands of wasted staples about. Other than that, the job was uneventful. Insulation makes a big visual difference. And talk about quiet. All of a sudden the building noises get absorbed and it doesn't sound like a herd of elephants when you walk across the floor. In addition, the fireplace and the furnace

started having an immediate impact on the temperature of the house. Insulating is one of those tasks out of which a small amount of work pays big dividends.

Wallboard

I call it wallboard. You may know it as sheet rock, gypsum, drywall, plasterboard, or just plain rock. It's the signal that you're coming down the home stretch.

You've probably heard stories about how a group of young gorillas show up on a truck one morning and, by that evening, the whole darn house is wallboarded. The next day, a couple of guys with "bazooka guns" show up to tape and putty the place. The third day sees a guy with a compressor and texturizer who finishes the ceilings and walls. Finally, the painter shows up, spray paints the entire house and puts sparkles on the ceiling if you want them. On the fifth day you move in and live happily ever after.

You can feel my envy spilling out into the typewriter. Even with help, it took me two months to do what the pros do in a week. I might seriously consider contracting the wallboard job if I build another house. Wallboard makes no structural difference; it's entirely for looks. And the professional job usually will look as good or better than my job in spite of the speed of its application. One good option might be to nail up the wallboard yourself and have a pro tape and putty it. A tip here, though, from an acquaintance who subcontracted this part: The standard price quote is based on square footage and assumes that everything will be texturized. If you want a smooth wall, which is necessary for wallpapered walls, the charge may be as much as $2,000 more. A smooth wall takes more time because there is no texturizing to hide all the sins. When you ask for a quote, be sure to specify if you want smooth or texturized walls and ceilings.

At any rate, with me it was a matter of stubborn pride. This house was my baby from stem to stern, and I wasn't about to share the glory with anyone. I didn't even bother getting quotes; I just charged ahead and learned a lot of things the hard way.

The first thing I learned was that the delivery charge for wallboard to my remote site was over a dollar per sheet. That was too steep for frugal Frank, so I decided to buy it twenty sheets at a time and haul it myself. My second lesson came when I tried to pick up a double sheet off the pile.

Holy cow, the stuff is heavy! It comes with two sheets taped together face to face and the 120 lb was more than I could handle; I had to take them apart and load them one at a time. The bed of my old panel truck isn't quite big enough to lay a 4 × 8 sheet flat, so I always put the sheets on edge and let them hang out the back door. That works a lot better with plywood than with fragile wallboard. I felt like I'd been through a workout with the Green Bay Packers by the time I got twenty sheets into my truck. The wallboard looked like the Packers had used it for turf.

As I sat in my truck recovering, a small flatbed truck pulled up. The gal had the fork lift load sixty sheets on her truck and was driving out the gate 5 minutes later with a smile on her face. There weren't any smiles in my truck. I was in pain and my truck creaked and groaned all the way home under the 1,200-lb load.

General Information

I did do some things right. I bought the right type of wallboard: ½-in tapered-edge gypsum wallboard in 4 × 8 sheets. The pros use 4 × 12 sheets so they will have fewer joints to tape and putty, but I don't know if I could handle a 90-lb sheet. I also bought the right type of nails: dimple head drywall nails, 1⅜-in or 1⅝-in. Fifty pounds last a long time.

To install the wallboard, the tools shown in Fig. 17-4 will be adequate. If you're still in good financial shape, a shop vacuum, a wallboard square and a floor lever (see Fig. 17-8) will be helpful also.

My procedure won't be hard for you to guess by now. I started in an out-of-the-way storeroom. A room that will be covered with wood paneling would also be a good choice. (Wallboard is a good fireproofing material, and often required even if the room will be paneled.) By wallboarding, taping, puttying, sanding, and painting the store-room, I got to practice where my errors would go undetected. The room turned out pretty well, though it took 2 weeks to complete. I progressed down the old learning curve as I went, eventually being able to do a room in about 5 days.

The ceiling wallboard is installed first, and it's the most difficult task. The sheets should be run perpendicular to the ceiling joists, if possible, but it is also alright if they run parallel to the joists since the sheets perform no structural function. The main consideration is to arrange the sheets so that each joint is backed by a joist (or by blocking installed specifically for the wallboard. Arrange the sheets so there are as few joints as possible. This often means using a full sheet rather than two smaller pieces, and results in throwing a lot of wallboard away. If you try to use small pieces, however, the room will take forever to tape and putty and will probably show all the joints in a few years.

Measuring and Marking Wallboard

Take the top sheet from the wallboard stack and turn it face up. Carefully measure the wall or the ceiling and transfer the measurements to the wallboard. To visualize and correctly mark the sheet can be tricky, particularly on ceilings. Remember, the good side faces down when the sheet is nailed on the ceiling. When marking the wallboard, the good side faces up. All cuts must be made on the face of the sheet. (See Fig. 17-5.)

Fig. 17-4 *Wallboard tool kit.*

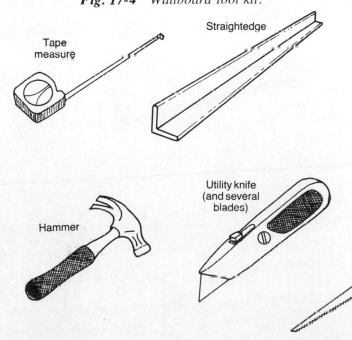

Tape measure

Straightedge

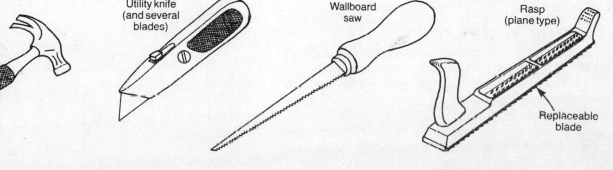

Hammer

Utility knife (and several blades)

Wallboard saw

Rasp (plane type)

Replaceable blade

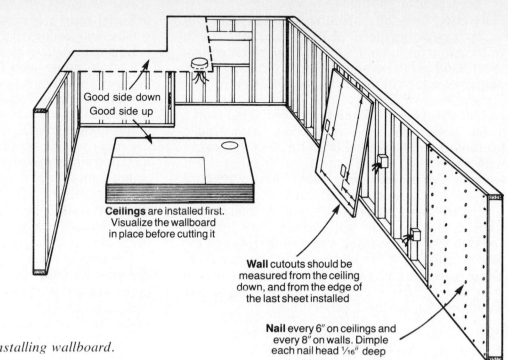

Good side down
Good side up

Ceilings are installed first.
Visualize the wallboard
in place before cutting it

Wall cutouts should be
measured from the ceiling
down, and from the edge of
the last sheet installed

Nail every 6″ on ceilings and
every 8″ on walls. Dimple
each nail head 1/16″ deep

Fig. 17-5   Measuring and installing wallboard.

I found that I made fewer mistakes with the walls if I always set the sheet up against the wall where it would be nailed and marked it right there. Always measure dimensions for cutouts (windows, electrical boxes, and the like) from the edge of the wallboard that is already installed. Also, always measure down from the ceiling, not up from the floor. There will be a gap along the floor which will be covered with the baseboard.

For windows and doors, mark the cutouts about 1/8 in larger than the opening. This gives you some leeway for aligning the sheet. The gap will be covered by the window or door trim.

For electrical boxes, use a spare box as a templet. The width of the pencil lead leaves about 1/16-in gap around the box, which will be covered by the cover plate or light fixture.

Mark long cuts with a straightedge. A piece of angle aluminum works well, or you can buy a wallboard T-square that is handy and quick to use.

Cutting Wallboard

The sheet must be face up. Lay it on the stack of wallboard or on 2 × 4's between sawhorses.

On long cuts, hold the straightedge firmly with one hand and one foot or knee. Run the utility knife along the straightedge slicing the paper facing and into the wallboard core about 1/8 in. Make a second pass with the knife to assure a clean cut.

Slide the sheet so the cut piece overhangs the stack of wallboard. Grasp the wallboard and break it backwards. Unless the wallboard is damp, it will break easily and cleanly when given a good solid snap. Then reach underneath the break and, supporting the hanging piece with one hand, slice the paper backing with the knife. (See Fig. 17-6.)

Fig. 17-6   Cutting wallboard.

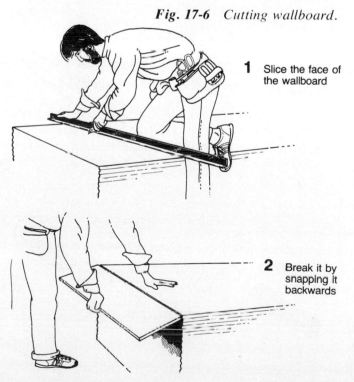

1 Slice the face of the wallboard

2 Break it by snapping it backwards

3 Reach underneath and slice the paper backing

As you get more proficient, you'll be able to make these cuts without the need for a "table."

On curved cuts and cutouts, use a wallboard saw. The previous technique only works on straight cuts where the end of the sheet can be snapped backwards. A wallboard saw works like any saw except that the tip can be punched through the wallboard by striking the handle with the palm of your hand.

When making a cutout, it's a good idea to back-cut the hole so the backside of the cutout is larger than the face side. This makes it easier to position and fit the sheet, particularly when there are several cutouts in one piece of wallboard.

Nailing the Ceiling Wallboard in Place

This would be easy if getting the sheet in position wasn't such a challenge. To start with, let's assume the more sensible approach of two people working together. You probably won't be as lucky as I was. Dad helped me, and he's installed a lot of wallboard. So let's take it step by step.

STEP-BY-STEP PROCEDURE

1. Turn off the electrical circuits to the room.

2. Check the ceiling and install any necessary blocking. The wallboard must be supported at least every 24 in to avoid sagging.

3. Mark the location of the ceiling joists on the top plate of the adjoining wall. When you're holding that wallboard up, you won't want to be dallying around looking for a joist to nail to. You may also wish to lightly mark the joist positions on the face of the wallboard. This makes it easy to find the right place to nail, but it takes some of the sport out of it.

4. Position a bench or some planks bridged between two saw horses under the center of the intended location of the sheet. The bench height should be such that your head reaches

the ceiling when you stand on it. Make sure you've got nails and a hammer in your apron.

OK, no more easy stuff. In some semblance of unison, both pick up the sheet, step up onto the bench, lift it over your heads, and place it tightly in the corner or against the last sheet that was installed. (See Fig. 17-7.)

As you stand there, check to see that the sheet is "tight to the wall" and/or "tight joint" to the last sheet installed. In a crisis, ¼ in might qualify as tight joint. Pull any wires for light fixtures through the cutouts. (The circuit is *off*, right?) Make sure there's enough joist showing so the next sheet can be nailed to it. By this time your arms are starting to shake, so push your head up against the wallboard and hold it tight against the ceiling. (A wool hat will cushion your head, if necessary.)

Fig. 17-7 *Installing wallboard on the ceiling.*

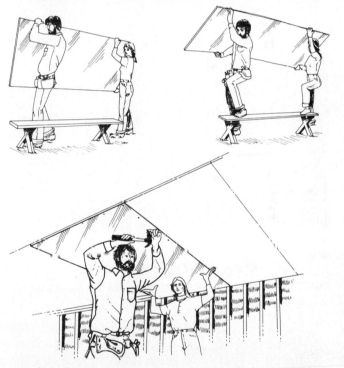

5. Holding the wallboard with your head, reach up and nail the sheet in several places. At each place, put two nails about 2 in apart. The challenge is to get about twenty well-placed nails in before your head and neck get tired.

Important: When nailing wallboard, sink the head of the nail just below the surface of the wallboard. Leave a smooth dimple about ¹/₁₆ in deep. The paper covering on the wallboard should not be ripped, just dimpled. It helps to practice ahead of time on a scrap of wallboard.

6. Fill in the nailing pattern. On ceilings, space the nails every 6 in. Don't skimp on the nails. Remove any nails that miss the joist by hooking them with the head of another nail. Dimple the hole now and putty it later.

7. When possible, railroad the wallboard, (as you did with the plywood sheathing and subflooring), so the end joints don't all land on the same ceiling joist.

   Applying ceilings alone is a tough job, unless you rent a wallboard jack. The sheet is set on the jack, rolled in place, and cranked up against the ceiling. Then it can be leisurely nailed in place. If you don't have a jack you have to nail a scab to the wall on one end, wrestle the sheet up into place, then hold it and nail it. Out in the center of the room a ceiling-high T-brace made with 2 × 4's is necessary, along with some scuffling, cursing, and heavy breathing.

Nailing the Wallboard to the Walls

Here we come to a decision point. Do we apply the wallboard horizontally or vertically? Horizontally seems slightly better, since the sheets can be railroaded to make the joints less likely to show if the house ever settles. But when working alone, wallboard is easier to apply vertically. I looked at a 25-year old house where the sheets were installed vertically and couldn't see any evidence of joints, so I decided to apply mine vertically also.

Step-by-Step Procedure

1. Turn off the electrical circuits to the room at the panel.

2. Measure and mark the wallboard with the sheet propped up against the wall as illustrated in Fig. 17-5. Remember to measure the cutouts from the ceiling down, not from the floor up. If necessary, add studs at the corners or elsewhere so all joints will be backed be a stud.

3. Cut the sheet to fit. If the walls were framed about 97 in high the sheets won't need to be cut heightwise.

4. Mark the stud locations on the floor and ceiling. Make sure that all pipes and electrical wires are safe from wallboard nails. Install nailplates where necessary.

5. Position the wallboard. Here's where one little trick will save oodles of time and sweat. There's a gizmo called a floor lever, shown in Figure 17-8, that lifts the sheet of wallboard and pushes it tightly against the ceiling. A scrap of plywood set on a 2 × 2 also makes an effective floor lever. (See Fig. 17-8.)

Fig. 17-8 *Using a floor lever.*

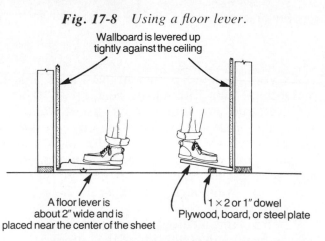

Wallboard is levered up tightly against the ceiling

A floor lever is about 2″ wide and is placed near the center of the sheet

1 × 2 or 1″ dowel
Plywood, board, or steel plate

Check the ceiling joint and the edge joint to be sure they are tight. You'll know from the practice room that joints that aren't tight are difficult to putty. Pull any wires through the electrical box cutouts. Make sure there's enough stud showing at the edge joint, (about ½ in) so the next sheet of wallboard can be nailed.

6. Nail the sheet in place. Usually six or eight nails will hold the sheet in place, so you can take your foot off the floor lever. Then fill in the nailing pattern with nails every 8 in. Dimple each nailhead into the wallboard so it can be covered with putty.

Once you have the basics down, cutting and hanging wallboard goes pretty fast. The first few sheets are terribly awkward to work with: to move, to cut, and to nail in place. If you watch a pro do it, it looks incredibly simple. Like everything else, soon you'll have made all the mistakes there are to make and will have added another skill to your bag of tricks.

Tape and Putty

Compared to how the inside of the house used to look, the place seems almost finished now. Yet if you painted now, it would look terrible, which brings us to the art of taping and puttying wallboard. Besides enhancing the appearance, the tape and putty also seals the draft from the ceiling joints and further fireproofs the house.

As I started talking around about how to tape and putty, it became apparent that everyone had a strong opinion on the subject. Some swore by self-stick tape. Others wouldn't touch the stuff. Some used a 6-in putty knife for all coats; others used three or four progressively wider knife sizes. Some folks were certain that two coats of putty were enough; others said it would take at least five coats. Pole sanders were big in one school of thought; damp sponging or hand sanding were espoused by other schools. On and on it went. Taped corners versus prefab reinforced corners; brand A putty versus brand B. Every tool and contraption from corner trowels to banjos to bazooka guns had their staunch proponents. Some folks said it was crazy for an amateur even to try it. I was getting confused.

Finally, I just bought some tape and putty and tackled the storeroom. I've done a couple of

Fig. 17-9 My tape and putty tool kit.

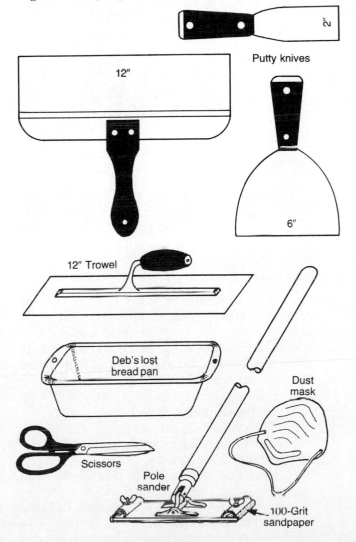

houses now and have developed my own opinions so I can jump into the fray and argue with the best of them. Figure 17-9 shows my tape and putty kit.

Applying the First Coat of Putty

Use embedding putty or all-purpose putty for the first coat. Start out in a closet or at least in a back bedroom.

STEP-BY-STEP PROCEDURE

1. Turn off the electrical circuits to the room at the panel. That putty knife is a great way to inadvertently connect a hot wire with a ground. The resulting flash is impressive, and possibly deadly.

2. Turn on the heat in the house to about 60°F to help dry the putty.

3. Tape the tapered joints first. (These are the joints between the tapered edges of two sheets of wallboard.) Apply self-sticking fiberglass tape with the flat of your hand or your fist as shown in Fig. 17-10.

4. Apply putty to the tapered joints. Use embedding or all-purpose putty (also called joint compound). Use a 6 in putty knife to fill your tray (or bread pan). Be careful to cover the putty container while you're working, and put a wet towel in over the putty. Otherwise the putty dries, forming little clinkers which will cause you extreme grief. You'll know what a clinker is when you make a long, smooth swoop with the putty knife and see a ditch in it where the clinker dragged through.

   Spread on the putty like peanut butter, covering the tape with putty and filling the recess formed by the tapered wallboard edges. When it's overfull, hold a trowel edgewise and strike the surface flat from ceiling to 3 in above the floor. Stay away from the floor; you'll just pick up more clinkers down there and it's all covered by baseboards eventually anyway.

   Some people use a 12-in putty knife for the strike-off, instead of the trowel. Whichever you use, be sure that the blade isn't bending or the surface will be slightly concave rather than flat as it should be.

   If there are minor imperfections in this first coat, don't worry about them. Tomorrow you'll be appalled at how much the putty

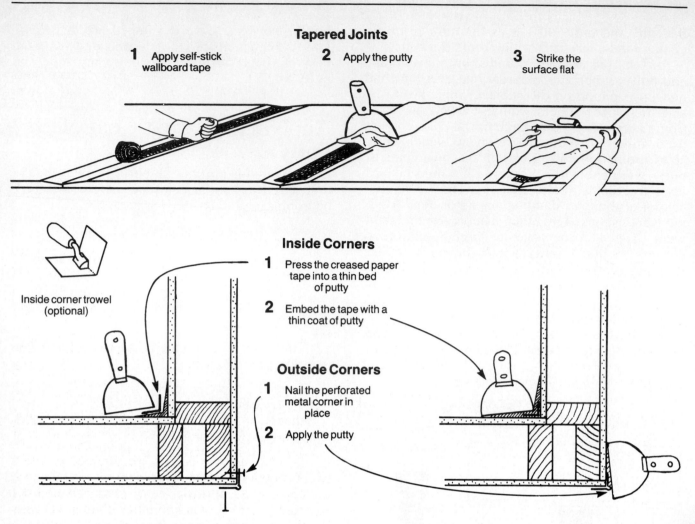

Tapered Joints

1 Apply self-stick wallboard tape

2 Apply the putty

3 Strike the surface flat

Inside corner trowel (optional)

Inside Corners

1 Press the creased paper tape into a thin bed of putty

2 Embed the tape with a thin coat of putty

Outside Corners

1 Nail the perforated metal corner in place

2 Apply the putty

Fig. 17-10   *Taping and puttying wallboard joints.*

shrank when it dried. It's the second and third coats where smoothness is important.

Continue this tape and putty process on every tapered joint in both the walls and the ceiling.

5. Tape and putty the inside corners, where one wall meets another. Leave the ceiling corners until later.

   a. Tear a piece of perforated paper tape to length (about 95 in) and fold it down the crease.

   b. Apply a *thin* coat of putty to both sides of the corner with a 6-in knife or with an inside-corner trowel.

   c. Press the creased paper tape into the putty with the 6-in knife or with the corner trowel. The putty should squeeze through the perforations in the tape.

   d. Apply a *thin* coat of putty covering the tape. Make this coat quite smooth. (See Fig. 17-10.)

The bugaboo here is to keep the putty thin. Most beginners slop on a bunch of putty because they think it'll be easier to smooth that way. It isn't. The more putty on a joint, the more it deviates from the flat wallboard surface and the lumpier the joint will appear.

6. It's easier to reinforce and putty the outside corners; unfortunately a house has about six inside corners for every outside corner.

   a. Cut the perforated metal corner piece to length with tin snips and nail it to the corner. Make sure it's on tight and square, then nail it every foot with wallboard nails. (See Fig. 17-10.)

   b. Glop putty on the length of the corner with a 6-in putty knife, embedding the prefab corner piece.

   c. Run the 6-in putty knife down both sides of the corner, striking off the excess putty. Stay 3 in above the floor to avoid picking up clinkers.

7. Putty the nailhead dimples. This is a breeze.
   a. Run the putty knife down a row of nail heads. If you hear any "chinning" sounds, locate the offending nail head and restrike it so it's below the surface.
   b. Fill a row of dimples with putty using a 4-in or a 6-in knife.
   c. Strike off the whole row of dimples with one long stroke. Some folks leave a thin coat of putty the length of the row; I strike off all of the excess putty as cleanly as possible to avoid any lumps.

8. Tape and putty the ceiling joints. Since the ceiling joints invariably overlap the tapered joints, the best approach for amateurs is to come back for these tomorrow. After the tapered joints are dry, do the ceiling corners just like the inside corners in step 5.

9. Tape and putty the butt joints. These are the joints where the ends of two sheets meet. They're well-named. After you've done a couple you'll be glad you bought tapered-edge wallboard, and will wonder why they don't make tapered ends on the wallboard.
   a. Apply self-stick fiberglass tape to the joint. Many pros use paper tape here; I prefer self-stick.
   b. Spread a 6-in band of putty over the tape. Feather the edges of the band to meet the wallboard. Sometimes bending the blade of the knife helps here.
   c. Smooth the putty as thinly as possible without exposing the tape.
   A butt joint is difficult because it can never be flat. There's always a gentle hump about $1/32$ in high. The challenge is to make the hump smooth and unnoticeable. Make the second and third coats of putty progressively wider, from a 6-in band to 8 in to 10 in. On the final coat it helps to grasp the 12-in putty knife at each side of the blade and bend it in a slight curve for that last smooth stroke.

General Notes

As you putty, you'll be constantly applying putty, cleaning the knife blade on the edge of the tray, then striking the putty smooth. The scraped off putty goes right back into the putty tray to be used again. If the putty starts getting too dry or gets a clinker in it, throw it away. Scrape it off, wipe it into a paper towel, fling it out the window, but *get rid of it*. Putty is cheap and your time is valuable.

If the putty won't go on smoothly, try wiping the blade with a paper towel frequently. Also, mixing some water with the putty in the tray will make it smoother.

Don't worry about getting glossy smoothness on the first coat. It'll be covered by two more coats anyway. On the other hand, do worry about any lumps or ridges. When finished, the surface must be absolutely flat wherever possible. If you think you're going to sand off the excess putty, there's an unpleasant surprise in store for you.

Applying the Second and Third Coats of Putty

The purpose of the *first coat* of putty was to bury the tape and nail heads without causing any ridges or bumps.

The *second coat* is to bring the surface up to final grade. The first coat must be dry before the second coat is applied or both coats will crack. Use finish putty for the second coat. While some people use an all-purpose putty for all coats, the finish compound has the advantage that its tan color helps to highlight high spots in the first coat, which appear light colored. Finish putty is wetter and spreads smoother than embedding putty. Use a 6-in putty knife to apply the putty and a trowel or 12-in blade for striking it off. Pretend you're a road grader and go for smoothness. Again, it'll look real good until the next morning when the nail head dimples will show and the taped joints will be slightly concave. If you can see or feel a low spot now, it will be visible after the surface is painted.

The *third coat* is the finish coat and smoothness is imperative. Use the finish putty again, but this time apply it with a 12-in knife, feathering all joints to meet the wallboard. Add water to the putty as necessary to make it smooth. The third coat can drive you crazy as you try to eliminate all of the small imperfections. Be patient and realize that a high degree of skill must be attained.

Sanding and Touchup

After the third coat of putty is dry (allow at least 24 hours), sand the walls and ceilings with 100-grit sandpaper. A pole sander (see Fig. 17-9) is a labor saver because it allows long strokes to be taken and avoids the need for constant bending over and standing up.

Compared to puttying, sanding goes fast. Wear a dust mask and try not to stand under the place you're sanding. Only light pressure is needed; the putty sands quite easily if there are no lumps or uneven spots. By changing the sandpaper every fifteen minutes or so you can keep the pressure light and the strokes long. If a piece of tape is exposed, stop sanding immediately. Be particularly careful at butt joints.

You'll find that this fine, light dust gets everywhere. Vacuum the dust up so it doesn't get blown all over when the furnace comes on. Some people use damp sponges instead of sandpaper to smooth the putty without creating dust.

Finally, look over the walls and ceilings for any spots that need a touchup. Hit them with a dab of putty (a fourth coat, if you will) and follow up with a light sanding after the putty dries. Usually the touchup is so thin that it dries in a couple of hours. Now you're ready to paint. Tape and puttying is slow, painstaking work at first, and reaching the painting stage is a happy milestone.

A Change in Perspective

You'll notice a change in the type of skills needed as you move into the finishing stages of the house. Aesthetics and craftmanship play a larger part; attention to details such as the application of the putty and the color of the paint becomes more important. While concrete is smoothed out rather roughly, wallboard putty is smoothed very carefully, the way Michelangelo would do it. Framing nails are sort of blasted in, leaving a hammer dent in the wood, while finish nails in interior trim are carefully driven in place, a nail set used to set the head slightly below the surface, and the divot puttied with wood putty. As I got to this stage, I started to think of myself more as a craftsman than an amateur builder.

Incidently, I now find myself writing to a totally different audience than I started with. At first, I geared the book to my memories of myself as a beginner, giving explanations of elementary things like wood grain, use of a skill saw, and how to make things square. Gradually I've shortened the explanations until now I'm writing to a person who has worked with a lot of building situations and a variety of materials: concrete, steel rebar, framing lumber, plywood, asphalt roofing, finish wood around windows and doors, glass, plastic and copper piping, electrical connections, sheet-metal duct work, fiberglass insulation, gypsum wallboard, putty, and probably a handful of other things. By this time, even if you've never done a task like painting or trim work before, I doubt that you'll need much explanation on how to do it. You'll know when detailed instructions would be helpful and where to find them.

Painting

Look on the back of the paint can. There they are, more instructions than you usually care to read.

I find that it helps to apply a **primer coat** before the final paint is applied. Primer is usually cheaper than regular paint and that new wallboard can really soak up the paint. If you try to paint with just one coat, often there will be a slight color variation between the putty spots and the wallboard.

Flat latex or **alkyd paint** is normally used on walls and ceilings. Flat paints are less likely to show imperfections than glossy paints. If the putty job didn't go so well, an application of texturizer may be in order or a different sort of lighting that doesn't shine across the ceilings at a low angle.

Semi-gloss or **glossy enamel paint** is normally used in bathrooms and kitchen where the walls are likely to be scrubbed. Flat paints, even if labeled as washable, won't hold up to a scrubbing. Enamels are often used on interior trim also.

Roller, Pad, Brush, or Spray Gun?

It's the cook's choice here. I used a long-handled roller for the bulk of the painting and found the long handle to be a major labor saver. As I came to a window, door, or corner I used a small paint pad to get better accuracy and less splatter. (See Fig. 17-11.)

Remember that there will be trim around the windows and doors so the paint job need not be super-accurate there. The gap is covered by the trim. The base of the wall will be covered by baseboard, so stay away from the sawdust and dirt on the floor with that roller.

Painting is surprisingly time-consuming. It takes me about a day to apply one coat to a medium-size room with a closet.

Fig. 17-11 *Painting.*

Alagabber

Nearly every homeowner has taken a shot at wallboarding and has some interesting stories to tell. The more gutsy folks tackle the job without any reading or help. My friend Dennis tells about cutting the wallboard with a skill saw and darn-near choking to death in the ensuing duststorm. I've heard of several people who bought tape and putty and vigorously proceeded, never dreaming that masking tape wasn't the type of tape everyone was referring to. Apparently the results look moderately horrid. Uncle Barney just applied putty and didn't bother with the tape. A week later, the house looked like it had been through a severe earthquake and Aunt Katharine was attempting to mend the cracks with Elmer's glue. Wrench's first whack at wallboarding included an attempt to tape and putty a joint that wasn't backed by a stud. After paving nearly 50 lb

of putty into one joint he reconsidered, tore out the wallboard, added a stud and started over.

Even when I knew what steps were supposed to be taken, I was constantly tempted to take shortcuts. Wallboarding was a step that took twice as long as I had planned, covered up all my previous work (of which I was quite proud), and made the house noisy again. I was zealously jealous of Steve, whose log house didn't require a single piece of wallboard. I started feeling better about the wallboard after some paint was applied and certain areas were covered with cedar. The combination of wood and wallboard came off well. When we moved in with our plants, rugs, and paraphernalia, the place quieted back down and finally felt like an honest-to-God home. Now I like the wallboard, the light ceilings, the flexibility of changing colors, the crisp angles and corners and I'm glad I took the time to do a good job on it.

I spent some time puzzling over the sequence of the finishing tasks. The normal order is wallboard, paint, paneling, cabinetry, plumbing fixtures, electrical devices, finish floors, and trim.

The cupboards and counters go in after the paint because it's time-consuming to mask and paint around them. The sinks fit into the counters so they come next. The electrical devices (outlets, switches, and lights) can be finished anytime after the paint is dry; the sequence isn't critical.

Finish floors go in nearly last so they aren't marred or stained by other activities. I once asked a knowledgeable fellow what wallboard putty is made from. He replied, "I don't rightly know, sonny, but it leaves the same kind of stain on oak floors that my cat leaves." I haven't learned anything to contradict that, so putty first, then do the floors. Baseboards can't go on until the finish floor or carpet is installed. The trim around the windows and doors can go on any time. Deb suggests I do it soon.

Well then, fellow adventurers, we're nearly to the end of the path through the woods that I know. Soon we'll be arriving at a trailhead from which a thousand paths lead. From the security of your soon-to-be functional home, a multitude of Topics (some friendly, some ferocious) can be pursued. Past the trailhead, travel as a group would be a waste of energy for everyone.

CHAPTER 18

Finishing Touches

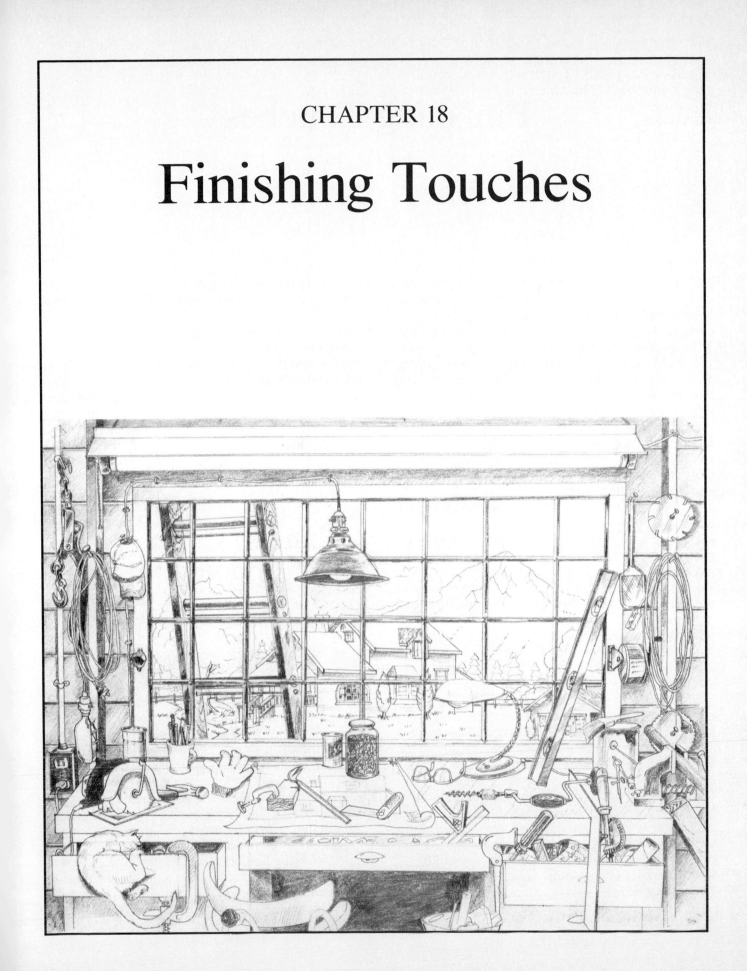

Finishing Touches

The finishing touches are those little expressions of imagination that give a house its character. Remember when you were a kid how there were certain houses that always seemed intriguing? Take yourself back in time to when you were six years old and your imagination was in its prime. When you look up at that dormer window, you wonder what it would be like if that were *your* bedroom. You can just see yourself standing up there looking out at everything below. What an exciting, cheery place it is with its window box brimful of bright, colorful flowers and the big front porch that beckons you to walk up the steps and come on in.

When you walk into those special houses, they make a lasting impression. The tile floor with the huge vase and bending rubber tree plant in the corner, the warm feeling of natural wood, the stairway enticing you to explore upstairs. The smooth, glossy bannister just fits in your hand as you go up. The window seat at the landing is one of those pleasant surprises that you expect to find. What a great place to be, just sitting there with the cat on your lap, letting the sun shine in and dreaming away.

This house was built for people to live in. What is it that makes it so appealing? Possibly it's that distinctive, personal touch of someone who loved to build. The people who live here may have a special place in your heart too, which adds to the warmth. Someday you'll live in a house like this one.

Maybe that day is finally drawing near. You've built a livable, functional house. Some of the special features are already started, the dormers, the arched windows, the upstairs, and probably some other things. These will give you something to work with as you start adding the house's charisma.

The charisma of a house doesn't happen overnight. It usually develops as the house matures.

Ma finally realizes that gaudy wallpaper will never make it in the dining room. Pa finally earns enough spare change to buy some hardwood flooring to cover up the plywood. Eventually those bare light bulbs, one by one, get replaced with distinctive light fixtures. Ma has a great plan for a sundeck and a gazebo where we can sit with friends, sipping cider and laying plans for new adventures. Slowly the house develops to fit the people living in it.

We're in that stage now. I expect it will take several years to finish. No rush, this is the best part of all.

Cabinets, Counters, and Cupboards

There's no rush unless you want a place to put your dishes, Deb reminds me. In which case you can go rustic with apple crates and beach planks, or go modern with prefabricated cabinets that are purchased ready to hang. Both can be installed in 2 or 3 days and are primarily an exercise in finding a stud in the right place and strategically placing the nails. The other option is to custom-build your own cupboards and cabinets. That's the tack I took but that's a ferocious topic with too many appendages and too much hair to get into here.

Planning the cabinets, counters, and cupboards is the key task, be they prefab or custom-built. In the kitchen, carefully arrange the available space to reflect the way the kitchen is used. Make a detailed drawing and be sure that both the cook and the dishwasher agree on the final design. Figure 18-1 shows a plan for the kitchen in the sample house.

Brochures are available for prefab cabinetry

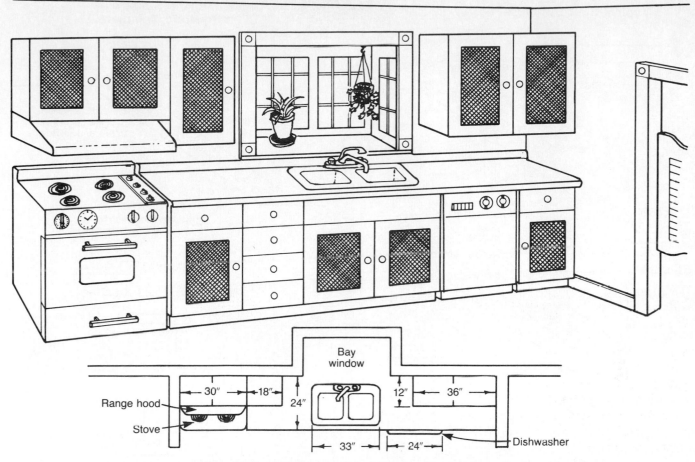

Fig. 18-1 *Kitchen cupboard and cabinet layout.*

showing the available units, dimensions, and suggested combinations of the units. Avoid cabinets made of particle board; recent studies have shown that the formaldehyde glue used in making the boards gives off unhealthy fumes.

Design custom cabinetry with the plywood in mind, remembering that the face grain should run vertically on the doors and drawer fronts. Work out the hinging arrangement carefully, observing the differences between regular and pivot-pin hinges. I found that pivot-pin hinges simplify the cabinet job and I like the fact that the hinges are hidden from view.

Although there is no pressing reason to build your counters at a standard height, kitchen counters are normally 36 in high and bathroom counters 31 in high. Deb (5 ft 5 in) and I (6 ft 5 in) find 38-in and 36-in counters more to our liking. We decided on prefabricated counters for the kitchen because they are easy to clean, easy to install, and the backsplash and raised front edge contain the inevitable milk spills. Most of our friends built beautiful plank, laminated wood, or tile counters. In terms of appearance, prefabs can't compete.

Stairways

Some of the design requirements for stairs were discussed in Chapter 5. The Uniform Building Code UBC Sec. 3305 covers stair requirements in more detail. The design must be well thought out or you'll get into trouble when you start to build.

Most methods of building stairs require stair stringers to be made. The stringers are the structural backbone of the stairs and determine the angle and placement of the steps. A smart thing to do is to make a plywood templet of the stringer and get it to fit perfectly. (Clench nail two long scraps of plywood together if necessary.) Then build the stringers to match the templet.

In the case of stairs with mid-flight landings, build the landing first at precisely the desired height. Don't forget to allow the normal ¾ in for the finish flooring (carpet, hardwood, linoleum over ⅝-in plywood, or whatever). The stringers are installed between the landing and the floor

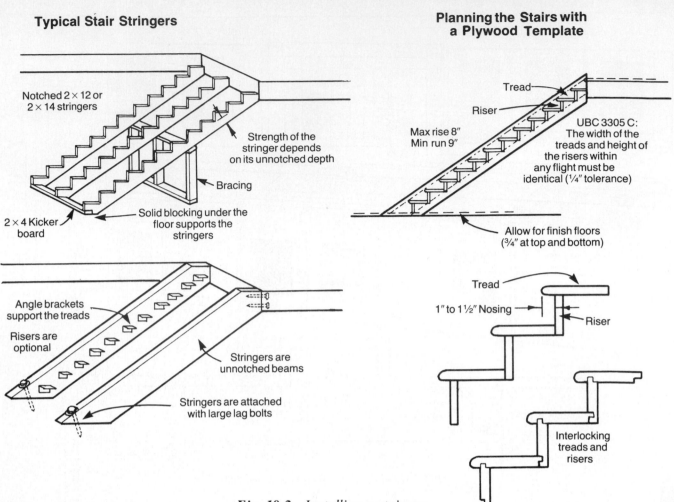

Typical Stair Stringers

Notched 2 × 12 or
2 × 14 stringers

Strength of the
stringer depends
on its unnotched depth

Bracing

2 × 4 Kicker
board

Solid blocking under the
floor supports the
stringers

Angle brackets
support the treads

Risers are
optional

Stringers are
unnotched beams

Stringers are attached
with large lag bolts

**Planning the Stairs with
a Plywood Template**

Tread

Riser

Max rise 8″
Min run 9″

UBC 3305 C:
The width of the
treads and height of
the risers within
any flight must be
identical (¼″ tolerance)

Allow for finish floors
(¾″ at top and bottom)

Tread

1″ to 1½″ Nosing

Riser

Interlocking
treads and
risers

Fig. 18-2   Installing a stairway.

and are attached with lag bolts or nails to the solid blocking under the floor. A kicker board may also be used to hold the stringers in place. Finish flooring can be fitted around the stringers later. (See Fig. 18-2.)

The attachment of the stair treads to the stringer is important. Stair treads will see a lot of action during the life of the house and generally will hold up better if screwed, rather than nailed. The old timers and few remaining craftsmen make stairs with interlocking treads and risers. It's a good way, but requires skill and time to painstakingly glue and nail everything together.

I used 3 × 12's for treads and attached them to 4 × 12 stringers with pieces of 2 × 2 angle iron. No risers were necessary. This was an easy way to get some nice looking stairs. Boy, was I glad to get those stairs in and finally be able to carry stuff upstairs without balancing on the ladder. "Should have built these stairs months ago," says my diary.

For more information on stairs see *Floors and Stairways* by Time-Life Books.

A Porch

The sample house we've built still doesn't look quite right. It needs a front porch with railings and perhaps a couple of rocking chairs out there. Although there are several ways the porch could be built, Fig. 18-3 shows how I would do it.

A Bay Window

Figure 18-4 shows one way to build a bay window. There are also greenhouse window units available that will fit right into a window opening without the need for any special framing.

This bay window structure lends itself nicely to making a "pop-out" window seat also. Another way to frame bay windows is to extend the floor joists and build the bay window from floor to ceiling.

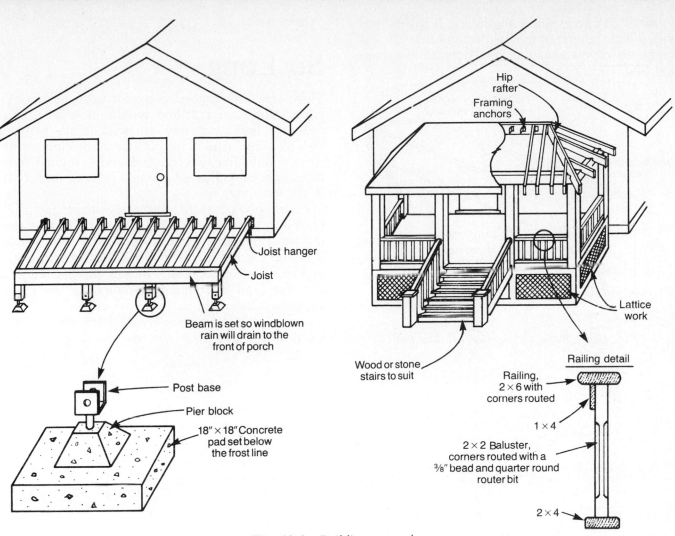

Fig. 18-3 *Building a porch.*

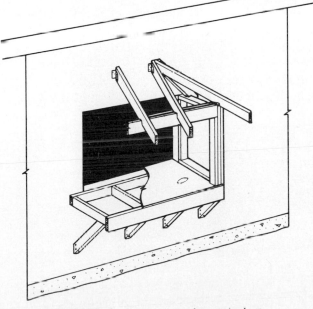

Fig. 18-4 *Framing a bay window.*

A Window Box

When you're interested in window boxes, suddenly they start appearing everywhere you go. While at a stop light on Roosevelt Way, I spied a window box on a cafe. Pulling over to the curb, I got out and took a look at it. I made a couple of improvements on the design and built ours as shown in Fig. 18-5.

It's a pretty window box and I was real proud of myself for building it. The glow didn't last long, though. All of Deb's flowers croaked soon after they were transplanted into it. Dear me, dear me, I probably shouldn't have used that strong wood-life preservative on it. The wood will last forever but the flowers won't last two days. We emptied the dirt and dead Impatiens out with stifled curses and complaints, lined the box with plastic, then put in new dirt and flowers. I'm happy to report that the new flowers are doing just fine.

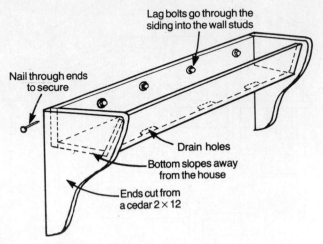

Lag bolts go through the siding into the wall studs

Nail through ends to secure

Drain holes

Bottom slopes away from the house

Ends cut from a cedar 2 × 12

Fig. 18-5 *Building a cedar window box.*

So Long

This seems like a good place for us to part company. You're a certified builder now and can probably figure out how to build things just as well as I can. And if we get started talking about elegant finishing touches and crafty special features in homes, I'm afraid we'll be here until two in the morning.

Every book has to end somewhere and this book has found the natural end of its trail. I see the sun is starting to peek out of the clouds and it's time for me to get to work on that sun deck and gazebo. So good luck to you, patient reader, and may you and your house enjoy many good years together. I hope that you will freely share your building experience with others, the way so many folks have done with me. So long!

SCRAPBOOK

Donna, Martin & Karey

Wrench

Chris, Spike & Nuclear

Dave's house — erecting a 60' wall

Marcia & Carl   Hartmut

Steve, Moki & Marie

Sis &
her practice project

Carl's addition

Trish

Dad

Guess who?

Mare & Steve

Deb, Garth, Frank & solar cat no. 3

APPENDIXES

APPENDIX A

Lumber and Plywood

TABLE A-1 Lumber Sizes and Their Grades*

Nominal size (inches)	Actual size (inches)	Typical grade progression
1 × 2	¾ × 1½	
1 × 3	¾ × 2½	
1 × 4	¾ × 3½	Clear
1 × 5	¾ × 4½	Standard
1 × 6	¾ × 5½	Utility
1 × 8	¾ × 7¼	
1 × 10	¾ × 9¼	
2 × 2	1½ × 1½	
2 × 3	1½ × 2½	
2 × 4	1½ × 3½	
2 × 6	1½ × 5½	
2 × 8	1½ × 7¼	
2 × 10	1½ × 9¼	
2 × 12	1½ × 11¼	
		Select Structural
		No. 1; Construction
4 × 4	3½ × 3½	No. 2; Standard, Stud
4 × 6	3½ × 5½	No. 3; Utility†
4 × 8	3½ × 7¼	No. 4; Economy†
4 × 10	3½ × 9¼	
4 × 12	3½ × 11¼	
6 × 6	5½ × 5½	
6 × 8	5½ × 7¼	
6 × 10	5½ × 9¼	
8 × 8	7¼ × 7¼	
8 × 10	7¼ × 9¼	

* For more information write to Western Wood Products, Yeon Building, Portland, Oregon 97204.

† Not suitable for structural framing.

TABLE A-2 Board Feet versus Lineal Feet

1 board foot = 1 in × 12 in
                = 2 in × 6 in* or equivalent
1 lineal foot = length as measured
Example: 10 lin ft of 2 × 12 = 20 bd ft
              60 lin ft of 1 × 4 = 20 bd ft

Note: Lumber is normally available in 2-ft increments from 6 to 20 ft.

* Actual size is 1½" × 5½".

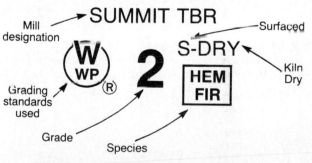

Fig. A-1 *Typical lumber grade stamp.*

TABLE A-3 Lumber Defects

Improper dimensions
Wetness (KD lumber stored in the rain)
Checks, splits, and surfacing defects
Warpage, twists, and bows
Pitch pockets, large or loose knots
Insect damage

TABLE A-4 Common Lumber Terms

STB	Lumber graded standard or better (most is standard, some is No. 1)
S4S	Surfaced on all four sides (most lumber is S4S)
S2S	Surfaced on two sides
S1S1E	Surfaced on one side and one edge
RL	Random length (refers to a sling of lumber of varying lengths)
T&G	Tongue and groove
KD	Kiln dried—controlled drying resulting in about 15 percent moisture content in the lumber
Green	Lumber with 20 percent or more moisture content; will usually split or warp if not dried properly
Rough cut	Unsurfaced lumber

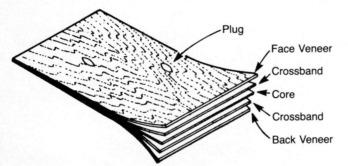

Fig. A-2 *Five-ply plywood. The grain of the crossbands runs perpendicular to the face veneer and core grain, and may have voids depending upon the kind of plywood.*

TABLE A-5 Plywood Surface Grades

N	Smooth, natural finish
A	Smooth, paintable (up to 18 plugs allowed)
B	Solid surface (1-in knots and minor splits allowed)
C	Passable strength, but not too pretty (1½-in knots and ⅛-in splits allowed)
D	Worst grade to be used normally (2½-in knots, knot holes and splits allowed)

TABLE A-6 Typical Plywood Types and Usage*

Underlayment	Used for subflooring. No voids are allowed in the crossbands. Can withstand occasional wetting during construction.
CDX	Used for roof and wall sheathing. Will withstand occasional wetting during construction.
AC exterior	Used for soffits and other exposed exterior surfaces.
T-111	Used for siding, either over sheathing or directly over the studs. It has a texturized surface that resembles siding boards.
AB	Used for exposed cabinet work.
Shop	Used for hidden cabinet work, shelves, and so on. The defects are filled with putty.

* More information is available in The Plywood "How To" Book, *which you can get by writing to American Plywood Association, 1119 A Street, Tacoma, Washington 98401.*

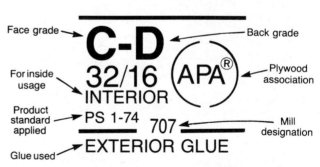

Fig. A-3 *Typical plywood grade stamp.*

APPENDIX B

Nails and Fasteners

TABLE B-1 Types of Nails and Their Uses

Nail		Typical usage
Common or sinker		Framing
Box		Toenailing
		Sheathing
		Subflooring*
Duplex head		Concrete forms
Annular ring		Subflooring
		Flooring
Siding		Siding
		Siding shingles†
Casing		Trim
Finish		Cabinets
		Fine woodwork
Roofing		Roofing‡
Drywall		Wallboard

Holds best if galvanized (hdg).
† Should be hdg or aluminum to avoid rust stains.
‡ Must be hdg.

TABLE B-2 Nail Sizes

Type	Length (inches)
2d	1
3d	1¼
4d	1½
5d	1¾
6d	2
7d	2¼
8d	2½
10d	3
12d	3¼
16d	3½
20d	4
40d	5
60d	6

TABLE B-3 Commonly Used Abbreviations

8d	Eight penny (a designation of nail size)
hdg	Hot dip galvanized (this is a zinc coating to prevent rust)
eg	Electro-galvanized (this is a thin zinc coating)

TABLE B-4 Types of Fasteners and Their Uses

Fastener		Typical usage
Wood screws		
Flat head		Cabinets Hinges*
Round head		Light metal to wood
Lag or hex head		Post-and-beam connections
Eye		Swings Safety lines
Sheet-metal screws		
Pan head		Metal to metal
Hex head or gimlet point		Ductwork Metal chimneys
Bolts		
Machine		General purpose Post-and-beam connections Wood to wood
Threaded rod		Wood to metal Metal to metal
Carriage		Wood to wood Metal to wood†
Stove		Cast iron stoves
Into concrete:		
Lag screw into expansion sleeve		Wood to concrete‡
Concrete nail		Wood to concrete
Into wallboard or hollow concrete block:		
Toggle bolt		Plant hooks in ceilings (heavy plants need a hook screwed into a joist)
Moly bolt or expansion bolt		Picture frames on walls (sleeve expands as bolt tightens)

* *Countersink needed.*
† *Square shank embeds in wood or metal cutout.*
‡ *Screw expands lead sleeve.*

APPENDIX C

Nailing Schedule

TABLE C-1 Required Nail Spacing for Typical Joints*

Connection	Nails required (at each joint)
Floors	
Floor header to mudsill	Toenail, one 8d box every 16 in
Floor joists to floor header	End nail, three 16d sinkers
Floor joists to mudsill	Toenail, three 8d box
Blocking between floor joists	Toenail, three 8d box
¾-in subfloor to joists	Face nail, 8d hdg box or 6d annular every 6 in along edges, every 8 in elsewhere
Walls	
Studs to sole plate and top plate	End nail, two 16d sinkers
Window and door headers, two planks nailed together	Face nail, one 16d every 16 in along edges
Window and door headers to stud	End nail, two or three 16d sinkers or toenail, four 8d box
½-in plywood sheathing to studs and plates	Face nail, 6d hdg box or 8d box every 6 in along edges, every 12 in elsewhere
Sole plate to floor framing	Face nail, 16d sinkers every 16 in
Second top plate to first top plate	Face nail, 16d sinkers every 16 in
Wall corners, stud to stud or stud to blocking	Face nail, 16d sinkers every 16 in
Wall corners, top plate overlap	Face nail, two 16d sinkers
Ceilings and rafters	
Ceiling joists to top plate	Toenail, three 8d box
Ceiling joists to rafters	Face nail, three 16d sinkers
Rafters to top plate	Toenail, three 8d box. In windy areas use metal connectors
½-in plywood sheathing to rafters	Face nail, 6d hdg box or 8d box every 6 in along edges, every 12 in elsewhere
Built-up girders and beams	
Two planks nailed together	Face nail, 16d sinkers every 16 in along edges
Three or four planks nailed together	Face nail, 20d common every 16 in along edges

*To avoid splitting, nails should be spaced at least 1½ in apart, except small nails, which can be spaced half their length apart. Also, avoid nailing within 1 in from the end of a board. Finally, avoid nailing two nails along the same grain line.

APPENDIX D

Structural Tables

TABLE D-1 Relative Bending Strengths of Wood Species*

Wood Species	Rating
Structural†	
Northern Douglas Fir, Larch	1.00
California Redwood	0.97
Southern Douglas Fir	0.93
Eastern Hemlock, Tamarack	0.89
Southern Pine	0.83
Mountain Hemlock	0.83
Hem-Fir (a mixture of species)	0.79
Northern Pine	0.75
Nonstructural	
Coast Sitka Spruce	0.72
Lodgepole Pine	0.72
Idaho White Pine	0.69
Eastern Spruce	0.69
Western Cedars	0.69
Ponderosa Pine, Sugar Pine	0.67
Balsam Fir, Red Pine	0.65
Western White Pine, Engelmann Spruce	0.63
Subalpine Fir (White Woods)	0.58
Northern White Cedar	0.57

* This table is for kiln-dried (KD) lumber from 2 × 6 to 2 × 12. All lumber must be graded No. 2 or better.

† For the purposes of this book, lumber is considered "structural" if it has a modulus of elasticity, E, of 1.0×10^6 psi or greater (stiffer) and is rated to withstand a bending stress of 1,100 psi or more after appropriate safety factors are applied. For more complete information, refer to the Uniform Building Code (Table 25-A-1) or to the Western Woods Use Book, which you can get by writing to the Western Wood Products Association, Yeon Building, Portland, Oregon 97204.

TABLE D-2 Floor Joists*

Nominal size (inches)	Maximum allowable span†
2 × 6	7 ft 3 in
2 × 8	9 ft 7 in
2 × 10	12 ft 2 in
2 × 12	14 ft 10 in

* This table is for wooden floor joists spaced 16 in on center. All lumber must be of a wood species designated "structural" in Table D-1, graded No. 2 or better, and kiln-dried.

† This table is based on an evenly distributed floor load of 60 lb or less per sq ft. The modulus of elasticity, E, of the lumber must be 1.0×10^6 psi or stiffer. These spans allow a maximum bending stress of 1,100 psi and a maximum deflection of span ÷ 360. Lesser load conditions or the use of stronger, stiffer wood species may allow longer spans. For more information, refer to the Uniform Building Code (Table 25-T-J-1) or the Western Woods Use Book.

TABLE D-3 Wooden Floor Girders*

Nominal size (inches)	Maximum allowable span (feet)†		
	Condition A	Condition B	Condition C
4 × 6	5	3½	—
4 × 8	6	4	3½
4 × 10	8	6	4½
6 × 8	8	6	4½
6 × 10	10	7	6
8 × 8	9	6½	5
8 × 10	12	8½	7

* This table is for wooden girders of a wood species designated as "structural" in Table D-1, graded No. 1 or better, kiln-dried, with no splits allowed. The maximum spans are based on loads evenly distributed along the girder, where the roof and floor(s) supported do not exceed a 15-ft span and the total load does not exceed the amounts indicated for each condition. The wood used for the girder must have a minimum E of 1.3×10^6 psi. These spans allow a maximum bending stress, F_b, of 1,000 psi and a maximum shear stress, F_v, of 70 psi under most conditions.

† Condition A is any situation where the girders support one floor and no more than 60 psf; condition B is when the girders support one floor, one ceiling, and one roof and no more than 120 psf; and condition C is when the girders support two floors, one ceiling, one roof, and 180 psf.

TABLE D-4 Wall Headers*

Nominal size	Maximum allowable span (feet)†	
	Condition A	Condition B
4 × 4 (or double 2 × 4's)	3	—
4 × 6 (or double 2 × 6's)	5	3½
4 × 8 (or double 2 × 8's)	7	4½
4 × 10 (or double 2 × 10's)	8	5½
4 × 12 (or double 2 × 12's)	10	7

* This table is for wooden headers of wood species designated as "structural" in Table D-1, graded No. 1 or better, kiln-dried, no splits, and used in walls framed according to instructions in Chapter 13. The recommendations are based on loads evenly distributed along the header, where the roof and floor(s) supported do not exceed a 15-ft span and the total loading does not exceed the amounts indicated for each condition. The wood used for the header must have a minimum E of 1.3×10^6 psi. These spans allow a maximum bending stress, F_b, of 1,000 psi and a maximum shear stress, F_v, 70 psi under most conditions.

† Condition A is where the headers support one roof, one ceiling, and no more than 60 psf; condition B is where the headers support one floor, one ceiling, one roof, and no more than 120 psf.

TABLE D-5 Ceiling Joists*

Nominal size	Maximum allowable span†
2 × 4	8 ft 4 in
2 × 6	13 ft 2 in
2 × 8	17 ft 5 in
2 × 10	22 ft

* This table is for wood ceiling joists spaced 16 in on center. All lumber must be of a wood species designated as "structural" in Table D-1, graded No. 2 or better, and kiln-dried.

† The recommendations are based on an evenly distributed ceiling load of 15 psf or less. The wood must have an E value of 1.0×10^6 psi or stiffer. These spans allow a maximum bending stress of 1,100 psi and a maximum deflection of span ÷ 240 under most conditions.

TABLE D-6 Rafters*

Nominal size	Maximum allowable span†	
	Low slope roofs (Less than 4 in 12)	High slope roofs (Steeper than 4 in 12)
2 × 4	—	5 ft 8 in
2 × 6	8 ft 10 in	9 ft
2 × 8	11 ft 8 in	11 ft 10 in
2 × 10	14 ft 10 in	15 ft 1 in
2 × 12	18 ft 1 in	18 ft 5 in

* This table is for wood rafters spaced 16 in on center. All lumber must be of a wood species designated as "structural" in Table D-1, graded No. 2 or better, and kiln-dried.

† This table is based on an evenly distributed roof load of 50 psf or less. The spans are measured horizontally as indicated in Fig. D-1. The wood must have an E value of 1.0×10^6 psi or stiffer. These spans allow a maximum bending stress of 1,100 psi and a maximum deflection of span ÷ 240 under most conditions.

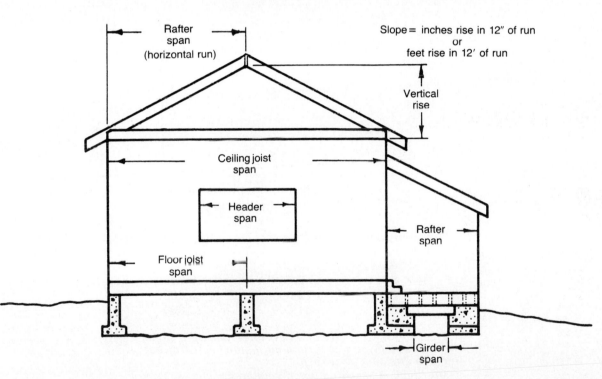

Fig. D-1 *Span diagram.*

TABLE D-7 Beams*

Nominal size (inches)	Actual size (inches)	Maximum allowable load (in pounds by span)[†]					
		10 ft	12 ft	14 ft	16 ft	18 ft	20 ft
4 × 10	3½ × 9 ¼	3,000	2,600	2,000	1,400	—	—
4 × 12	3½ × 11¼	3,600	3,600	3,400	2,500	—	—
6 × 10	5½ × 9¼	4,700	4,200	3,400	2,500	—	—
6 × 12	5½ × 11¼	5,700	5,700	5,300	4,000	3,100	—
6 × 14	5½ × 13¼	6,800	6,800	6,800	6,400	5,100	4,100
8 × 10	7¼ × 9¼	6,200	5,500	4,600	3,500	2,700	—
8 × 12	7¼ × 11¼	7,600	7,600	7,000	5,500	4,400	3,500
8 × 14	7¼ × 13¼	8,900	8,900	8,900	8,000	6,300	5,000

* This table is for solid or built-up wood beams of Douglas Fir, Larch, California Redwood, and Hem-Fir, graded No. 1 or better, kiln-dried and seasoned.

† This table is based on a load that is evenly distributed over the length of the beam. To determine the load, total the area of all floors supported and multiply by 60 lb per sq ft, the area of roofs supported multiplied by 50 psf, and the area of ceilings supported multiplied by 10 psf (assuming normal conditions in all three cases). Add any known additional loads. The wood must have an E value of 1.3×10^6 psi or stiffer. These loads and spans allow a maximum bending stress of 1,000 psi and a maximum deflection of span ÷ 360 under most conditions. Store beams level and straight in a dry, shady, well-ventilated place for 2 to 6 months to season them and minimize splitting.

TABLE D-8 Posts*

Nominal size	Maximum allowable load (in pounds)[†]	
	8-ft posts	10-ft posts
4 × 4	5,300	3,400
4 × 6	8,400	5,300
6 × 6	24,000	20,000
6 × 8	31,000	27,000

* This table is for solid wood posts of Douglas Fir, Larch, California Redwood, and Hem-Fir, graded No. 1 or better, kiln-dried and seasoned.

† This table is based on wood with an E value of 1.2×10^6 psi or stiffer. These loads allow a maximum compressive stress of 800 psi. Posts must be securely attached in the vertical position at the top and bottom and braced laterally. Season the posts by storing them in a dry, well-ventilated, shady place for 2 to 6 months to reduce splitting. Protect the base of the post from moisture. Refer to the Uniform Building Code Sec. 2507, 2607, 2703, 2705, and 2706 for more information on wood, steel, and concrete posts (columns).

TABLE D-9 Holding Power of Bolts*

Pounds per bolt[†]

Bolt diameter (inches)	Thickness of beam or post (whichever is smaller)				
	2½ in	3 in	3½ in	4 in	5½ in
½	300	350	—	—	—
5⁄8	350	400	500	600	—
¾	400	500	600	700	—
1	600	700	800	900	1,300

* This table is for steel bolts installed horizontally in wood species designated as "structural" in Table D-1. All lumber must be graded No. 1 or better, kiln-dried, and seasoned.

[†] These loads allow a maximum compressive stress of 245 psi on the wood under most conditions. Bolt holes must be drilled the same diameter as the bolt and spaced 3 in or more from each other and from the ends and edges of the wood. Always use washers so the wood doesn't split when the bolt is tightened. Under some conditions bolts will hold more than these loads. Refer to the Uniform Building Code Sec. 2510 for more information.

INDEX